ISFRAM 2014

Sahol Hamid Abu Bakar • Wardah Tahir •
Marfiah Ab. Wahid • Siti Rashidah Mohd Nasir •
Rohana Hassan
Editors

ISFRAM 2014

Proceedings of the International Symposium on Flood Research and Management

Springer

Editors
Sahol Hamid Abu Bakar
Universiti Teknologi MARA
Selangor, Malaysia

Marfiah Ab. Wahid
Universiti Teknologi MARA
Selangor, Malaysia

Rohana Hassan
Universiti Teknologi MARA
Selangor, Malaysia

Wardah Tahir
Universiti Teknologi MARA
Selangor, Malaysia

Siti Rashidah Mohd Nasir
Universiti Teknologi MARA
Selangor, Malaysia

ISBN 978-981-287-364-4 ISBN 978-981-287-365-1 (eBook)
DOI 10.1007/978-981-287-365-1

Library of Congress Control Number: 2015937600

Springer Singapore Heidelberg New York Dordrecht London

Printed on acid-free paper

Springer Science+Business Media Singapore Pte Ltd. is part of Springer Science+Business Media (www.springer.com)

Preface

The world is frequently being devastated by unpredictable disasters. Years of civilization of a nation can simply be shattered by a disaster such as tsunami in a matter of hours. Flooding has been one of the most disastrous natural hazards striking many parts of the world. The increasing trend in flood disasters has resulted from the combined impacts of several factors including global warming effects (e.g., increasing frequency of intense rain, glacier melting, and sea-level rise), land-use changes, and growing population in flood-prone areas.

International Symposium on Flood Research and Management 2014 (ISFRAM2014) is organized by the Flood Control Research Center (FCRC), Faculty of Civil Engineering, Universiti Teknologi MARA, to promote advances in flood research and management in finding solution toward reducing flood disasters. The objective of ISFRAM2014 is to provide a forum to researchers, scientists, and engineers to share and exchange their views, experiences, and researches in flood-related areas and sustainable management in Malaysia and worldwide. The symposium presents innovative work and best practices in managing flood and recommendation of flood solutions.

The full paper submissions were reviewed by national and international panel of reviewers before final acceptance. The selected papers were evaluated based on originality, research content, and relevance to contributions. Papers selected cover the fundamentals and latest advancements in related areas to flood research and management. The book proceedings to be published by Springer will serve as the source of knowledge and state-of-the art technology in managing flood for the betterment of the quality of life.

This symposium is supported by IEEE Malaysia IE/IA Joint Chapter and FCRC collaborators—Colorado State University, IIT Roorkee, AIT Thailand, Stuttgart University, National Hydraulic Research Institute, and the Drainage and Irrigation Department. I would like to express my deepest gratitude to all committee members, panel of reviewers, authors, chairpersons, delegates, and everyone who had contributed to make ISFRAM2014 a success. I wish all participants a beneficial symposium, valuable experiences, and a pleasant stay at Sabah.

Selangor, Malaysia Wardah Tahir

About FCRC

Flood Control Research Center was founded by the Vice Chancellor of Universiti Teknologi Mara, Tan Sri Dato' Sri Prof. Ir. Dr. Sahol Hamid Abu Bakar, in collaboration with Stuttgart University, Colorado State University, Asian Institute of Technology (AIT), and Indian Institute of Technology (IIT) Roorkee as a five-university collaboration. The objectives of the center are:

- To identify important issues on flood
- To provide the solutions for flood problems in Malaysia
- To train young researchers and staff
- To share the pool of expertise
- To go beyond Malaysian border
- To get national and international research grants

The national research collaborators include the National Hydraulic Research Institute Malaysia (NAHRIM), Drainage and Irrigation Department (DID), Lembaga Urus Air Selangor (LUAS), Malaysian Meteorological Department (MMD), Department of Environment (DOE), National Security Council, and Local City Councils. These agencies would contribute:

- To share resources, data, and expertise
- To act as the pushing factor to the government for implementation of the proposed flood solution
- Joint research and supervision
- Joint publication
- Co-organizers of events and activities

Current and future activities of FCRC include:
Establishing a Flood Library which has:

- Library of models (e.g., HEC-RAS, HEC-HMS, Infoworks, Mike-11)
- Collection of structural solutions and key methodologies for flood problem

- Repositories of technical papers
- Open access to libraries especially to the five universities

Defining flood master planning methodology

- Flood risk and damage assessment
- Flood plan to incorporate entire flood hydrological regime
- Economic cutoff point for extending flood plan by using nonstructural method
- Nonstructural flood planning

Providing training program

- To increase capacity building for those involved in flood control and management
- To enable access to experts in the consortium of the five-university collaboration

Organizing annual meeting and symposium

ISFRAM2014 Organizing Committee

Patron
Y. Bhg. Tan Sri Dato' Sri Prof. Ir. Dr. Sahol Hamid Abu Bakar FASc

UiTM Advisors
Prof. Dr. Azmi Ibrahim
Mustafar Kamal Hamzah, IEEE Malaysia IE/IA Joint Chapter

International Advisors
Dr. Daniel Gunaratnam (USA)
Prof. Dr. Pierre Y. Julien (Colorado State University)
Prof. Dr. Silke Wieprecht (Stuttgart University, Germany)
Prof. Dr. M. Perumal (Indian Institute of Technology (IIT) Roorkce, India)
Prof. Dr. Mukand Singh Babel (AIT Asian Institute of Technology, Thailand)

Industry Advisor
Ir. Dr. Nasehir Khan (National Hydraulic Research Institute Malaysia)

General Chair
Assoc. Prof. Dr. Wardah Tahir

General Co-chair
Dr. Siti Rashidah Mohd Nasir

Technical Program Chair
Dr. Marfiah Ab Wahid
Dr. Mohd Fozi Ali
Dr. Janmaizatulriah Jani
Dr. Lee Wei Koon
Ir. Turahim Abd Hamid
Assoc. Prof. Dr. Ramlah Tajuddin

Publication Chair
Dr. Rohana Hassan

Finance Chair
Ros Mini Ismail

Secretariat
Azinoor Azida Abu Bakar
Suzana Ramli
Dr. Jazuri Abdullah
Sh Nurul Huda Sy Yahya
Mohd Shahir Abd Rahman

Information and Website Chair
Khasiah Zakaria
Ahmad Amiri Mohamad
Bahroni Moojee

Reviewers

Dr. Mokhtar Guizani	SET King Saud University, Saudi Arabia	g_mokh@yahoo.fr
Dr. Milad Jajarmizadeh	UTM	milad_jajarmi@yahoo.com
Assoc. Prof. Dr. Marlinda Abdul Malik	UNITEN	marlinda@uniten.edu.my
Assoc. Prof. Dr. Wardah Tahir	UiTM	wardah_tahir@yahoo.com
Mrs. Azlinda Saadon	UNISEL	azlinda303@gmail.com
Prof. Dr. Abdul Halim Ghazali	UPM	abdhalim@eng.upm.edu.my
Mr. Fauzi Baharudin	UiTM	fauzi1956@yahoo.com
Prof. Dr. Faridah Othman	UM	faridahothman@um.edu.my
Mrs. Azinoor Azida Abu Bakar	UiTM	azinoor@salam.uitm.edu.my
Prof. Dr. Ahmad Khairi Abd Wahab	UTM	drakaw@gmail.com
Dr. Marfiah Ab Wahid	UiTM	ce_marfiah@yahoo.com
Dr. Mohd Fozi Ali	UiTM	mohdfozi@salam.uitm.edu.my
Mrs. Suzana Ramli	UiTM	suzana799@salam.uitm.edu.my
Prof. Dr. Ahmad Yahya	USM	ceshukri@usm.my
Assoc. Prof. Dr. Ramlah Tajuddin	UiTM	reviewramlah2014@gmail.com
Dr. May Raksmey	UiTM	may_raksmey@salam.uitm.edu.my
Dr. Lee Wei Koon	UiTM	leewei994@salam.uitm.edu.my
Mrs. Nuryazmeen Farhan Haron	UiTM	neemzay@yahoo.com
Prof. Dr. Muthiah Perumal	IIT Roorkee, India	p_erumal@yahoo.com
Dr. Sucharit Koontanakulvong	Chulalongkorn University, Thailand	sucharit.K@chula.ac.th
Mrs. Caroline Peter Diman	UiTM	carolinepeter968@ppinang.uitm.edu.my
Dr. Jazuri Abdullah	UiTM	jazuri.abdullah@yahoo.com

(continued)

Dr. Chia Chay Tay	UiTM	taychiay@perlis.uitm.edu.my
Prof. Krzysztof S. Kulpa	Warsaw University of Technology, Poland	K.Kulpa@elka.pw.edu.pl
Prof. Dr. Md Azlin Md Said	USM	azlin@eng.usm.my
Assoc. Prof. Dr. Sayang Mohd Deni	UiTM	sayang@tmsk.uitm.edu.my
Sharifah Nurul Huda Syed Yahya	UiTM	sharifahnurulhuda@hotmail.com

Keynote Speaker

Pierre Y. Julien

Dr. Julien is professor of Civil and Environmental Engineering at Colorado State University. As Professional Engineer, he has completed projects for 50 different agencies including UNESCO and the World Bank. Dr. Julien authored more than 500 scientific contributions including two textbooks, 20 lecture manuals and book chapters, 170 refereed journal articles including 90 full papers in scientific journals, 150 professional presentations, 190 conference papers, and 120 technical reports. He supported and guided more than 100 graduate students (including 37 Ph.D.) to complete engineering degrees. He delivered 15 keynote addresses at international conferences. He received the H.A. Einstein Award for his research on sedimentation and river mechanics.

Contents

Part V Weather and Climate

Keynote: Analysis of Extreme Floods in Malaysia

Pierre Y. Julien, Jazuri Abdullah, and Nur Shazwani Muhammad

Abstract This article reviews some of the recent advances in the analysis of extreme flood events in Malaysia. First, a detailed analysis of daily rainfall precipitation measurements leads to new understanding regarding Malaysian monsoons: the conditional probability of rainfall steadily increases as a function of the number of successive rainy days. The probability of multiday rainfall events has also been analyzed using stochastic models like DARMA(1,1) to demonstrate lower periods of returns of large precipitation amounts for rainfall events between 4 and 12 days. Advances in numerical modeling of surface runoff using the TREX model allowed improved simulations of large floods when considering rainfall amounts between the 2- and 100-year events and the PMP for extreme floods on both small to large watersheds in Malaysia. Examples on Lui, Semenyih, and Kota Tinggi have also been possible with GIS data at 30–90 m resolution. The recent floods of the Kota Tinggi and Muda River are also briefly discussed. Finally, a brief overview of the DID River Management Manual is also presented.

Keywords Monsoon precipitation • Extreme floods • Flashflood modeling • River management

1 Introduction

Southeast Asia has long experienced a monsoon climate with dry and wet seasons. With a mean annual rainfall precipitation around 2,500 mm and locally in excess of 5,000 mm, the very intense rainstorms in the steep mountains of Malaysia have caused frequent and devastating flash floods. In the valleys, floodwaters spread over very wide flood plains developed for agriculture, predominantly rice paddies and oil

P.Y. Julien (✉)
Department of Civil and Environmental Engineering, Colorado State University, Fort Collins, CO 80523, USA
e-mail: pierre@engr.colostate.edu

J. Abdullah
Flood Control Research Center, Universiti Teknologi MARA, Shah Alam, Selangor, Malaysia

N.S. Muhammad
Department of Civil Engineering, Universiti Kebangsaan Malaysia, Bangi, Selangor, Malaysia

S.H. Abu Bakar et al. (eds.), *ISFRAM 2014*, DOI 10.1007/978-981-287-365-1_1

palm. Urbanization and deforestation exacerbate flooding problems due to the increased runoff from impervious areas. The industrial developments fostered a new way of urban life, and flood control in Malaysia is undergoing significant changes.

The objective of this article is to provide a brief description of recent developments in the analysis of new engineering methods for the analysis of extreme floods. The first objective is to describe developments in the analysis of daily rainfall precipitation data under monsoon climates. The second objective is to share some of the developments in hydrologic modeling of extreme surface runoff from exceptional rainstorms on small to large watersheds. The third objective is to illustrate some of the implications in terms of direct applications to recent flood events in Malaysia and specifically on the Muda River and near Kota Tinggi. Finally, a brief overview of the DID manual of River Management will be presented.

2 Extreme Rainfall Precipitation

2.1 Analysis of Daily Rainfall Precipitation

Muhammad [1] recently reviewed the daily rainfall precipitation data at Subang Airport from 1960 to 2011. During this period of 18,993 days, there were 10,092 rainy days with more than 0.1 mm of precipitation. The average daily rainfall is 13 mm and standard deviation 17 mm. She demonstrated that the distribution of rainfall precipitation followed a gamma distribution. The equation of the probability density function can be approximated as

$$f(x,t) \cong \frac{1}{|24.0|\Gamma(0.6t)}\left(\frac{x}{24.0}\right)^{0.6t-1}\exp\left(-\frac{x}{24.0}\right) \tag{1}$$

where x is the daily rainfall depth in mm, and t is the number of consecutive rainy days. The cumulative distribution function is the integral of Eq. (1). As shown in Fig. 1, there is a 37 % probability that the total precipitation from six consecutive rainy days will exceed 100 mm. It is interesting to note that the NE and SW monsoons produced fairly similar rainfall distributions at that location.

One of the main findings from her research was that the conditional probability of rainy days increased with the number of consecutive rainy days as shown in Fig. 2. Monsoon rainfall events cannot be considered to be independent.

Muhammad [1] then developed a detailed DARMA(1,1) model for the simulation of long sequences of wet and dry days including the amount of daily precipitation. The main results of this analysis have demonstrated that the periods of return of the amount of precipitation from multiple rainy days vary with the number of rainy days as shown in Fig. 3. The agreement between the DARMA(1,1) model

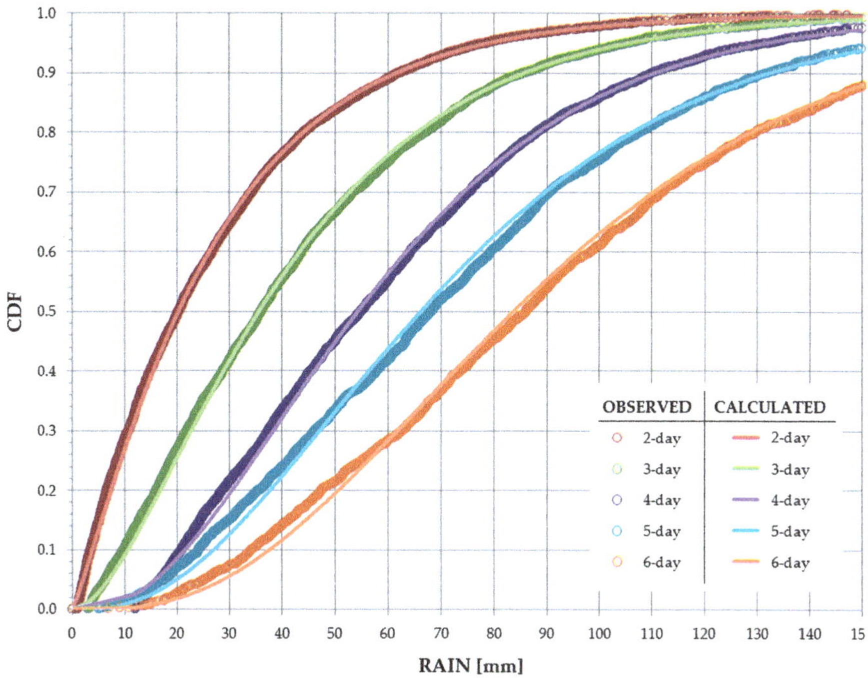

Fig. 1 Multiday cumulative distribution functions for the daily rainfall data at Subang Airport from 1960 to 2011, from Muhammad [1]

simulations and the field measurements was excellent. For instance, the accumulation of 120 mm of rainfall in a single day has a period of return of approximately 20 years; however, when accumulated over eight consecutive rainy days, the period of return is now reduced to approximately 2 years.

Daily rainfall simulation sequences up to 1,000,000 days (i.e., ~2,700 years) are readily possible as shown in Fig. 4. For instance, it now becomes possible to predict that a 250 mm rainfall in 4 days will have a period of return of about 500 years. The extension of daily precipitation analysis to rare and extreme events can now be better investigated using this methodology. The practical implications of this research are most important for the analysis of floods on large watersheds, i.e., larger than 1,000 km^2.

2.2 *Return Periods and Probable Maximum Precipitation*

Abdullah [2] recently examined the frequency distribution of rainfall precipitation as a function of storm duration for the State of Selangor. His analysis of data from several sources led to Fig. 5 where large events are comprised between the two lines with a period of return of 2–100 years. The world maximum precipitation (WMP)

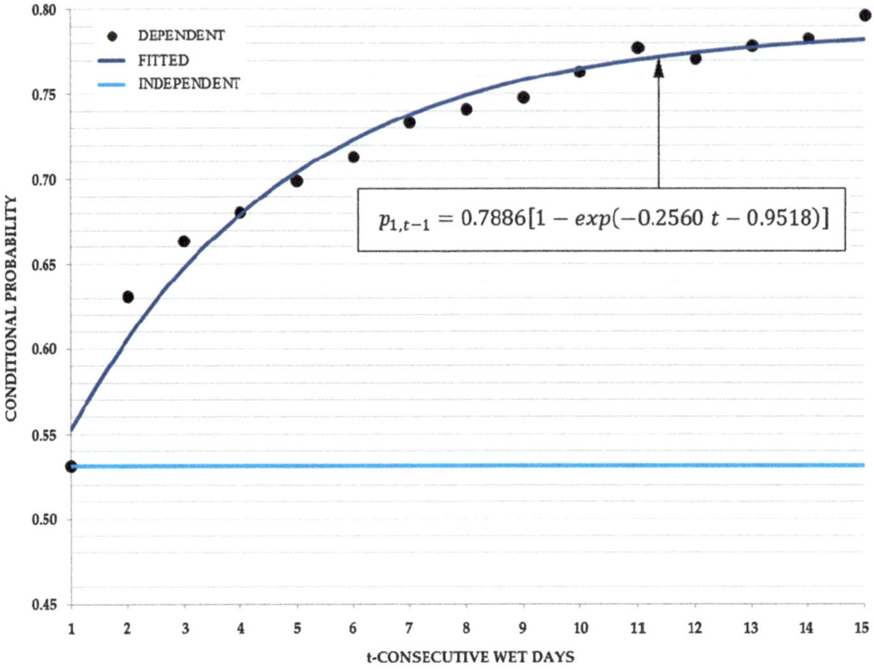

Fig. 2 Increase in conditional probability of rainfall at Subang Airport, from Muhammad [1]

events can be found on the top line with the probable maximum precipitation (PMP) for Selangor between the WMP and the 100 year rainfall precipitation. It is observed that the 100-year precipitation is approximately two times the 2-year rainfall precipitation event, and the rainfall depth increases approximately with the square root of rainfall duration. This means that the average rainfall intensity gradually decreases with rainfall duration. The PMP for the State of Selangor is approximately three times the 100-year rainfall depth and half the world maximum precipitation.

3 Large Flood Simulation with TREX

Abdullah [2] successfully applied the fully distributed two-dimensional TREX model to the simulation of infiltration, overland runoff, and channel flow during extreme rainfall events on small and large watersheds in Malaysia. There are four main processes in the TREX hydrological sub-model developed by Velleux et al. [3, 4]: (1) precipitation and interception; (2) infiltration and transmission loss; (3) depression storage; and (4) overland and channel flow.

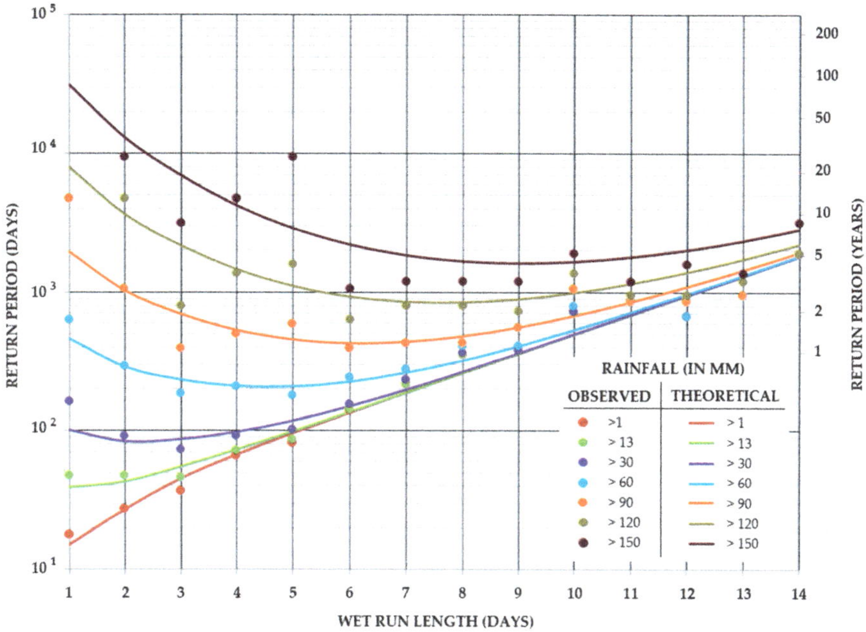

Fig. 3 Return period of multiple rainy days as a function of the cumulative rainfall precipitation, from Muhammad [1]

3.1 *Precipitation and Interception*

The precipitation volume reaching the near surface can be written as the product of rainfall intensity and surface area. The presence of forests or other vegetation influences the distribution pattern of the net rainfall precipitation. Some of the precipitation is intercepted and retained by the leaves and other parts of the canopy, and then eventually returned to the atmosphere in the form of evaporation.

3.2 *Infiltration and Transmission Losses*

In the TREX model, the infiltration rate is calculated using the well-known Green and Ampt equation. Transmission losses describe the water reaching the groundwater, overbank flow onto floodplains, wetlands and billabongs, and water never returning to the river. The rate of transmission may be affected by several factors, particularly the soil hydraulic conductivity.

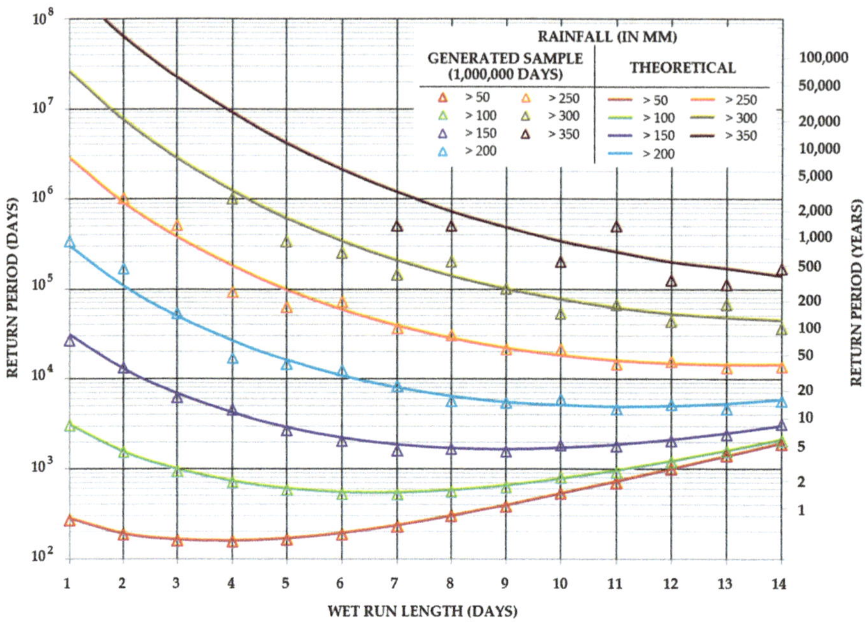

Fig. 4 Extension of the return period analysis of multiple rainy day precipitation by using the DARMA(1,1) model, from Muhammad [1]

3.3 *Depression Storage*

Precipitation retained in small surface depressions is called the depression storage, which may be conceptualized as a depth when normalized by the surface area. When the water depth is below the depression storage threshold, overland flow is zero. Note that water in depression storage is still subject to infiltration and evaporation. Similar to depression storage in overland areas, water in channels may be stored in depressions in the stream bed, which are caused when the channel water depth falls below some critical level, flow is zero and the water surface has discontinuities, but individual pools of water remain. This mechanism is termed dead storage. Note that the water in dead storage is still subjected to transmission losses and evaporation. For single storm events, the recovery of depression storage volume by evaporation can be neglected. Similarly, the recovery of dead storage volume by evaporation can also be neglected for single storm events.

3.4 *Overland and Channel Flow*

Overland flow occurs when the water depth of the overland plane exceeds the depression storage threshold. Overland flow is governed by the conservation of

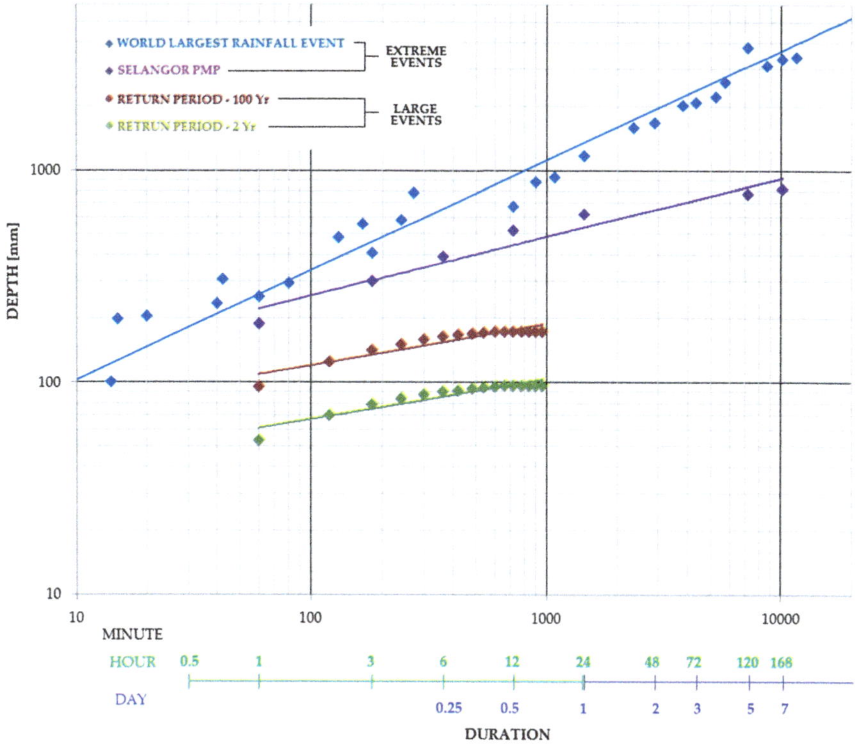

Fig. 5 Rainfall depth vs. duration for the 2-year, 100-year, the PMP-Malaysia and the world maximum precipitation for small and medium watersheds in Selangor, from Abdullah [2]

mass (continuity) and conservation of momentum. The two-dimensional vertically integrated continuity equation for gradually varied flow over a plane in rectangular coordinates is used in the TREX model.

The application of momentum equations (Saint-Venant equations) for two-dimensional runoff calculations are derived in terms of net forces per unit mass or acceleration. Five hydraulic variables must be defined in terms of depth-discharge relationship to describe flow resistance before the overland flow equations can be solved.

One-dimensional channel flow (along the channel in the down-gradient direction which laterally and vertically integrated) is also governed by conservation of mass (continuity) and momentum. The method suggested by Julien [5] is applied for gradually varied flow. To solve the channel flow equations, from the momentum equation (by neglecting the local and convective terms), the diffusive wave approximation may be used for the friction slope. The Manning relationship is used to describe resistance to flow.

3.5 Model Parameters

The TREX model simulates infiltration, overland runoff, and channel flow during extreme rainfall events. Input data were prepared using ArcGIS 9.3 and converted into a text file. The surface topography of the watershed was discretized at a 230 by 230 m scale. The grid size was used to delineate these watersheds. The DEM was downloaded at a 90 m resolution from the ASTER GDEM website. The watershed was described with a total of 31,000 active grid cells within a matrix of 292 rows and 292 columns. The total river length of the large watershed was ~250 km (1,081 nodes and 42 links).

Calibrated model parameters and modeling details can be found in Abdullah [2]. A sensitivity analysis showed that the hydraulic conductivity, K_h, and Manning, n, are the most sensitive parameters during calibration. These values were adjusted to achieve very good agreement between observed and simulated discharges. The antecedent moisture condition for the watershed was assumed to be dry at the beginning of simulation. Rainfall was generally sufficiently abundant to neglect interception and detention storage.

The TREX model provides illustrations of the evolution of flow depth with time during a flood event. It can provide maps of the distribution of flow depth at different times. For instance, Fig. 6 illustrates the simulation of an extreme event on the Semenyih watershed by Abdullah [2].

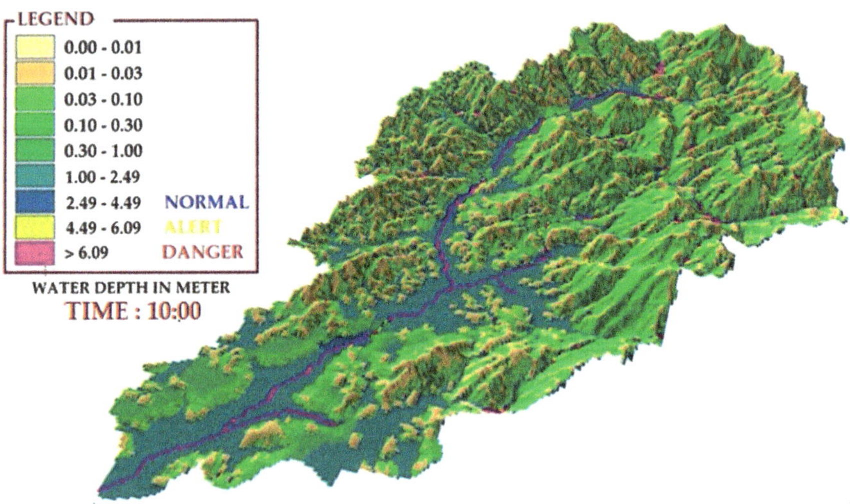

Fig. 6 Illustration of the distribution in flow depth on the Semenyih watershed, from Abdullah [2]

4 Kota Tinggi Flood

Shafie [6] compiled the information relevant to the extreme rainfall precipitation leading to the Kota Tinggi flood. During the calibration, in the TREX model, Abdullah [2] was able to simulate the hydrological conditions of the Kota Tinggi Flood with reasonable accuracy, as shown in Fig. 7.

The validation process was performed using stage data from December 14, 2006 to January 25, 2007 as shown in Figs. 8 and 9.

Figure 8 shows a detailed water depth distribution on the Kota Tinggi watershed from the TREX model at the time when the water reached the alert level on December 19, 2006. The stage continued to increase and easily passed the alert and danger levels as a result of the continuous rainfall. Figure 9 shows the TREX model results in terms of the flooding areas on the Kota Tinggi watershed on December 21, 2006. The maximum stage was reached on December 22, 2006, i.e., 2 days after the rainfall stopped.

The model gave very good estimates of the peak discharge and total volume with average overestimation of about 0.8 % and 1.5 %, respectively. The hydrological modeling results presented here give a physical representation of the flooding at Kota Tinggi. The results further prove that the multiday rainfall events are the main causes of severe flooding on this large watershed.

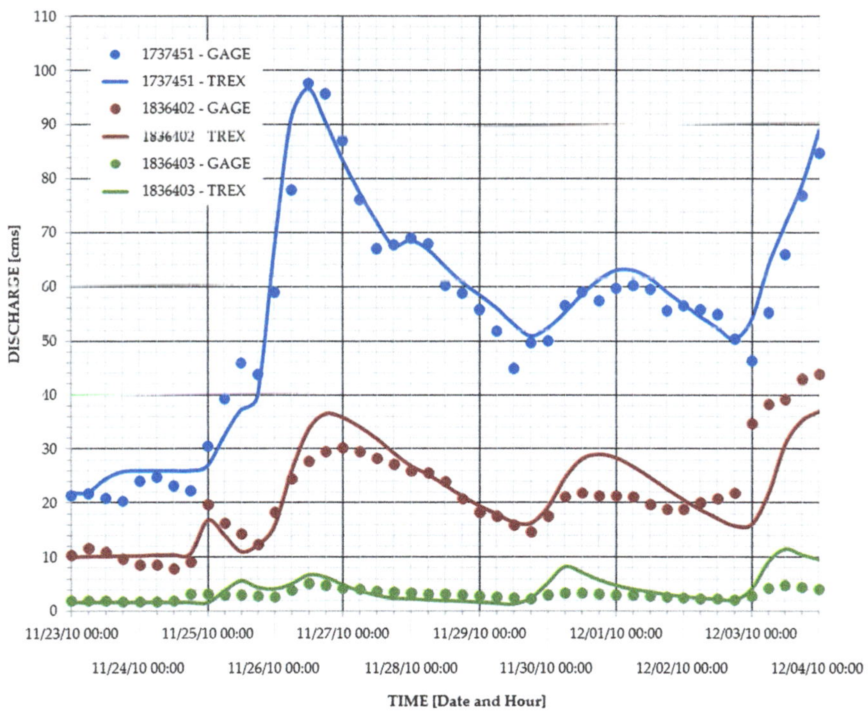

Fig. 7 Calibration results of the TREX model applied to the Kota Tinggi watershed, from Abdullah [2]

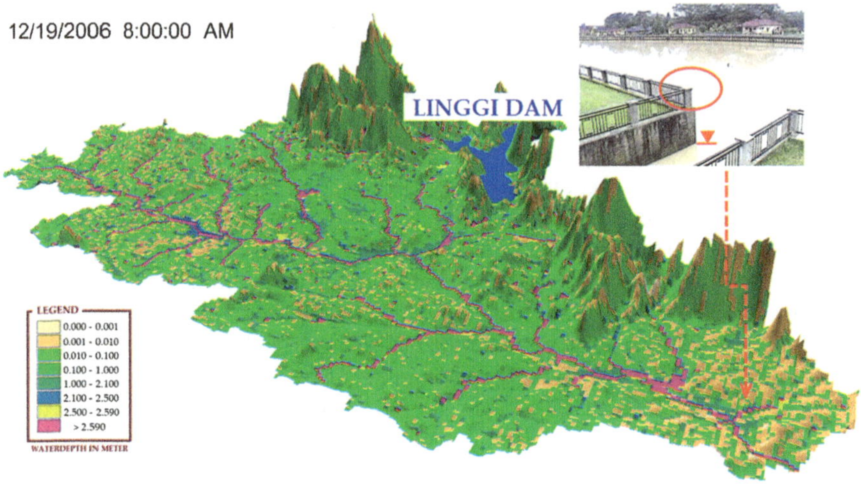

Fig. 8 TREX simulation of the Kota Tinggi Flood on December 19, 2006, from Abdullah [2]

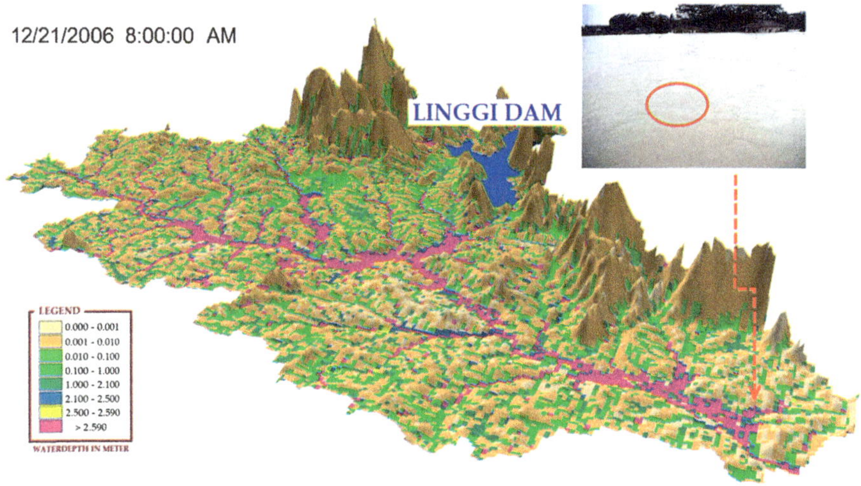

Fig. 9 TREX simulation of the Kota Tinggi Flood on December 21, 2006, from Abdullah [2]

Figure 10 illustrates the relationship between the estimated flood thresholds, return periods, and flood thresholds. A return period of 220 years (upper value) is the flood threshold for 1 day of rainfall. Overall, the return period estimated for the multiday rainfall is significantly lower than a single day event. For example, the return period to reach the flood threshold in four consecutive rainy days is only 24 years.

These results are useful in determining the design rainfall for a flood mitigation structure on a large watershed like the case of the Kota Tinggi flood.

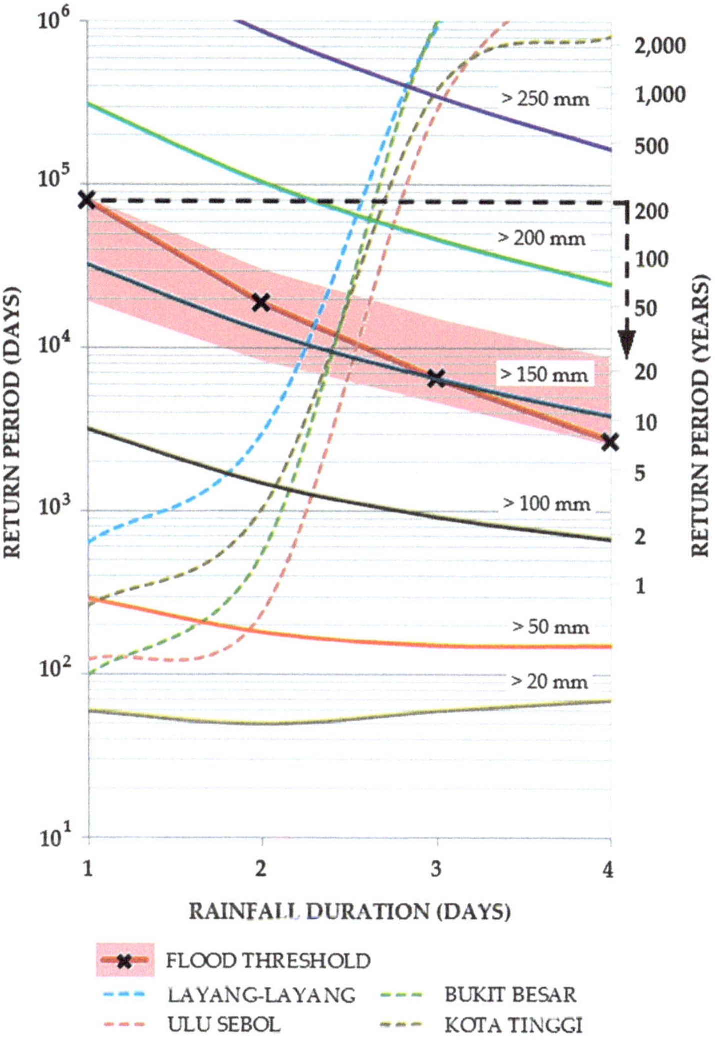

Fig. 10 Return period of multiple rainy days as a function of the cumulative rainfall precipitation and flooding threshold, from Muhammad [1]

5 Muda River Flood

The Muda River in Malaysia experiences floods every year, and the floods of 1996, 1998, and 1999 were particularly high. The Department of Irrigation and Drainage (DID) in Malaysia (Jabatan Pengairan dan Saliran Malaysia is also known as JPS) enacted a Flood Control Remediation Plan. Figure 11 illustrates the aerial extent of this flood, which adversely impacted 45,000 people in the State of Kedah.

Fig. 11 Aerial extent of flooding of the Muda River, from Julien et al. [8]

Julien et al. [7, 8] reviewed the flood control remediation plan for the Muda River. This highlights several important points in the design of flood remediation countermeasures against the frequent and intense floods during the monsoons. For instance, the analysis of measured daily discharge records can produce a more reliable 50-year peak discharge than hydrologic models. There was a 25 % difference in flood frequency analysis between the field measurements (1,340 m^3/s) and hydrologic model results (1,815 m^3/s) as shown in Fig. 12.

Sand and gravel mining operations have caused major problems associated with riverbed degradation. The riverbed degradation has exposed the foundations of numerous bridges and caused additional stability problems to riverine structures like bridges and pumping stations as shown in Fig. 13. Also, the river longitudinal profile shown in Fig. 14 indicates that the bed elevation remained below the mean sea level up to some 25 km upstream of the river mouth. There is always a significant risk of having saltwater intrusion in the river for long distances, which requires the construction of estuary barrages besides pumping requirements at irrigation canal intakes.

It is thus recommended to replace in-stream sand and gravel mining operations with off-stream mining operations within the flood plain corridor at a minimum distance of 50 m from the river banks.

The concept of river corridor has been explored, as shown in Fig. 15, and the proposed levee height of the Muda River could have been based on the 2003 flood plus a 1-m freeboard.

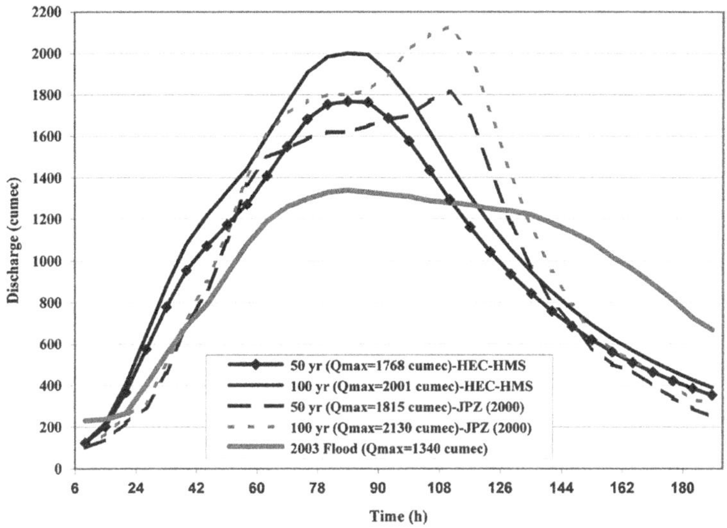

Fig. 12 Design hydrographs and measurements of the Muda River at Ladang Victoria, from Julien et al. [8]

Fig. 13 Impact of river bed degradation near bridge crossings, from Julien et al. [8]

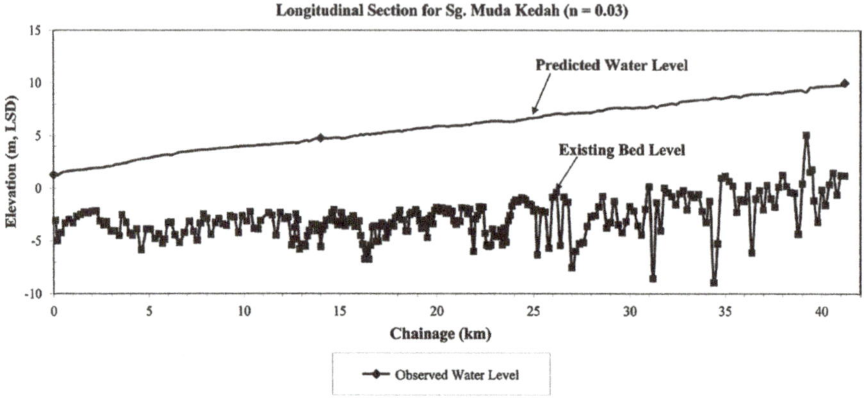

Fig. 14 Longitudinal profile of the Muda River, from Julien et al. [7]

Fig. 15 Concept of river corridor on the Muda River near Merdeka Bridge, from Julien et al. [7]

6 River Management Manual

This new River Management Manual of the Malaysia Department of Irrigation and Drainage [9] has been a tremendous effort for the development of rivers in Malaysia. The manual contains more than 600 pages of technically sound river management practices; see Fig. 16.

The first chapter of the manual contains a statement of river management issues as well as a clear definition of the responsibilities of the Federal, State, and Local

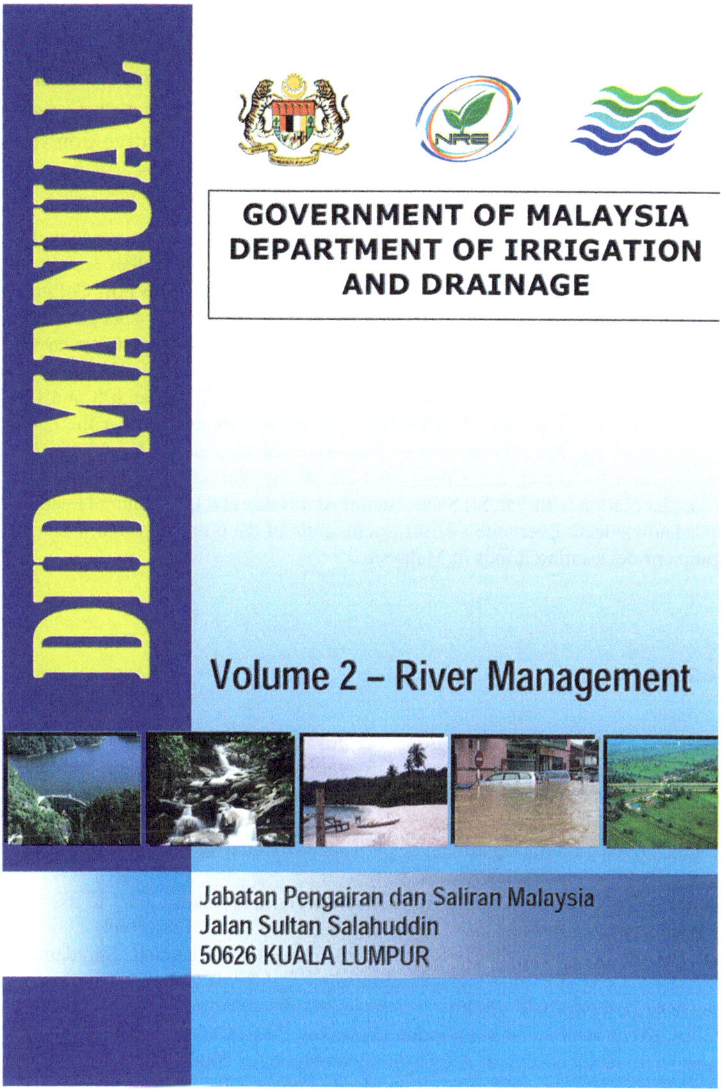

Fig. 16 Cover page of the DID River Management Manual

authorities in Malaysia. Chapter 2 discusses the concepts of Integrated River Basin Management (IRBM) and describes the main river characteristics and fluvial geomorphology followed with a description of river ecology and river health. Chapter 3 provides an ample discussion on river restoration, river rehabilitation, and river engineering. There is also a discussion of the recovery of disturbed river systems and on rehabilitation monitoring and management. Chapter 4 focuses on the concept of river corridor. Issues relative to the management of river corridors, riparian land, and floodplains are covered in detail. Chapter 6 broaches other relevant topics like sand/gravel mining, water quality improvement, and solid waste management. There is also a brief description of climate change, research and development, as well as periodic reviews. Throughout the manual, there are numerous figures, illustrations, worksheets, and examples to assist the engineer and scientist with their river work. This is one of the most innovative contributions to river engineering management in Malaysia.

Acknowledgment The author is most grateful for the opportunity to work with numerous engineers, scientists, and professionals in Malaysia. Drs. Jazuri Abdullah and Shazwani Nur Muhammad first come to mind since they spend several years with us in Fort Collins, CO. Most of the rainfall precipitation and TREX modeling results presented in this paper stem from their Ph. D. research at Colorado State University. The flood application on the Muda River has been carried out at REDAC (Universiti Sains Malaysia) in Nibong Tebal under the leadership of Drs. Azazi Zakaria and Aminuddin Ab. Ghani. Collaboration with Atikah Shafie at JPS and Dr. Junaidah Ariffin at the Universiti Teknologi MARA has been greatly appreciated for the discussion and detailed analysis of the Kota Tinggi flood. The River Management Manual has been mostly conducted by Chop Ai Kuang, Mr. Cheng, and Dr. Wong Wai Sam at Dr. Nik and Associates. Continued collaboration with Tan Sri Sahol Hamid Abu Bakar and Dato Ahmad Fuad Embi is also gratefully acknowledged. Everyone's effort is a measure of the promising future developments in the mitigation of devastating floods in Malaysia.

References

1. N.S. Muhammad, Probability structure and return period calculations for multi-day monsoon rainfall events at Subang, Malaysia, Ph.D. Dissertation, Department of Civil and Environmental Engineering, Colorado State University, Fort Collins, CO, 2013, 198 p
2. J. Abdullah, Distributed runoff simulation of extreme monsoon rainstorms in Malaysia using TREX, Ph.D. Dissertation, Department of Civil and Environmental Engineering, Colorado State University, Fort Collins, CO, 2013, 226 p
3. M. Velleux, J. England, P.Y. Julien, TREX: Spatially distributed model to assess watershed contaminant transport and fate. Sci. Total Environ. **404**(1), 113–128 (2008)
4. M. Velleux, P.Y. Julien, R. Rojas-Sanchez, W. Clements, J. England, Simulation of metals transport and toxicity at a mine-impacted watershed: California Gulch, Colorado. Environ. Sci. Technol. **40**(22), 6996–7004 (2006)
5. P.Y. Julien, *River Mechanics* (Cambridge University Press, Cambridge, 2002). 434 p
6. A. Shafie, Extreme Flood Event: A Case Study on Floods of 2006 and 2007 in Johor, Malaysia. Technical Report, Department of Civil and Environmental Engineering, Colorado State University, Fort Collins, CO, 2009, 82 p
7. P.Y.A. Julien, A. Ab. Ghani, N.A. Zakaria, R. Abdullah, C.C. Kiat, R. Ramli, J. Dinor, A. Manap, F. Yusof, Design Option of the Flood Mitigation Plan of Sg. Muda, Sungai Muda, Kedah, Malaysia. Final Report prepared at REDAC, Universiti Sains Malaysia, Nibong Tebal, for DID Malaysia, 2006, 149 p
8. P.Y. Julien, A. Ab. Ghani, N.A. Zakaria, R. Abdullah, C.K. Chang, Case-study: Flood mitigation of the Muda River, Malaysia. J. Hydraul. Eng. ASCE **136**(4), 251–261 (2010)
9. Government of Malaysia, Department of Irrigation and Drainage, *Volume 2—River Management Manual*. Jabatan Pengairan dan Saliran Malaysia, Jalan Sultan Salahuddin, Kuala Lumpur Malaysia, 2009, 612 p

Part I
Flood Management

Trend Analysis of Publications on Watershed Sustainability Indicators in Popular Academic Databases

Siti Mariam Akilah Mohd Yusoff and Noorul Hassan Zardari

Abstract This chapter presents trend analysis of publications on watershed sustainability and other related topics. We selected keywords related to watershed sustainability from an extensive literature review. Following keywords were used in database search: "WSI & watershed," "WSI & catchment," "watershed sustainability," "WSI & river basin," "river basin sustainability," "watershed sustainability index," "catchment sustainability," and "sustainable use of watershed." The trend analysis was performed on the number of retrievals from seven popular databases such as Scopus, Science direct, Wiley online library, ProQuest, Springer link, Taylor Francis online, and Web of Knowledge. The search period was divided into different periods to see the trend of publications. The trend analysis shows that the number of publications for the recent period (say 2010–2014) was much higher than older periods for all databases. The popular keyword on which most of the publications were belonging was "WSI & watershed." The most popular database where higher numbers of publications on watershed sustainability was retrieved was Scopus. More trend analysis results are shown in the body of this chapter.

Keywords Watershed sustainability index • Sustainable development • Sustainability • Water resources management

1 Introduction

A watershed (or catchment basin) is an area of land that captures water in diverse forms such as rain, snow, or dew and drains it to a common water body, i.e., stream, river, or lake [1]. A catchment basin generally comprises areas with a high population density due to favorable living conditions such as water for vegetation,

S.M.A.M. Yusoff
Department of Hydraulics and Hydrology, Faculty of Civil Engineering, Universiti Teknologi Malaysia, Skudai, Johor, Malaysia

N.H. Zardari (✉)
Institute of Environmental and Water Resources Management (IPASA), Faculty of Civil Engineering, Universiti Teknologi Malaysia, Skudai, Johor, Malaysia
e-mail: noorulhassan@utm.my

S.H. Abu Bakar et al. (eds.), *ISFRAM 2014*, DOI 10.1007/978-981-287-365-1_2

irrigation, industrial, and drinking purposes. In Malaysia, 97 % of the water supply is from surface water such as river while the remaining is from groundwater. However, the demand for water has been increased due to population growth, urbanization, and industrialization and these gave rise to water shortage [2]. Thus, it is vital to keep our watersheds healthy and sustainable by developing sustainability indices for major watersheds.

Sustainability assessment and development, which meets the need of the present generation without compromising the ability of future generations to meet their needs [3], is acknowledged as a major objective of watershed management. Now it is generally recognized that the sustainability of water resources in a watershed is directly related to its hydrology, environmental, life, and policy parameters. However, only few attempts have been made to integrate them in one single and comparable manner [4]. For last few decades, integrated indicators are being used for survey and planning purposes. The United Nations have been using the Human Development Index (HDI) [5] for several years. It integrates education, life expectancy, and income information from municipalities, states, and countries. The HDI is simple to use and applied worldwide.

One approach to achieve sustainable and integrated water management is through the application of the indicator-based approach [6]. This approach has been previously used to develop water sustainability indices, namely Water Poverty Index (WPI) by [7], Canadian Water Sustainability Index (CWSI) by the Policy Research Initiative [8], and Watershed Sustainability Index (WSI) by [4]. All these three indices have common objectives to provide information on current conditions of water resources, provide inputs to decision makers, and prioritize water-related issues [9].

The WSI, which attempted to integrate hydrologic, environmental, life, and policy issues, has shown advantages, both in the process of its development as well as in the implementation. In the process of its development, the WSI has provided decision makers with a clear and concise framework of water sustainability. During implementation, it has helped policy makers to improve water resources policies and minimize sewage pollution [4].

Reference [10] applied Water Quality Index (WQI) to different Malaysian rivers and found that the water quality of the Skudai River was declining, and the river was slightly polluted and [11] applied WQI to nine monitoring stations along Skudai River and found that all are classified as seriously polluted. More recently, [12] assessed water quality in the Skudai River and reached the same conclusions which were previously made by [10]. However, the WQI is not an enough yardstick for determining the sustainability of water resources in a particular river or basin. More indicators (e.g., hydrologic, environmental, life, and policy) should be considered while determining the sustainability level of watershed resources (especially land and water).

Therefore, it is important to have an overview of the research currently being conducted in the field of sustainable management of watersheds. This chapter presents a review of papers published in some popular databases. The search keywords were decided from the literature review and with the authors' consultation.

2 Methodology

We used seven databases subscribed by the Universiti Teknologi Malaysia (UTM) library. These databases were Springer Link, Scopus, Science Direct, Web of Knowledge, ProQuest, Wiley Online Library, and Tailor Francis Online.

2.1 *Brief Overview of the Databases*

Springer is a global publishing company that publishes books, e-books, and peer-reviewed journals in science, technical and medical. Scopus is a bibliographic database containing abstracts and citations for academic journal articles. It covers nearly 21,000 titles from over 5,000 publishers, of which 20,000 are peer-reviewed journals in the scientific, technical, medical, and social sciences (including arts and humanities). Science Direct has over 15 million researchers, healthcare professionals, teachers, students, and information professionals around the globe who rely on Science Direct as a trusted source of nearly 2,200 journals and close to 26,000 book titles. Web of Knowledge (formerly known as ISI Web of Knowledge) is an academic citation indexing and search service, which covers sciences, social sciences, arts, and humanities. It provides bibliographic content and tools to access, analyze, and manage research information. ProQuest consists over 125 billion digital pages. Content is accessed most commonly through library and internet gateways. John Wiley & Sons, Inc., also referred to as Wiley, is a global publishing company that specializes in academic publishing. The company produces books, journals, and encyclopedias, in print and electronically. Taylor Francis Online is one of the world's leading publishers of scholarly journals, books, e-books, reference works, and databases. Taylor & Francis publishes more than 1,000 journals and over 1,800 new books each year, with a backlist of over 20,000 titles.

2.2 *Keywords Selection and Database Search Criteria*

Search in these databases was made for these keywords related to sustainable management of watersheds: "watershed sustainability," "watershed sustainability index," "catchment sustainability," "river basin sustainability," "sustainable use of watershed," WSI & watershed, WSI & "River basin," and WSI & catchment. Although we tried to cover all sustainability issues related to a watershed in the selection of keywords, the scope of watershed sustainability may not be fully covered in the selected keywords.

We selected only journal, review papers, and conference publications. Other publications (e.g., e-book chapters) were excluded. We also include all years of publications. Later, we divided these publications into two periods (older than 2004 and 2004–present) to see whether the number of publications on watershed sustainability was increasing in the more recent past compared to the period more than 10 years before.

3 Results and Discussion

3.1 Identification of Popular Keywords

The search by keywords was made database to database. Exact phrases of the keywords were used for searching each of the databases. Keyword retrievals from e-books were excluded, and only journals and conference papers were counted for each of the keywords. Table 1 presents the number of papers retrieved from each database.

A search of SpringerLink shows that "WSI & Watershed" was the most popular keyword with 37 retrievals. On the other hand, "sustainable use of watershed" was the least popular keyword as no paper was found in SpringerLink database. For the search of SCOPUS database, we found that "watershed sustainability" has the most retrieval (i.e., 69 publications) and should be the most popular keyword compared to the rest of the keywords. Similar interpretation for other databases can be made from the data shown in Table 1. The most popular keyword for all the seven databases was "WSI & watershed" as it has the highest number of papers ($n = 344$) in combined retrieval of all databases. On the other hand, "sustainable use of watershed" was the keyword for which the lowest number of publications ($n = 8$) was retrieved from all seven databases. The search data analysis also reveals that the SCOPUS has more publications related to watershed sustainability compared to other databases. A total of 358 papers were found on the applied keywords. On the other hand, Web of Knowledge database retrieves the lowest number of keywords ($n = 31$). Ranking of other databases and keywords can be seen in Table 1.

3.2 Trend Analysis of Publications

The whole search period was divided into two periods: (1) older than 2004 and (2) from 2004 to present. This type of analysis was made to see if there is any visible trend of publications on watershed sustainability keywords for these two periods.

Table 1 Retrievals of keywords from different databases

Keywords exact phrase (in full text)	Abbreviation	Scopus	Science Direct	Wiley Online Library	ProQuest	Springer Link	Taylor Francis Online	Web of Knowledge	Column total	Percent (column-wise)	Rank
WSI & watershed	WSIW	63	91	59	80	37	9	5	344	24.7	1
WSI & catchment	WSIC	48	73	29	48	28	85	0	311	22.3	2
"Watershed sustainability"	WS	69	21	123	33	12	6	13	277	19.9	3
WSI & "river basin"	WSIRB	53	63	33	59	34	9	5	256	18.4	4
"River basin sustainability"	RBS	66	6	1	3	1	1	0	78	5.6	4
"Watershed sustainability index"	WSI	31	8	4	14	6	0	4	67	4.8	5
"Catchment sustainability"	CS	25	13	1	2	5	3	3	52	3.7	6
"Sustainable use of watershed"	SUW	3	0	1	0	0	0	1	8	0.6	7
Row total		358	278	251	239	123	113	31	1,393	100.0	
Percent (row-wise)		25.7	20.0	18.0	17.2	8.8	8.1	2.2	100.0		
Rank		1	2	3	4	5	6	7			

Fig. 1 Percentage of papers on each keyword in Scopus

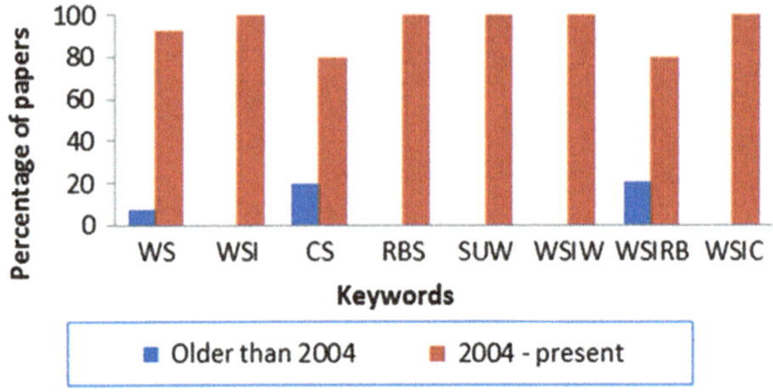

Fig. 2 Percentage of papers on each keyword in ScienceDirect

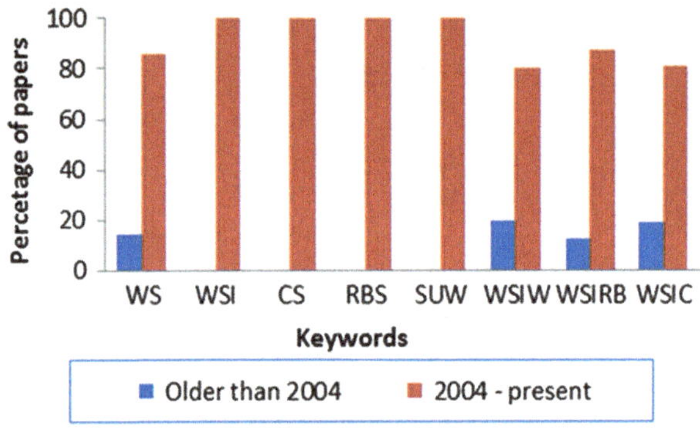

Fig. 3 Percentage of papers on each keyword in Wiley Online Library

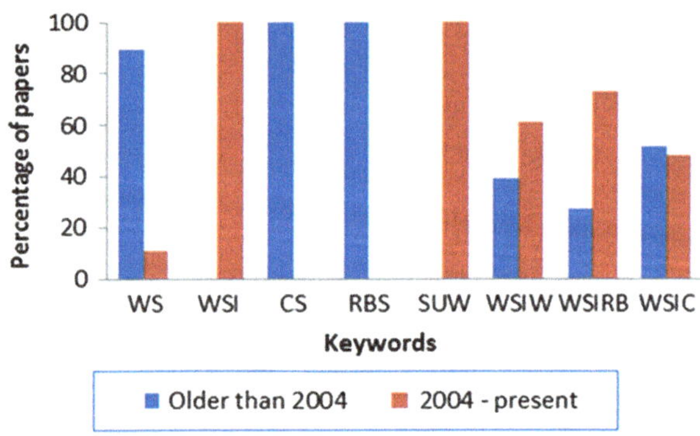

Last 10 years were assumed to be the recent period and the period before 2004 was considered as old period. Here, we emphasize that some databases are new, so the number of retrievals for those databases may not be showing the exact number of papers published during the selected periods. The search results for these two periods for different keywords and databases are shown in Figs. 1, 2, 3, 4, 5, 6, and 7. We can see that much percent of papers for each keyword was published in the last 10 years compared to the years before 2004. This clearly indicates that the awareness of sustainable management of watersheds is growing very fast.

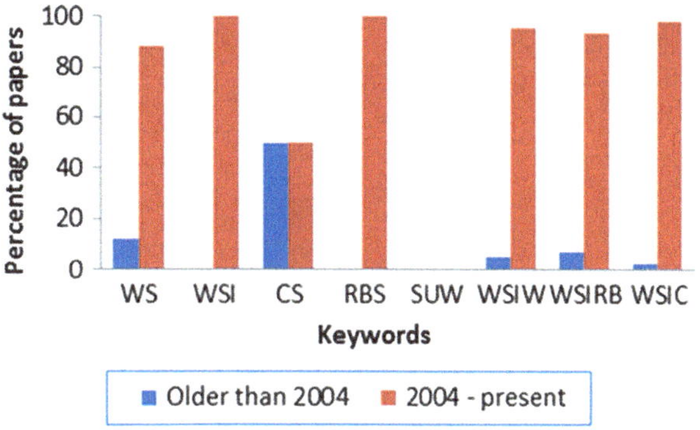

Fig. 4 Percentage of papers on each keyword in ProQuest

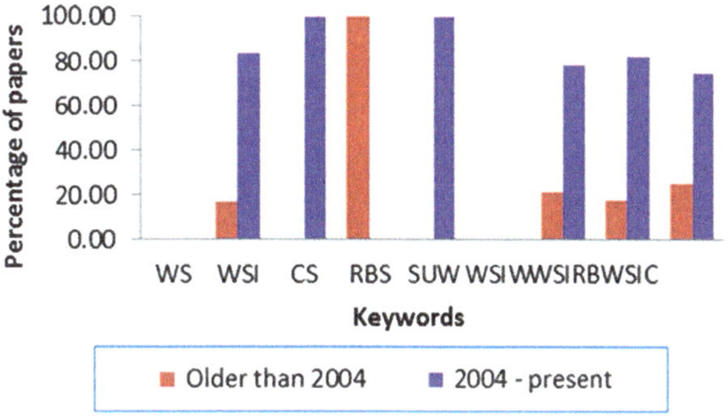

Fig. 5 Percentage of papers on each keyword in SpringerLink

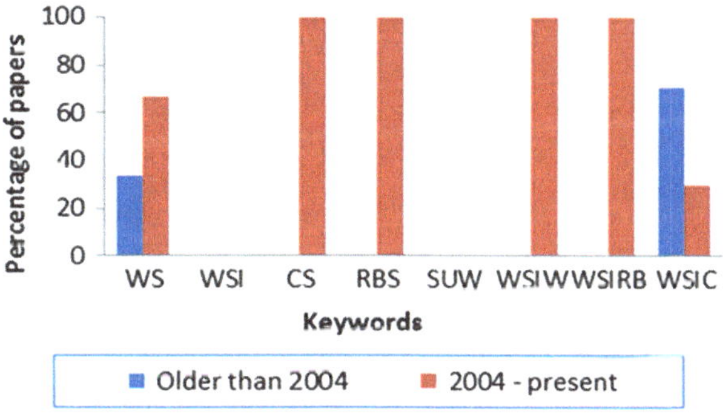

Fig. 6 Percentage of papers on each keyword in Taylor Francis Online

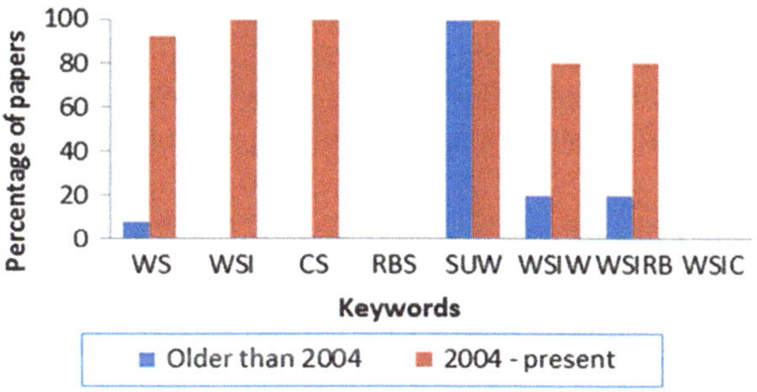

Fig. 7 Percentage of papers on each keyword in Web of Knowledge

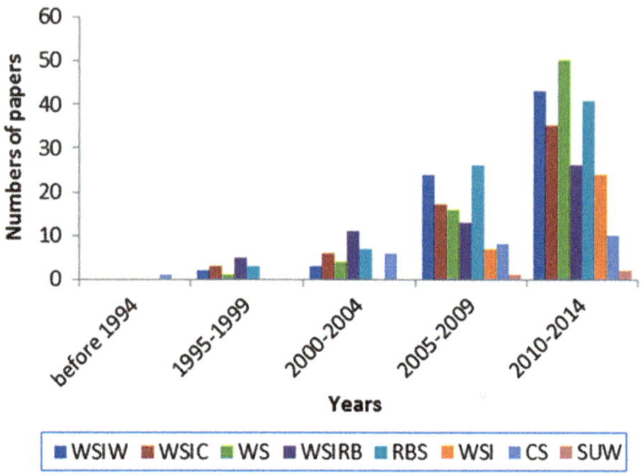

Fig. 8 Number of papers retrieved from Scopus for different search periods and keywords

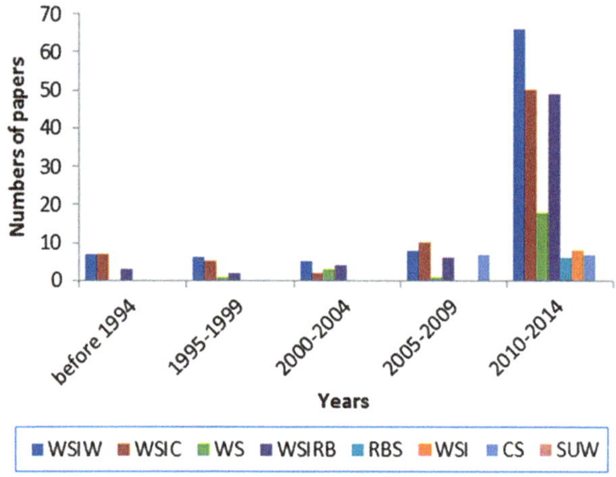

Fig. 9 Number of papers retrieved from ScienceDirect for different search periods and keywords

In the second phase of publication trend analysis for different keywords and databases, we divided whole period of search into five different periods (say 1991–1995, 1996–2000, etc.) with each period spanning 5 years. This analysis was done to check whether the trend of publications for small span periods was similar to the earlier analysis in which two periods (before 2004 and 2004–present) were used. The findings of this type of analysis are shown in Figs. 8, 9, 10, 11, 12, 13, and 14. This clearly shows that most of the publications on watershed sustainability were found for the latest period (i.e., 2010–2014). All keywords on watershed sustainability indicators had a higher number of retrievals for this period. Not surprisingly, the number of publications on watershed sustainability for the older period (before 1994) was very low, which indicates that not much research was conducted on watershed sustainability before 1994.

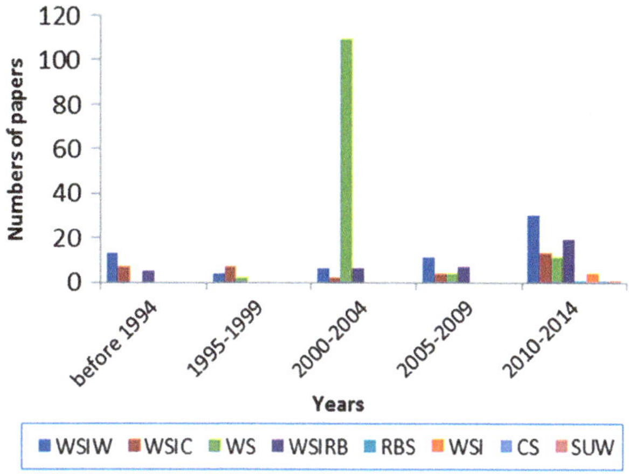

Fig. 10 Number of papers retrieved from Wiley Online for different search periods and keywords

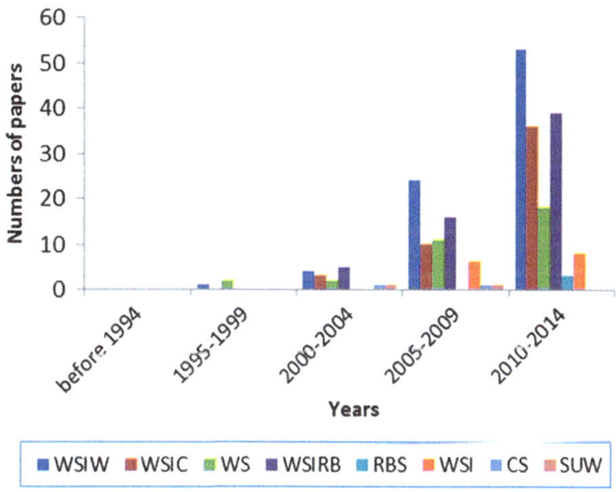

Fig. 11 Number of papers retrieved from ProQuest for different search periods and keywords

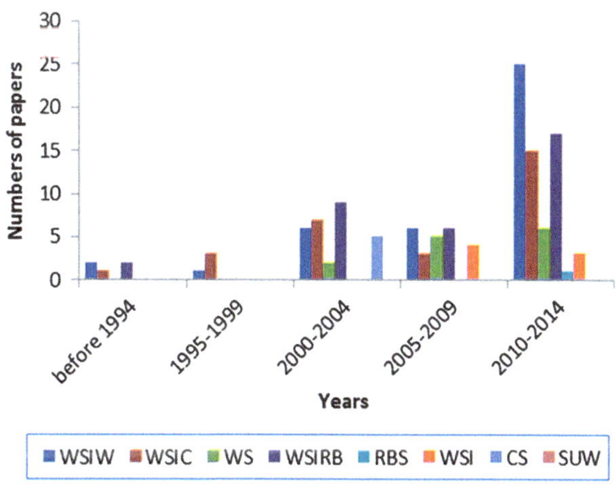

Fig. 12 Number of papers retrieved from SpringerLink for different search periods and keywords

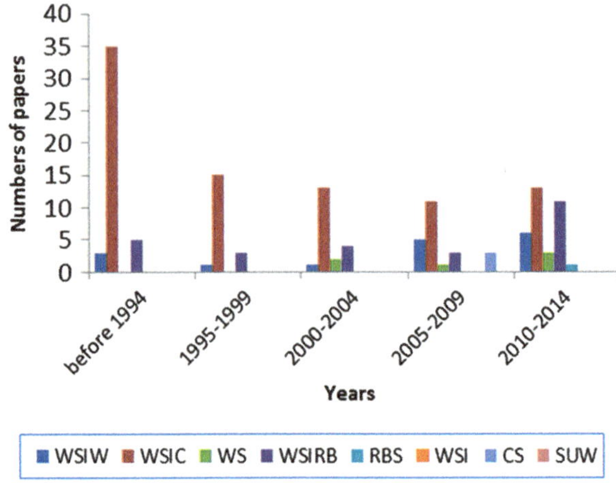

Fig. 13 Number of papers published in different years in Taylor Francis

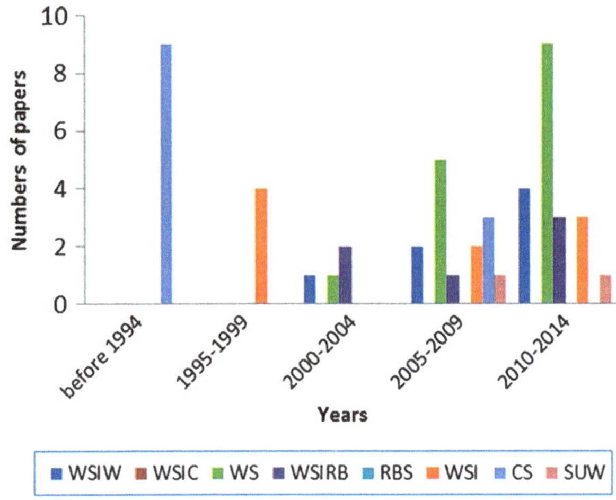

Fig. 14 Number of papers retrieved from Web of Knowledge for different search periods and keywords

4 Conclusion

The database research revealed that the trend of publications on watershed sustainability has increased sharply for the last 10 years compared to the period before 2004. We found that "WSI & watershed" was the most popular keyword in all databases. The second most important keyword was "WSI & catchment." On the other hand, the least popular keyword was "sustainable use of watershed" for all the searched databases. The most popular database that publishing more papers on watershed sustainability was SCOPUS followed by Science Direct. On the other hand, the least popular database publishing least number of papers on watershed sustainability was Web of Knowledge as only 31 papers were retrieved for all the keywords. We emphasize that the paper provides insight into the research work that has been completed on managing watersheds on a sustainable basis. However, it does not give any definite conclusions on the management of watersheds as only seven databases were searched for only a few keywords. In future, the number of keywords and databases may be increased to get complete research outputs and trends in watershed sustainability publications.

Acknowledgment This study was partly funded by the Ministry of Higher Education (MOHE), Malaysia, and the Universiti Teknologi Malaysia (UTM) under GUP Grant Tier-1 with Vot No. 08H43 and FRGS Grant Vot No. 4 F539. This study was also supported by the Asian Core Program of the Japanese Society for the Promotion of Science (JSPS).

References

1. P.A. DeBarry, *Watersheds: Processes, Assessments and Management* (Wiley, Hoboken, NJ, 2004)
2. K. Abdullah, Towards realising integrated river basin management in Malaysia. Integrated River Basin Management Report. Department of Irrigation and Drainage, Malaysia (2011). http://www.water.gov.my/component/content/article/19-river-management/20-integrated-river-basin-management-report?lang=en. Accessed 30 Mar 2014
3. World Commission on Environment and Development (WCED), *Our Common Future* (Oxford University Press, Oxford, 1987)
4. H.L. Chaves, S. Alipaz, An integrated indicator based on basin hydrology, environment, life, and policy: The watershed sustainability index. Water Resour. Manag. **21**, 883–895 (2007)
5. UNDP, *Human Development Report*, vol. 228, Lisbon, 1998
6. A.J. Jakeman, R.A. Letcher, S. Rojanasoonthon, S. Cuddy, *Integrating Knowledge for River Basin Management: Progress in Thailand* (Australia Centre for International Agricultural Research, Canberra, 2005)
7. C. Sullivan, Calculating a water poverty index. World Dev. **30**(7), 1195–1210 (2002)
8. Policy Research Initiative, Sustainable development briefing note. Canadian water sustainability index (2007), http://publications.gc.ca/collections/Collection/PH2-1-14-2007E.pdf. Accessed 30 Dec 2013
9. P. Lawrence, J.R. Meigh, C.A. Sullivan, *The Water Poverty Index: An International Comparison*. Keele Economic Research Papers 2003/18 and Center for Ecology and Hydrology (CEH), Wallingford, 2003
10. L.F. Ling, Comparison of Water Quality Index (WQI) between DOE Method and Harkin's Index, Master Thesis, Universiti Teknologi Malaysia, 2007
11. Z.M. Ali, F.A. Shahabuddin, Y.S. Shee, N.J.Z. Mamat, Water quality study at Skudai River 2002–2006, in *World Congress on Nature and Biologically Inspired Computing (NaBIC 2009)*, 2009, pp. 600–606
12. I. Mohamed, Water quality assessment of Sungai Skudai, B.Sc. Thesis, Universiti Teknologi Malaysia, 2011

Flood Management to Reduce Flood Hazards of Gumti River Using Mathematical Modelling

A.F.M. Afzal Hossain

Abstract The Gumti River originates from India, flows into the territory of Bangladesh and ultimately falls into the river Meghna. Each year several flash floods occur in this river during monsoon. Due to high flood level and heavy rainfall during monsoon, in the recent year serious flood and breaches occurred in the embankment of this river. Accordingly, the objective of the study was to investigate different scenarios for finding the most feasible option for reducing flood hazard and improving flood management in the Gumti basin using mathematical model. A total of 13 options have been studied. Model results of all options have been demonstrated to the field officials and stakeholders for social acceptability. Finally, one option has been recommended considering technical, social, economical and environmental aspects. Implementation of the recommended option will reduce flood hazards and flood damages in the Gumti Basin.

Keywords Flood management • Flashy River • Mathematical model • Technical alternatives

1 Introduction

The Gumti River originates from the Tipperan hilly region of India and flows into territory of Bangladesh through Katak Bazar border and ultimately falls into the river Meghna at Daudkandi (Fig. 1). In Bangladesh, the length of meandering Gumti River is approximately 83 km. The length of the embankment on left bank is about 91 km from Indian border to Eliotganj Bazar and 64 km of right bank (RB) embankment from Indian border to Dashkandi. In the remaining part of the right embankment, there is submersible embankment (low height) at places. The catchment area of the Gumti River is 2,173 km^2. Most of it lies in India, comprises of 44 % hilly area and remaining area are plain lands studded with low land and lakes.

A.F.M. Afzal Hossain (✉)
Institute of Water Modelling (IWM), Dhaka 1206, Bangladesh
e-mail: afh@iwmbd.org

S.H. Abu Bakar et al. (eds.), *ISFRAM 2014*, DOI 10.1007/978-981-287-365-1_3

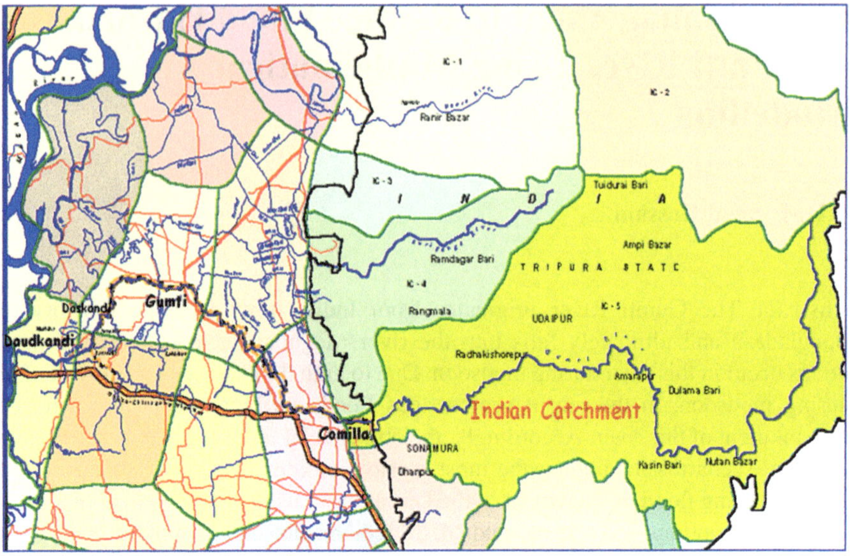

Fig. 1 Gumti River and its catchment area

Gumti is a flashy river in nature. Each year several flash floods occur in this river during monsoon, i.e. from May to August. During flash floods both discharge and velocity of the flowing water are tremendous, and water level rises very rapidly. Maximum discharge of the Gumti River is about 1,059 m^3/s with water level at Comilla is 13.59 m PWD and minimum discharge is about 8.00 m^3/s. It also carries a lot of sediments with it [1, 2].

2 Problem Statement

Severe flood occurred in Gumti basin in 1954, 1956, 1961, 1966, 1968, 1970, 1974, 1983, 1987 1993, 1997 and 1999. Total 120 numbers of breaches of embankments occurred since 1954 as per available record, out of which 85 occurred in Right Bank and 35 in Left Bank. Loop cutting, strengthening and protection works have now reduced the frequency of breach from 2 to 3 times per year before 1970 to once after 2–3 years.

In the recent year serious flood and breaches occurred in the embankment due to high flood level and heavy rainfall. Due to breach of the embankment in 1999, the inundated area was about 890 km^2 and crop damages area was about 350 km^2. The total estimated damage was about US$12.988 million. Now breach occurrence places are concentrated mainly in 20–30 km of right embankment. Breaches of embankments occur mainly due to sudden rise of water level in the river and heavy seepage through weaker sections of the embankment. However, major causes of breaches are as follows: (1) Trees, bushes, homesteads, sugar cane, paddy cultivation and crossroads in the flood plains obstruct the flood flow of the river. (2) Movement of the thalweg of the river very near to the embankment causes erosion of the embankment and consequently breach occurs. (3) Embankment was constructed with the silts and sandy materials and was not compacted onto the designed level. (4) Spot holes due to heavy local rainfall and rat holes in the embankment at the riverside are potential places for breach in the embankment [3–5].

3 Objective

The flood management of the Gumti River is very complex, which requires a knowledge-based solution. It is widely acclaimed that modelling efforts could provide a logical and amicable solution to these long-standing problems. A tool capable of explaining the flows of the Gumti system is of prime need in designing appropriate measures for improving the flood management of the area. Accordingly, the main objective of the study was to investigate different scenarios for finding the most feasible option for reducing flood hazard and improving flood management in the Gumti basin using mathematical model.

4 Study Components

Study activities involved in representing the physics of the project area into the model and then proceeding with model simulations in order to simulate different options with the model. For conducting this study, an approach has been adapted with due emphasise on data collection. Data collections are needed for model development, calibration and validation. However, different components of the study have been enumerated below: (1) Hydrometric, hydro-meteorological and cross section data collection. (2) Development of digital elevation model (DEM) based on secondary data. (3) Model development, calibration and validation. (4) Application of model and analysis of model result. (5) Impact assessment considering socio-economic and environmental aspects. (6) Formulation of a flood management plan [5–7].

An intensive hydrometric and hydro-meteorological data collection programme was carried out for the monsoon (June–August) 2002 and 2003 for model development, calibration and validation. Series of cross sections, channel connectivities, structure locations, etc. were procured for model development. The land topography was also collected from 1:16,000 water development maps prepared by FINN Map authorities in 1989 based upon aerial photographs and field surveys done in 1987. DEM for the study area has been developed from this topographic data. Stage-discharge rating also has been developed for the station Comilla at Chainage 5.59 km of Gumti River using measured discharge from 1997 to 2003. Frequency analysis of the available observed data at Comilla has been carried out to select the design year for different return period. Fifty-year flood event has been selected as design event. In this study, year 1983 has been determined as design event year.

5 Application of Model

Various technical alternatives have been formulated in consultation with stakeholders to resolve the existing flooding problems of Gumti River. The development alternatives were studied using calibrated and verified model against the monsoon of 1983, i.e. 50-year flood event since this year has been determined as the design event year. Total 13 nos. of options have been studied and are given in Table 1. In each option re-sectioning of existing embankments to some extent has been considered to protect the embankments from overtopping at 50-year design flood.

Table 1 Options for flood management of Gumti River

Sl. no.	Option	Description of option
1	Option 0	Base condition, i.e. existing situation: without any intervention
2	Option 1	Increase conveyance capacity by re-excavation of Gumti River: (1) Dredging up to 45 km. (2) Dredging depth: 1.5–2 m and design width: 100–125 m
3	Option 1A	Increase conveyance capacity by optimised re-excavation of Gumti River: (1) Dredging from 20 to 30 km. (2) Dredging depth: 1.5–2 m
4	Option 2	Increase conveyance capacity by reducing roughness of the floodplain of Gumti River
5	Option 3	Loop cut at three locations of the Gumti River: (1) Loop cut at Balibari, Ch. 37.70 km. (2) Loop cut at Muradnagar, Ch. 49.30 km. (3) Loop cut at Dorirampur Ch 56.50
6	Option 4	Without existing bridges of the Gumti River
7	Option 5	Release of excess flow from Gumti to flood plain by emergency diversion at 4 locations: (1) Spillway 1 at Ch. 9.5 km, sill level; 10.5 m, depth = 1.0 m, width = 15.0 m. (2) Spillway 2 at Ch. 13.5 km, sill level; 10.75 m, depth = 1.0 m, width = 20.0 m. (3) Spillway 3 at Ch. 18.28 km, sill level; 9.85 m, depth = 1.0 m, width = 15.0 m. (4) Spillway 4 at Ch. 26.9 km, sill level; 9.5 m, depth = 1.0 m, width = 15.0 m
8	Option 6	Release of excess flow from Gumti in Ghungur and Buri Nadi through bypass: (1) 64.8 m^3/s passes through Ghungur River. (2) 93.7 m^3/s passes through Buri River
9	Option 6A	Release of excess flow from Gumti in Ghungur and re-excavation of Ghungur river: (1) 102.8 m^3/s passes through Ghungur River
10	Option 7	Combination of Option 1A and Option 3
11	Option 8	Combination of Option 1A, Option 3 and Option 6A
12	Option 9	Combination of Option 3 and Option 6A
13	Option 10	Expansion of the flood width between the embankments of the Gumti River

6 Analysis of Model Results

Results of model simulations under different options have been analysed in three different ways as follows: (1) Hydrographs of water levels. (2) Long Profile. (3) Flood maps under different options. The hydrographs indicate the river stages and flow volume of key locations throughout the period, while long profile provides river stages for the whole reach at a specific time. Flood maps provide qualitative assessment of depth of flooding at a specific time. Comparison of hydrographs indicates the changes in respect of water level/discharge throughout the period of interest, while the flood maps under base and option condition indicate the extent of relief of flooding in the study area. MIKE11-GIS facilities with land level data from the DEM have been applied to develop the flood maps.

From the results of the model simulation of the above options, it has been observed that in each option re-sectioning of existing embankments to some extent is required to protect the embankments from overtopping at 50-year design flood.

Out of the above options, strengthening and raising of embankments, Loop Cuts and a diversion from Gumti to Ghungur (Option 0, Option 3, Option 6A and Option 9) are recommended for further study on engineering, social, economic and environmental aspects to select the best option to be recommended for implementation.

However, the existing and design crest level for the different reaches of the Gumti River for Option 9 is given in Table 2, and long profile for this option is given in Fig. 2.

Table 2 Existing and design crest level of Gumti River for Option 9

	Existing crest level				Diff of crest level (design-existing)	
Chainage (km)	Left embank	Right embank	Water level (m)	Design crest level (WL+ 0.9 m freeboard)	Left embank	Right embank
6.59	15.023	14.92	14.373	15.273	0.250	0.353
10.26	14.589	14.3	13.989	14.889	0.300	0.589
16.1	13.589	13.588	13.57	14.47	0.881	0.882
20.5	13.175	12.886	12.984	13.884	0.709	0.998
30.31	12.090	12.42	11.486	12.386	0.296	−0.034
40.26	10.850	10.212	10.002	10.902	0.052	0.690
50.35	9.331	9.611	8.308	9.208	−0.123	−0.403
55.35	9.281	8.880	8.134	9.034	−0.247	0.154

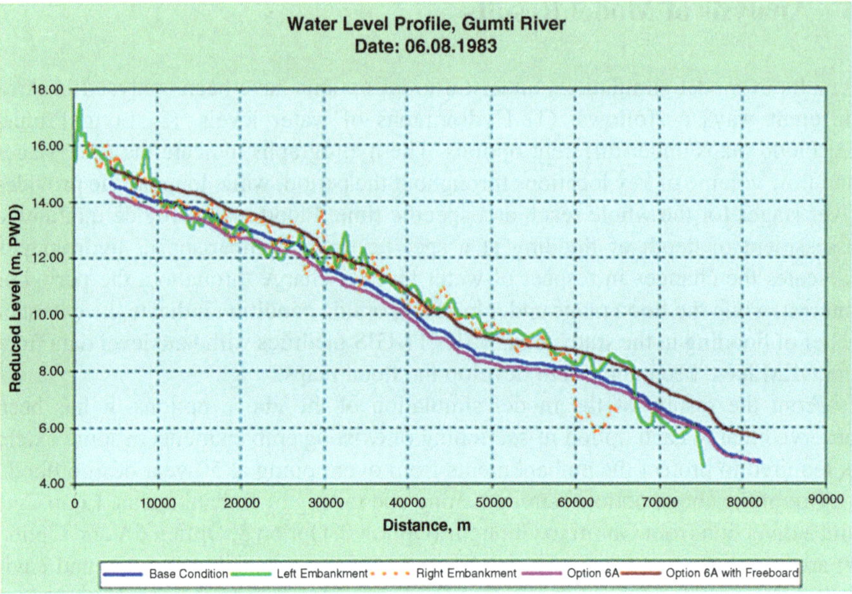

Fig. 2 Water level profile of Gumti River for Option 9

7 Impact Assessment

7.1 Socio-environmental Aspects

A socio-economic survey has been conducted to assess social impacts of different options. The people identified various options of development. People rank re-excavation/dredging of Gumti River as first followed by raising and strengthening of both banks of Gumti River. Other than these methods, most of the people suggested for loop cutting during survey and focus group discussion. The above activities, if executed properly, are supposed to bring overall positive benefits on environment.

7.2 Engineering Aspects

Costs of several alternatives have been estimated. Comparative statement of the four recommended feasible options is presented in Table 3.

Table 3 Comparative statement of feasible options

Sl. no	Options	Cost in million Tk. (1$=77.56 Tk.)	Impact		
			Economical	Social	Environmental
1	Option 0 (base condition)	442.83	Economically viable	Socially accepted	No adverse impact is anticipated
2	Option 3A	467.74	Economically viable	Socially accepted	No major adverse impact is anticipated
3	Option 6A	510.40	Economically viable	Socially accepted	No major adverse impact is anticipated
4	Option 9	635.61	Economically viable	Socially accepted	No major adverse impact is anticipated

8 Formulation of Flood Management Plan

Considering technical, social, economic and environmental aspects, **Option 9** (Fig. 3) is finally selected for implementation. Cost of Option 9 though is higher than other recommended options, it will provide extra safety in case of flood exceeding the design flood.

The regulator at diversion will be used to divert floodwater in excess of design flood. Although it has been suggested to operate the bypass in extreme emergency situation, bypass can also be used for supplementary irrigation during or at the end of monsoon season at the time of draught or when necessary. In addition to the implementation of finally recommended option, following issues also need to be considered for reducing flood hazard and improving flood management.

8.1 Flood Monitoring

Flood monitoring is an essential part of a better flood management programme. At present, BWDB maintains only three water level gauge stations in Gumti river: one gauge is at Comilla and other two are at Kangshanagar and at Daudkandi. Pier of four bridges between Comilla and Companiganj should be used as permanent water level gauges. It has been observed that almost every breach occurred when water level at Comilla is higher than 11.75 m, PWD. This level can be taken as critical level; as above this level, risk of embankment failure is increased. Discharge at different water level at Comilla can be computed using stage-discharge rating. The critical levels of other four locations are as follows: (1) Chandpur bridge at 7.60 km, Critical level: 11.55 m. (2) Palpara bridge at 10.80 km, Critical level: 11.30 m. (3) Kalikapur bridge at 28.90 km, Critical level: 8.95 m. (4) Companiganj bridge at 43.35 km, Critical level: 6.65 m.

When water level approaches towards the critical level, gauge reader will frequently inform the office. BWDB staff from office will warn the local people

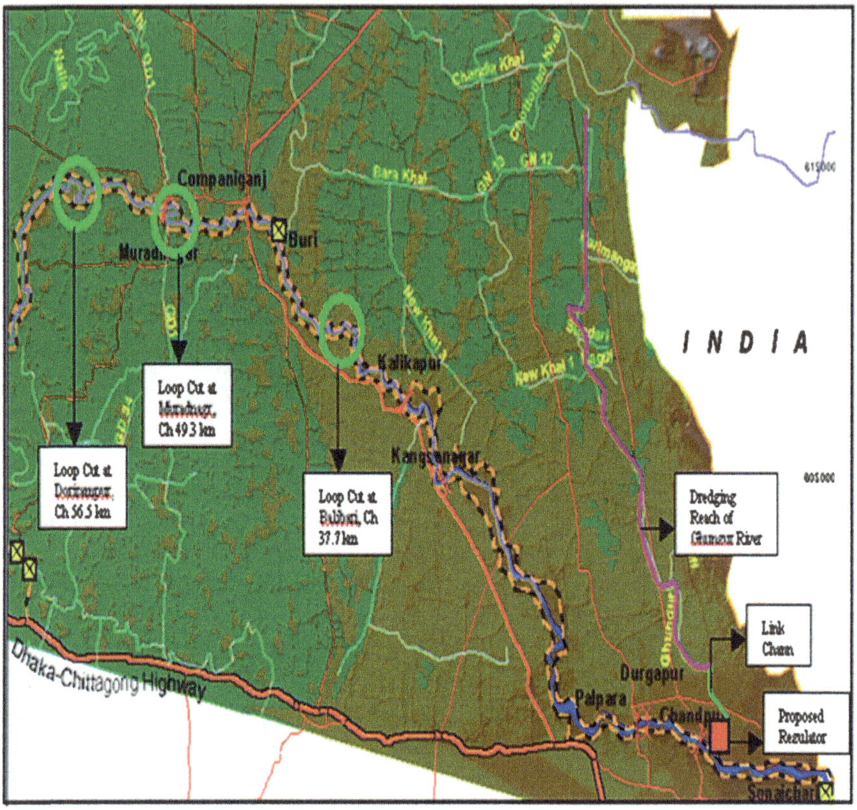

Fig. 3 Final recommended option

and local administration, who may take emergency actions including evacuation of the local inhabitants from flood-prone areas. Local embankment protection vigilances committee including local people, BWDB, Army and public representative should be formed. The Committee will be able to timely detect the weaker portion of the embankment and identify the works to be done for repair of the embankments during emergency situation [6–8].

9 Operation and Maintenance of Embankments

The maintenance of embankments is proposed in two different routines, one for yearly maintenance and one for periodical re-sectioning. The principal maintenance activities can be summarised as follows:

- Maintenance of the embankment as per designed level.

- Operation and maintenance of existing and proposed regulators on the embankments of both banks as per design criteria and needs (Three regulators on the left embankment and two regulators on the right embankment).
- Maintenance of existing and proposed 60 km road pavement on the embankments on both banks.
- Maintenance of the protective works for river erosion.
- Operation and maintenance of irrigation inlets and rams. Seepage under embankment is one of the major causes for breach. Seepage occurs through the irrigation inlet pipes of LLP poorly done by a group of farmers and loosely filled up after use in dry months. These activities must be stopped. Removal of earth by public cut in the countryside of the embankments for fish culture and other purpose is other major cause. These activities must also be restricted.

Removal of trees, bushes, homesteads, etc. on the embankments as well as in the floodplain of the Gumti River is also recommended. Plantation of trees should be followed as per national guideline.

10 Real-Time Flood Forecasting Model

Real-time flood forecasting model would help in attaining more confident decisions during flood period. In the long run, integration of flood forecasting with existing flood management model is also necessary. Based on the flood forecasting model, flood forecasts are made for a sufficient time period in advance in order to enable the local authorities to warn for possible evacuation of the local inhabitants of flood-prone areas at the basin, which will minimise the loss of life and property due to flood.

11 Conclusion

Flood management models can be used effectively for people's participatory-based flood management studies to reduce flood hazards. It was observed that through adopting a series of options and demonstrating their effective outcome to the beneficiaries save a lot of efforts in resolving conflicting issues and complex situations regarding flood management. It is very beneficial for the decision makers to observe the impact of any intervention or modification in the project infrastructure through the use of models.

References

1. Bangladesh Water Development Board and Netherlands Consultant, Feasibility Study on Gumti South Project (1964)
2. BWDB, Flood Damage Report of the Breach at Etbarpur (1999)
3. A.F.M. Afzal Hossain, Management of irrigation and drainage systems using mathematical modelling. J. Appl. Water Eng. Res. **1**(2), 129–136 (2013)
4. Danish Hydraulic Institute, MIKE11—A Modelling System for Rivers and Channels, Version 4.0 for Windows 95 and Windows NT, August 2001
5. Howard Humphrey's & Partners in association with Hydraulic Research Limited, Hydrological and Morphological Studies, Gumti–Titas & Atrai Basins, April 1990
6. Institute of Water Modelling, Final Report, Feasibility and Model Study for Rehabilitation of Flood Control Embankment on Both Banks of Gumti River, June 2004
7. Mott Macdonald Limited in association with Nippon Koei Company Ltd., House of Consultant Ltd. and Desh Upodesh Ltd. Gumti Phase-II Sub Project Feasibility Study (1993)
8. UNDP, World Bank, BWDB, South East Region Water Resources Development Programme (FAP-5), Planning Unit 10, Gumti Phase-1, August 1993

Community Awareness and Preparedness Towards Flood in Kuantan, Pahang

J. Jani, S.R. Mohd Nasir, and N.S. Md Zawawi

Abstract Flood occurrence has become a national issue as it threatens life and property and disrupts social and economic activities. In Malaysia, more than 85 water basins were identified as prone to flood. Although flood occurs frequently and is the most severe of all disasters in Malaysia, the awareness and preparedness of communities in facing flood are still low. Often it brings greatest damage to the flood victims as they may not be well prepared in facing flood. Questionnaire survey was conducted targeting flood victims in area of Kuantan. This study aims to analyse the level of awareness and preparedness among communities towards recent flood in Kuantan, Pahang. The successfully collected questionnaire is 31 numbers with a response rate of 62 %. This chapter also discusses the effectiveness of existing National Flood Disaster Relief in managing relief operations before, during and after flood events.

Keywords Flood occurrence • Awareness and preparedness • National flood disaster relief

1 Introduction

Flood is a major natural disaster in Malaysia. It is due to regular and heavy rainfall from October to March every year. It is also due to the increased number of urbanised areas with inadequate drainage that cannot support the amount of flow during the heavy rainfall [1]. The combination of natural and human factors has produced different types of floods such as monsoon, flash and tidal flood [2].

In Malaysia, National Security Council (NSC) and other agencies such as Department of Irrigation and Drainage (DID) are the responsible parties for coordinating all relief operations before, during and after flood events. Flood forecasting and warning system such as Short Messaging System (SMS), fax and online

J. Jani (✉) • S.R. Mohd Nasir
Flood Control Research Center, Faculty of Civil Engineering, UiTM, Shah Alam, Malaysia
e-mail: janmaizatulriah@salam.uitm.edu.my

N.S. Md Zawawi
Faculty of Civil Engineering, UiTM, Shah Alam, Malaysia

S.H. Abu Bakar et al. (eds.), *ISFRAM 2014*, DOI 10.1007/978-981-287-365-1_4

website were used for early warning system of upcoming flood. In addition, government also organised awareness programmes to increase awareness and preparedness among community towards flood disaster. These programmes would assist in raising levels of preparedness, aid response and recovery activities. Moreover, it could also reduce the impacts and trauma associated with flooding.

In this study, assessment of community awareness and preparedness in facing flood event as well as effectiveness of existing flood management committee in coordinating relief operations is carried out. The findings will give some overview of flood victims' preparedness level and effectiveness of flood management approach.

2 Study Area

Sungai Isap in Kuantan, Pahang, was selected for this research area due to its historical data of flood recurrence, its rainfall recorded and its recent flood disaster. During the end of 2013, the worst flood occurred at the East Coast of Malaysia that left Kuantan, Pahang, almost paralysed. More than 38,000 people were evacuated [3]. Among other states in Malaysia, Kuantan records the highest number of evacuees. Moreover, the extensive rainfall in high intensity is the main reason of flooding in Kuantan. The maximum rainfall amount was recorded at Kuantan rainfall station of 175 mm/day. The average amount of rainfall in the flood affected area during December 2013 to January 2014 exceeded 800 mm [7].

3 Methods

Questionnaire was designed as five sections. Section 1 is the demographic profile of the respondent; section 2 is the awareness and knowledge on flood, section 3 is on community preparedness; section 4 is on the government and insurance compensation and section 5 is on evacuation process and satisfaction on flood relief centre. Various scales were used in the questionnaire such as Likert Scale, Yes/No answer and multiple choice of answer. In section 3, Likert scales of five levels were used to obtain respondents' view on their preparedness. The scales are as follows: (1) unexpected, (2) no plan, (3) planned, (4) halfway executed plan, (5) fully executed plans. While in section 5, the following scales were used: (1) Very Dissatisfied, (2) Dissatisfied, (3) Neutral, (4) Satisfied and (5) Very Satisfied. Fifty Questionnaires were distributed to flood victims who resided in Sungai Isap, Kuantan, Pahang from 14 to 19 April 2014. Thirty-one questionnaires were successfully collected and the respond rate is 62 %.

Analysis on findings was conducted using descriptive analysis that describes the perception against the demographic profile.

4 Results and Discussion

From the results gathered, Fig. 1 shows that 90 % of the respondents are living in one-storey houses, while 6 % in two-storey houses. Eighty percent of the respondents' houses are made of concrete and bricks as shown in Fig. 2. In Fig. 3, 41 % of the respondents gained income of about RM 1,001–3,000 per month, and 38 % of the respondents gained RM 3,001–5,000 per month. Figure 4 shows that 32 % of the respondents had lived in that area for about 10–19 years and 26 % had lived there for more than 20 years. This shows that most of the residents are from a lower income group and are permanent residents.

In this study, the results were divided in two parts which include: (1) awareness and preparedness of the community and (2) community's perception on flood relief operations.

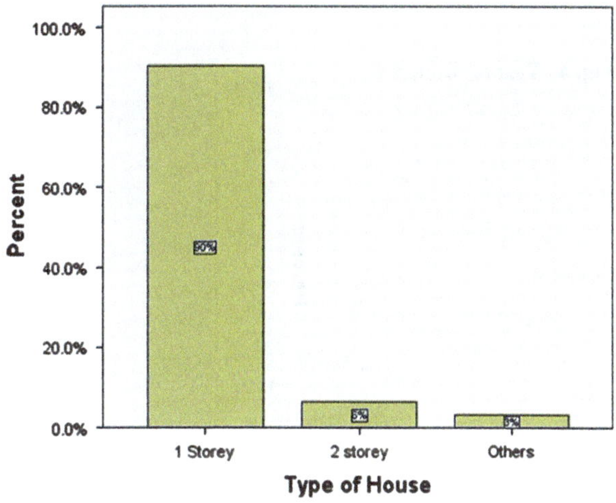

Fig. 1 Type of house

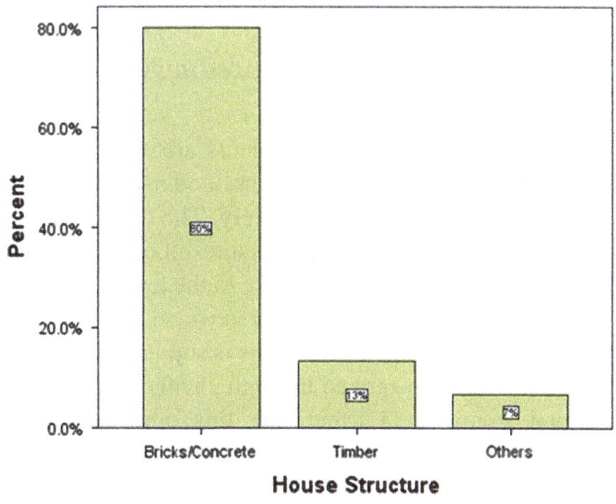

Fig. 2 House structure

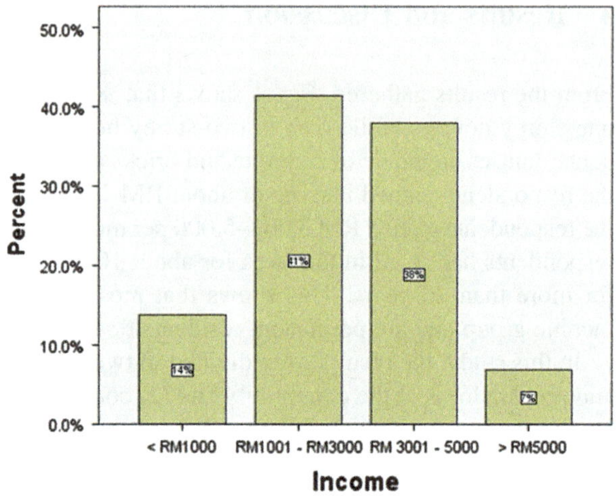

Fig. 3 Respondents' income

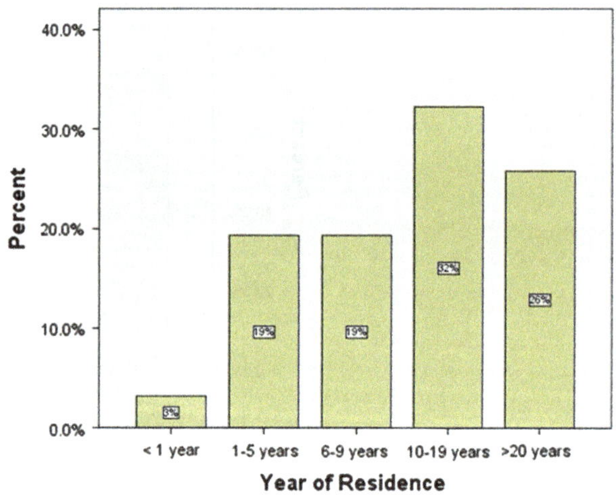

Fig. 4 Year of residence

4.1 Awareness and Preparedness of the Community

In general, 81 % of the respondents are aware that their residential area is flood prone area and 94 % are aware that flood reoccurred five times for the past 5 years as shown in Figs. 5 and 6, respectively. The flood is due to the increase in the number of residents in Kuantan, which is a result of the increase in the number of developments. On top of that, the increase in the level of sea water results in the insufficient level of river basin. To make it even worse, the blocked drainage obstructs the water from flowing to nearest river. This is supported by [1] the hydrological processes in basins where water moved through changes due to development will be the reason of flood occurrence. Furthermore, this indicates that the residents choose to stay at that area even if they knew it is a flood prone area.

Fig. 5 Community awareness on flood risk

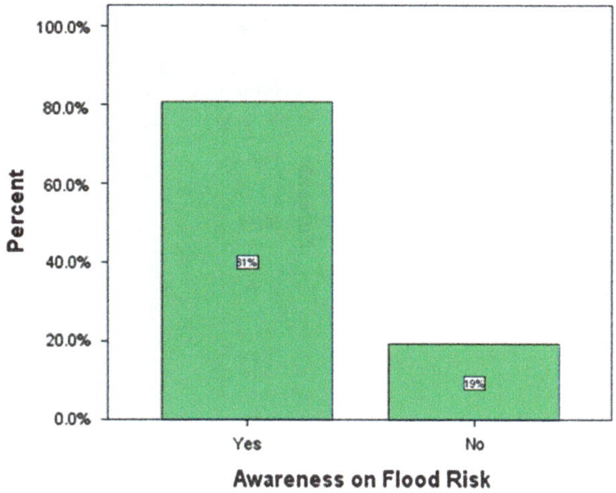

Fig. 6 Flood occurrence

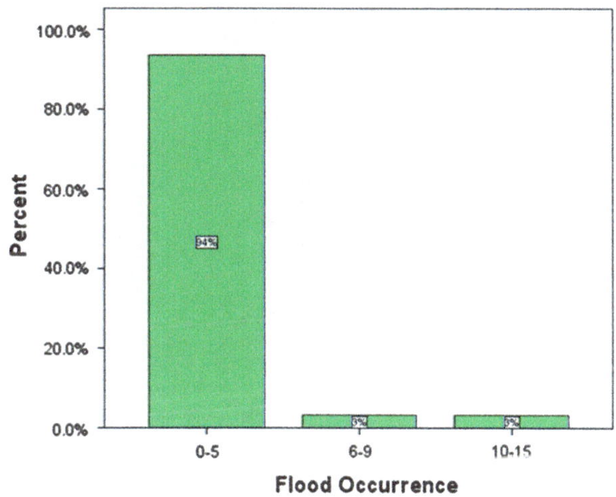

Flood forecasting and warning system has been developed by Department of Irrigation and Drainage Malaysia (DID). Two units of flood forecasting models known as Linear Transfer Function Model (LTFM) have been applied at Sungai Pahang and Sungai Kelantan, while the dissemination system such as warning siren, SMS, telephone, fax and online website will help in informing the sign of upcoming disaster [4]. Flood warning system is classified as warning given through media, siren, local authority/rescuers, JKKK and others. Figure 7 shows that 55 % of the respondents are aware of the existence of flood warning system in their area. This shows that there is lack of information on the flood warning system in that area. The community should be made aware of the warning system by the responsible parties. In Malaysia, the main organisations that are responsible for declaring early warning are Malaysian Meteorological Department (MMD), Malaysian Department of

Fig. 7 Flood warning system

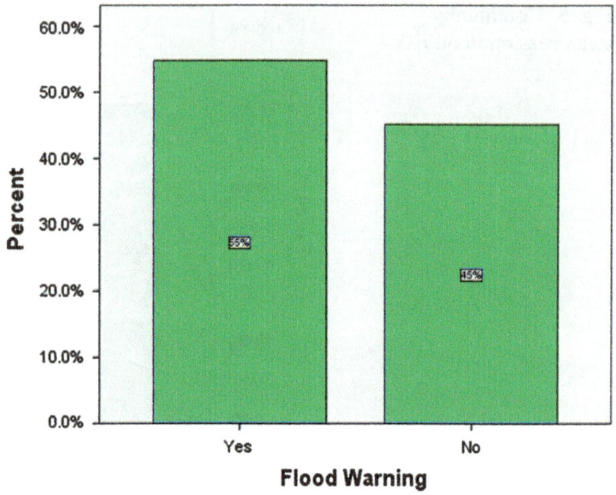

Fig. 8 Way(s) of receiving flood warning

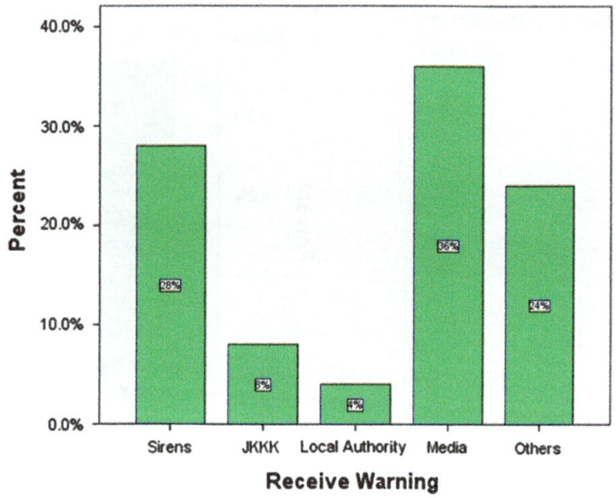

Irrigation and Drainage (DID) and Malaysian Centre for Remote Sensing (MACRES). MMD has specialty in weather forecasting, DID in Telemetry System, which includes inventing Flood Forecasting Models and Info Banjir, while MACRES has a unit named National Disaster Data and Information Management System (NADDI) that will give information about disaster [4]. As shown in Fig. 8, 36 % of the respondents received information on occurrence of flood through the media, whereas 28 % were warned by the siren.

The awareness of community on the occurrence of flood will reflect their level of preparedness. The high level of awareness will increase victims' level of preparedness in facing flood events and may respond to flood warnings promptly and efficiently. It also can reduce the potential for damage and loss of life. Meanwhile, low degree of flood awareness may lead to not fully appreciate the importance of

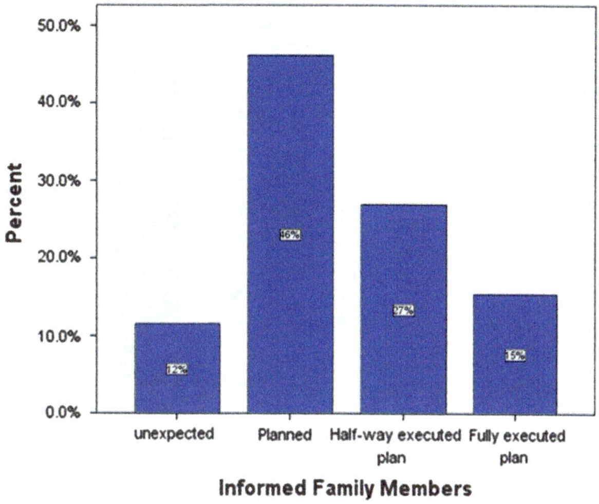

Fig. 9 Informed family members on flood preparedness plan

flood warnings and flood preparedness and consequently suffer greater personal and economic losses [5].

Awareness programmes and activities need to be conducted in order to increase the level of disaster awareness and disaster preparedness among citizens. The activities and programmes will educate citizens and let them realise how important it is to be prepared during disaster events [4]. General preparedness includes preparing emergency planning documents and contacts. Making some checklists will help in creating emergency plan. Other than that, gathering basic medical information about each family member will help in getting medical assistance during emergency events. Make a checklist of important items that are needed in order to get ready for an emergency case [6].

Figure 9 shows 46 % of the respondents has planned for flood preparedness and also has informed their family members of the plan. However, only 20 % of the respondents fully executed their plan, whereas 30 % of the respondents do not have any flood emergency plan as shown in Fig. 10. This indicates that the flood victims have poor awareness on flood preparedness and, thus, require more awareness programmes and activities from the government.

Disaster kits such as first aid kit, blanket, water and food are among the important needs during the flood occurrence. In Fig. 11, it shows that 12 % of the respondents managed to execute their emergency plan, whereas 35 % of the respondents do not have any plan in preparing flood disaster kit. This also shows that the flood victims have lack of planning for flood disaster. Thus, most of them rely on the flood evacuation team and the flood relief centre.

The negative impacts of flood are it can threaten lives, disrupt social and economic activities and also destroy properties. Other than that, it can cause distress, and recovery can be costly either to individuals or government [7]. This is supported by [8] who in their research indicated that impacts of flood such as loss

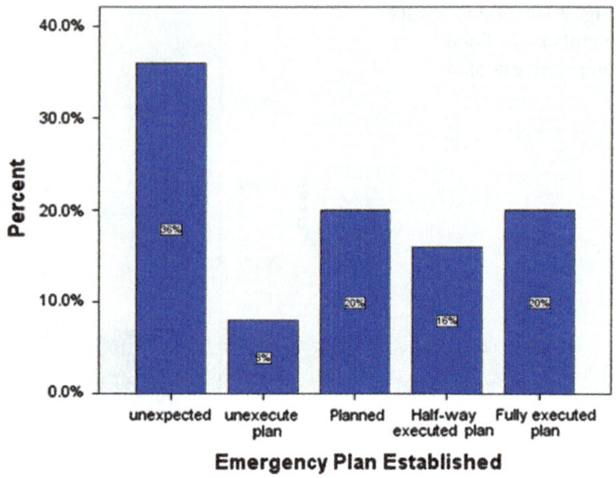

Fig. 10 Establishment of flood emergency plan

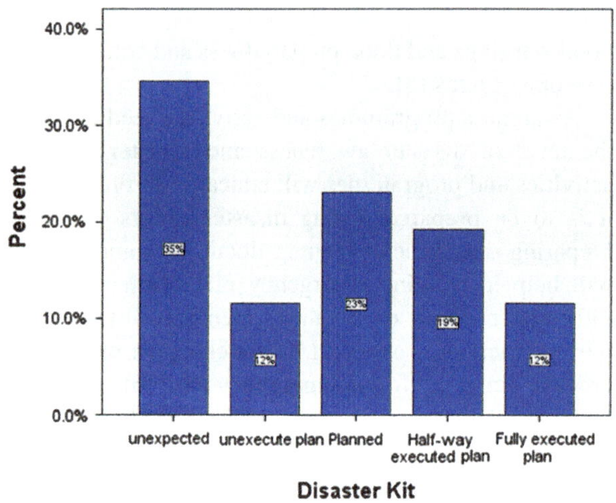

Fig. 11 Flood disaster kit

of life, damage of physical infrastructures (houses, schools, road and culvert) and damages on productivity of agriculture land will put a hold to victims' livelihood earnings. In this study, it shows that 24 % of the respondents have life insurance, 38 % of the respondents have insured their properties and 38 % have insured other valuable assets (Fig. 12). This indicates that most of the flood victims had prepared themselves in terms of financial, property and asset coverage.

It was recorded that more than 30,000 people were evacuated to 38 relief centres in Kuantan during the recent flood in December 2013 [7]. Figures 13 and 14 show that 93 % of the respondents are aware of the flood relief centres' location and 91 % of the respondents communicated using mobile phone during the evacuation process in order to get the latest information on the evacuation operations, respectively.

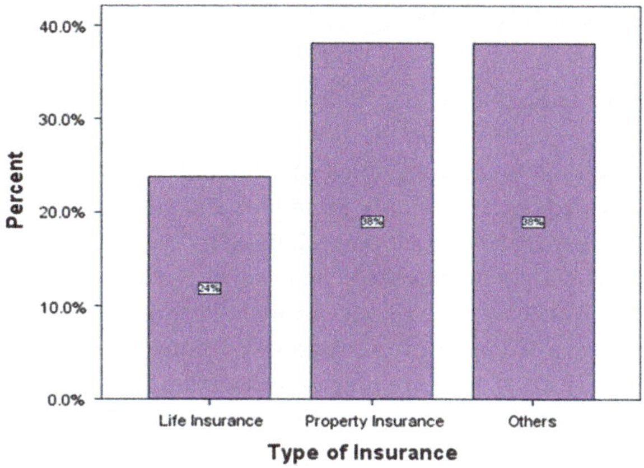

Fig. 12 Type of insurances

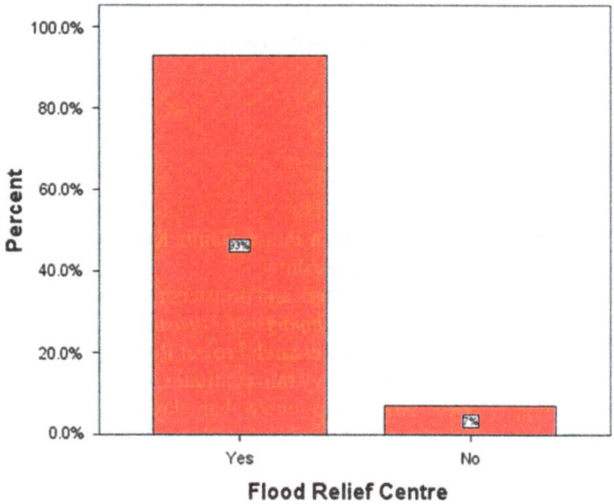

Fig. 13 Flood relief centre location

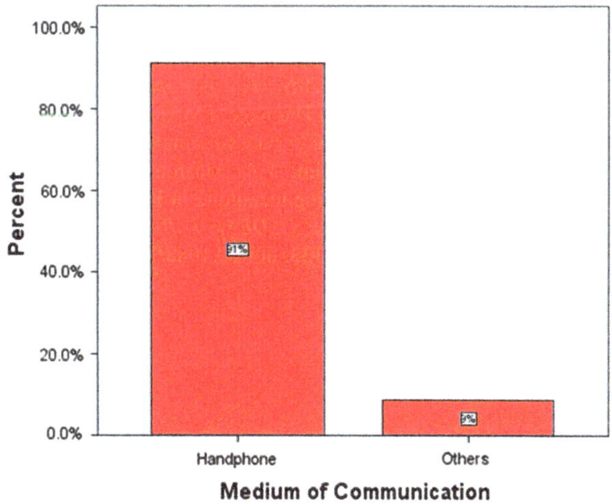

Fig. 14 Medium of communication

5 Conclusions

The findings conclude that the communities are still lacking of awareness and preparedness during the flood occurrence. In addition, the facilities at the flood relief centre are considered inadequate and require improvement. The findings of this study on community awareness and preparedness level towards flood and level of satisfaction on the flood relief operations could assist the responsible authorities in planning more comprehensive preparedness strategies towards future flood occurrence. Various flood awareness programmes can be performed in order to increase the level of preparedness among various communities. Therefore, community could be more prepared especially in dealing with before, during and after flood event.

Acknowledgement The authors gratefully acknowledge the support from the Water Resources and Environmental System (WRES) Division and Flood Control Research Centre (FCRC) Faculty of Civil Engineering, UiTM, Shah Alam.

References

1. A. Yunus, Air Pasang Bukan Punca Banjir Kuantan. Astro Awani, 7 Dec 2013, http://www.astroawani.com/budget/news/show
2. N.W. Chan, Impact of disasters and disasters risk management in Malaysia: The case of floods, in *Economic and Welfare Impacts of Disasters in East Asia and Policy Responses*, ed. by Y. Sawada, S. Oum. ERIA Research Project Report 2011-8 (ERIA, Jakarta, 2012), pp. 503–551
3. The Malay Mail Online, Heavy rain continues, over 38,000 flood victims throng relief centres in Pahang, 7 Dec 2013, http://www.themalaymailonline.com/malaysia/article/heavy-rain-continues-over-38k-flood-victims-throng-relief-centres-in-pahang#sthash.2Cv4US7p.dpuf
4. C.M. Umar, *National Seminar on Social Economic Impacts of Extreme Weather and Climate Change*. Rapporteur report, Ministry of Science, Technology and Innovation, 2007
5. M. Jempson, Johnstone River flood Study Final Report, vol. 1 of 2 (2003), http://www.cassowarycoast.qld.gov.au/c/document_library/get_file?uuid=b8cb70fa-3111-4fc8-91db-02a64504ea96&groupId=1422210
6. M. Dorasamy, M. Raman, S. Muthaiyah, M. Kaliannan, Disaster preparedness in Malaysia: An exploratory study, in *Recent Advances in Management, Marketing, Finances*, 2009, pp. 19–29
7. Department of Irrigation and Drainage, Flood management programmes and activities, 18 Mar 2014, http://www.gov.my/our-services-mainmenu-252/flood-mitigation-mainmenu
8. A.A. Ainullotfi, A.L. Ibrahim, T.A. Masron, Study on integrated community based flood mitigation with remote sensing technique in Kota Bharu, Kelantan, in *8th International Symposium of the Digital Earth (ISDE8). IOP Conference Series: Earth and Environmental Science*, vol. 18 (2014), 012133. doi:10.1088/1755-1315/18/1/012133

Perspective of Stakeholders on Flash Flood in Kuala Lumpur

Siti Rashidah Mohd Nasir and Sya'ari Othman

Abstract This chapter describes the perspective of stakeholders on flash flood in Kuala Lumpur. This chapter includes how to identify major causes of flash flood, major effects of flash flood, and suitable mitigation options to reduce flash flood in Kuala Lumpur. The questionnaire survey has been used to collect data from targeted respondents. The targeted respondents are government, consultant, and academic institutions who are involved in flood projects. The result shows that the major causes of flash flood in Kuala Lumpur include blocking of drainage channel, increase of surface runoff, uncontrolled urbanization, poor drainage maintenance, and inadequate drainage system. The results also show the major effects of flash flood in Kuala Lumpur are damage to property, damage to business premises, damage to roads, interruption to the utility services, and interruption to the transportation services. All three groups of the respondents agreed that the suitable mitigation includes frequent maintenance of drainage system, construction of flood diversion, control of land use activities, installation of flood bypass, limitation of deforestation, and enhanced and stringent acts on flash flood. The findings of this study will assist the local authorities to select suitable mitigation plans.

Keywords Flash flood • Flood mitigation • Kuala Lumpur

1 Introduction

Urbanization is a process that cannot be separated from the development of a country because it is like a catalyst for positive development. Unplanned development and rapid urbanization process can cause certain area to undergo a change of originality of earth and can lead to disasters. One of the disasters mentioned is flooding. Throughout the countries around the world, especially developing countries experience the flooding problem. For an example, Malaysia is one of the developing countries that constantly face the flooding problem. Flooding is the most important natural disaster, which impacts 4.9 million people and inflicts

S.R. Mohd Nasir (✉) • S. Othman
Flood Control Research Center, Faculty of Civil Engineering, Universiti Teknologi MARA, Shah Alam, Malaysia
e-mail: sitir015@salam.uitm.edu.my

S.H. Abu Bakar et al. (eds.), *ISFRAM 2014*, DOI 10.1007/978-981-287-365-1_5

damage worth several millions every year. In late 2006 and early 2007, Malaysia was rocked by severe flooding [1]. The Northeast Monsoon brought heavy rain through series of extreme and persistent storm causing devastating floods in southern region of Peninsular Malaysia especially to Kota Tinggi in December 2006 and January 2007 [2]. This unexpected type of disaster gives serious impact when there is loss of life. In addition, two students became victims of the flood that occurred in Machang, Kelantan, in early year of 2013 [3].

Flash flood has been a frequent disaster occurring in Kuala Lumpur. Flash flood mostly occurred in the areas of Sungai Klang, especially Kampung Dato Keramat, Jalan Yap Kwan Seng, Kampung Bharu, Jalan Dang Wangi, Jalan Tun Perak, Jalan Melaka, and Jalan Tun Razak in the center of the city as reported by Department of Irrigation and Drainage. As reported through the news, thousands of road users were stuck in the traffic jams on several areas in Kuala Lumpur, which were submerged by a flash flood when they were traveling back to home [4]. Besides that, there were massive traffic jams reported from Sungei Besi to Kuala Lumpur as a result of flash flood in Jalan Tun Abdul Razak [5]. This flood phenomenon is due to several factors.

Thus, this chapter studies the causes and effects of flash floods in Kuala Lumpur. This study also enables to reduce the occurrence and effects of flash flood in Kuala Lumpur by identifying the suitable mitigation plans. In addition, this study is conducted to identify the causes of flash flood even though there are flood mitigations implemented in Kuala Lumpur since 2003, such as Stormwater Management and Road Tunnel (SMART) Project.

2 Literature Review

2.1 Flood

Flood can be categorized as one of the hydrological disasters that occur worldwide. Department of Irrigation and Drainage Malaysia defines flood as a water body that rises and overflows into the land that is not usually covered [6]. They also defined flood as a stream or drainage that overflows into land as effects of channel obstruction. Flood also can be defined as an overflow of the immense quantity of water that submerges land [1]. Malaysia can be categorized as one of the developing countries and one of the "Tiger Economy" of Asia that is rapidly emerging in industrialization and urbanization [7].

Flash Flood A flash flood is a direct response to rainfall with a very high intensity. Flash flood is characterized by the occurrence of fast-flowing water usually after heavy and localized high-intensity rain [8]. This ability of water to move large objects such as cars and trees has a great impact and can cause losses [9]. This type of flood caused more deaths compared with other flood types [10].

2.2 Causes of Flood

Flooding can be caused by various factors. There are lots of researches and study has been done to identify the causes of flood disaster. The causes of flooding can be by natural factor, man-made, or any combination of these two factors [11]. In Malaysia, as reported by Department of Irrigation and Drainage Malaysia, there are two major categories of causes of flood in Malaysia, which are natural causes and human activities [6]. In the first category of causes, which is natural causes, the causes of flooding are heavy rainfall caused by flash floods and heavy rains, resulting in stagnant water [6]. Besides that, Department of Irrigation and Drainage also categorized the second category of causes of flood, which is man-made category, including dumping of solid waste into the river, sediment from land clearance and construction, increased impervious area, and obstacles in the river [6].

Causes of Flood due to Urbanization Human life cannot get away from the process of development and urbanization as people always want advancement and want to enjoy surviving. Based on previous study, the occurrence of the flooding is influenced by commercial development and urbanization of many areas [12]. The process of urbanization affects the size and probability of floods to occur and may expose flood hazard to public [13].

Causes of Flood due to Land Use Sustainable development project will result in the physical alteration of natural land to impermeable layer or a low rate of water infiltration into the soil [14]. Changes in land use will change the infiltration rate used and the design of the drainage system in the first place. In addition, the new infiltration rate, previously drainage system, may not be able to accommodate the new flow due to increased runoff [14]. In addition, Konrad [13] also stated that increased land usage also can increase the peak floods discharge.

Causes of Flood due to Inadequate Drainage System Drainage design that does not take into account the future developments in the environment aspect will not be able to function properly in the event of an increase in surface runoff due to development activities, and this will cause an overflow of water from drainage system and subsequent flooding, which means the drainage system is not perfect or is filled with debris, and suspended plant will affect the functioning of the drainage system that acts as a water flow path [14]. With these constraints, the speed and direction of flow rates through it will decrease and cause flooding [14].

2.3 Effects of Flood

The effects of flooding can be categorized as tangible and intangible effects [15]. Tangible effects can be referred as effects of flooding that can cause monetary loss [16]. In other words, tangible effects can be described as direct effects of flood that contact with human and property [17]. This means that the effects have an impact on the physical assets such as human and infrastructure. Meanwhile, intangible effects refer to the flood impact that cannot cause monetary loss [18]. Indirect effects are like interruption of production of goods and services that happen due to interruption of transport, utility services, and markets [17].

Effect to Structures During flooding, many structures have been totally submerged by flood water. Many structures that have been built for living and businesses are destroyed. Based on previous study, the study has identified the structures that are destroyed including flood wall and bridges [2]. There is acceleration to damage of building and structures if the water of flooding remains stagnant for a couple of days [19].

Effect to Roads Flood disaster also can give a serious impact to roads infrastructures [17]. Based on previous study, the effect of flood is damage and destruction of roads, railways, and bridges [20]. This is because water, the first enemy to bitumen, penetrates into the deeper layers of bitumen roads and causes disturbed subgrade layer [21]. Therefore, the incident can result in broken roads and potholes that can be seen along the roads [2].

Effect to Human Lives Loss of life including human lives is another critical impact that can happen due to flood disaster. In the last decade of the twentieth century, floods have caused the death of about 100,000 people, and more than 1.4 billion people are killed every year over the world [22]. Based on previous researches, flood impact that annually occurs in Malaysia including loss of life and property damage [16]. The loss of human and animal life was higher when high velocity of floodwater submerged the area [19].

Economic Effects The economy losses are considered as the second major impact of flooding in Nigeria [23]. Economy aspect flooding can affect the gross domestic product (GDP) and gross national product (GNP) [7]. They also stated that when faced with flood disaster, Malaysia reduced its GDP and the annual development budget for the country [7].

2.4 Flood Mitigations

In Malaysia, flood management solution focuses on the need to provide immediate flood relief work and to implement flood mitigation projects to catch up with the sharp rise of urban development and sustainable development [6]. Based on previous study, the study has mentioned about flood management measures that are divided into two measures which are structural and non-structural measures [24].

Structural Measures Structural measures are basically through physical implementation. In Malaysia, the type of structural measure strategies in flood mitigation that has been implemented is shown in Table 1.

Table 1 Structural measures of flood mitigations

Creative measures	Structural components
Regulate water level	Barrage
	Tidal gate
	Flap gate
	Lock
Store and controlled release of water	Dam
	Reservoir
	Detention and retention ponds
	Bund
	Inlet and outlet storage
	Spillway
Improve flow efficiency and controlled flows	River channelization
	Improvement works
	Flood wall
	Weir
	Control gate
	Culverts
Re-route the flood flow	River diversion
	Diversion channels
	Flood bypass (open channel and tunnel) Intake structures
	Bridge and culvert
Forced removal (non-gravity flow) of excess water	Pumps and pump house
	Inlet and outlet structures (DID, 2013)
Delineation and separation	Polder, ring bund, linear bund

Non-structural Measures Generally, a non-structural measure is about administering for flood mitigation. Non-structural measures are all about planning, organizing, policy-setting, coordinating, simplifying, creating awareness, supporting, and strengthening the community to face the threat and effects of flooding [2]. This type of management includes flood forecasting, flood warning, and flood plain zoning [25]. In Malaysia, the types of non-structural measures are:

- Compliance with laws, acts, and guidelines
- Awareness campaign and education programs
- Flood forecasting and warning
- Catchment management
- Flood proofing
- Flood hazard mapping
- Relocation

3 Methodology

Questionnaire is constructed in order to obtain the perception and opinion about the study. In order to develop the questionnaire, previous studies on flood disaster were reviewed. Three sets of Likert scales consisting of five points were used as follows: (1) very low contributing to (5) very high contributing; (1) seldom to (5) always; and (1) strongly disagree to (5) strongly agree.

Questionnaire Design The questionnaire has been divided into four sections as:

Section A: Background of respondent
Section B: Causes of flash flood
Section C: Flash flood effect to society, assets, and utilities
Section D: Mitigate flash flood in Kuala Lumpur

This study covers the areas that include Cheras, Segambut, Bandar Tun Razak, Lembah Pantai, Seputeh, Kepong, Batu, Wangsa Maju, Titiwangsa, and Setiawangsa district. The target respondents are engineers, authorities, government agencies, consultants, and academic institutions. The questionnaires were distributed by hand, postal, and online survey.

Sample Size Sampling size is important for a researcher to conduct a study because the better the sampling, the most accurate the data. In determining the sample as a representative of the opinions, table for determining sample size [26] is used. Sample sizes for this study are as follows: government (30), consultants (55), and academic institutions (30).

Survey Return Rate The researcher has managed to distribute 130 sets of questionnaires to all the targeted respondents; however, only 124 were successfully returned within 3 weeks and another six questionnaires were unreturned due to misplacement of questionnaire. From 124 questionnaires, two have been discarded for data analysis due to incomplete data. Thus, only 122 questionnaires were used for this study with the respond rate of 94 %.

Data Analysis Relative Important Index (RII) [27] will be conducted in order to interpret a set of data from Sections B and C of the questionnaire. Using RII assists in the ranking the importance level of the findings. Thus, the major causes and effects of flash flood can be determined. Besides that, Chi-squared test was used to analyze data in Section D in order to determine the suitable mitigation option to reduce flash flood in Kuala Lumpur according to the type of organization which are government, consultant, and academic institutions. The data collected were analyzed and considered as a normally distributed data.

4 Results and Discussions

4.1 Causes of Flash Flood

Table 2 shows the ranking of causes, where RII was used to identify the significant causes of flash flood in Kuala Lumpur. The causes of flash flood were ranked based on the values of RII. From the ranks assigned, the five major causes of flash flood as perceived by respondents are as follows: (1) blocking of drainage channel by rubbish (RII = 0.910), which is the highest frequency of very high contributing; (2) increase of surface runoff due to urbanization (RII = 0.910), which is the second highest frequency of very high contributing; (3) uncontrolled urbanization in squatters area (RII = 0.890); (4) low drainage maintenance (RII = 4); and (5) inadequate drainage system (RII = 0.848).

From these findings, it shows that blocking of drainage channel by rubbish that contributed to the cause of flash flood in Kuala Lumpur represents the major causes of flash flood. Lack of regular clearance of drainage system was listed as the main cause for flood disaster [1]. This shows that in Kuala Lumpur and cities in Malaysia, there is lack of regular clearance of drainage which influences the blocking of drainage channel with rubbish [1].

Increase of surface runoff due to urbanization is ranked second among the 15 causes of flash flood in Kuala Lumpur. Increase in surface runoff is mainly due to urbanization [28].

Table 2 also shows that uncontrolled urbanization ranked third by the respondents. This is due to the rapid and uncontrolled urbanization which has significantly caused flood risk [16].

Table 2 Ranking of causes of flash flood

Causes of flash flood in Kuala Lumpur	RII	Rank
Uncontrolled urbanization (squatters area)	0.890	3
Increase of surface runoff due to urbanization	0.910	2
Heavy rainfall	0.764	13
Prolonged precipitation	0.793	8
Blocking of drainage channel by rubbish	0.910	1
Inadequate drainage system	0.848	5
Poor drainage maintenance	0.854	4
Increasing of improper land use	0.797	7
Increasing of deforestation (low infiltration)	0.836	6
Upstream development did not take into account the downstream drainage capacity	0.789	10
Loss of flood storage	0.779	11
Inadequate river capacity cause river overflow	0.766	12
Obstruction in the river flow	0.790	9
Improper management of guidelines and control	0.736	14
Inadequate flood response and awareness	0.708	15

From Table 2, it shows that poor drainage maintenance is ranked as fourth, and inadequate drainage system is ranked fifth among the 15 causes of flash flood in Kuala Lumpur. These findings are supported by other studies [1] that identified poor drainage system as an issue which contributes to flood disaster. If the drainage system that is filled with debris and suspended plant is not properly and regularly maintained, this leads to blocked drainage, where surface runoff may not properly be discharged and may cause flash flood.

4.2 *Effects of Flash Flood*

Based on the ranking as in Table 3, the five important effects of flash flood in Kuala Lumpur as perceived by respondents are as follows: (1) damage to property (RII = 0.946); (2) damage to business premises (RII = 0.882); (3) damage of roads (RII = 0.836); (4) interruption to the utility services (RII = 0.811); and (5) interruption to the transportation services (RII = 0.769).

Table 3 shows that damage to property is the most frequent one faced by the society among the effect factors. Based on previous study [20], the loss of property is the perennial problem of flooding. Damages to business premises were ranked by respondents as second, where flash floods have effects on physical infrastructure, which involves shops and houses [29].

From Table 3, damage of roads is ranked third of flood effects in Kuala Lumpur. According to previous studies [2], the identified flood impacts and damages include damages to road, which can be clearly visible after the flood occurred. In addition, the water as the first enemy to bitumen penetrates into the deeper layers of bitumen roads and causes disturbed subgrade layer [21].

Interruption of the utility services was ranked as fourth among the flash flood effects in Kuala Lumpur. This finding was supported by previous studies [17] which have identified the interruption of utility services as one of the flood effects and as

Table 3 Ranking for effects of flash flood

S. No	Effects of flash flood in Kuala Lumpur	RII	Rank
1	Damage to property	0.946	1
2	Losses of life	0.672	10
3	Damage to business premises	0.882	2
4	Lead to economy losses	0.762	6
5	Interrupt the utility services	0.811	4
6	Interrupt the transportation services	0.769	5
7	Damage of roads	0.836	3
8	Disruption of communication	0.762	6
9	Cause diseases	0.692	9
10	Effects production of goods	0.695	8

an indirect category. This includes utility services such as electricity, water supply, and network affected during and after flash flood [30]. Lastly, the fifth rank of flash flood effects is the interruption of transportation. This leads to massive traffic congestion and is supported by previous studies [4], where there were thousands of road users stuck in traffic congestions on several areas in Kuala Lumpur, which were submerged by a flash flood when they were traveling back home. It is also stated that traffic congestion which affects daily life is also one of the effects of flood [30].

4.3 Flash Flood Mitigation Measures

Chi-squared test was used to see the significance between the mitigation options to reduce flash flood and type of organization. By using Chi-squared test, the null hypothesis for this section is constructed automatically which is the distribution the mitigation options to reduce flash flood is the same across categories of the type of organization.

If p-value is below 0.05, the null hypothesis for this significant value is rejected [31]. On the other hand, the variables which have a significant value of more than 0.05 indicate that the three types of organizations agreed that the mitigation options can reduce flash flood.

From Table 4, frequent maintenance of drainage system ($p > 0.05$) received similar perception from all three groups. This suggested that maintenance schedule of drainage system should be emphasized as a non-structural measure and include as a way to reduce flash flood in Kuala Lumpur. This is supported by previous study [2], which stated that the types of non-structural measures include planning, organizing, policy-setting, and coordination.

Table 4 shows that construction of flood diversion ($p > 0.05$) also received similar perception across the groups. This mitigation option reflected in the study states that the construction of flood diversion is feasible only if the intended diversion channel is not too long [16].

Table 4 shows that all three groups agreed that controlling land use activities ($p > 0.05$) is one method in flood mitigation plan. In this case, Land Conservation Act 1960 was introduced by federal government that focuses on flood control by regulation of land use.

However, the control of land use is not strictly followed by state government [16].

Enhancement and strengthening of the Act on flash flood also received similar perception across all groups ($p > 0.05$). This indicates that Land Conservation Act requires attention from all parties involved by taking into consideration the physical, social, and environment factors in reducing flood [32].

Installing flood bypass is another mitigation option ($p > 0.05$) which received similar perception among the three groups.

Table 4 Perspective of stakeholders on flash flood mitigation option

Flash flood mitigation option	Type of organization	Mean rank	Chi-square	Sig. (p-value)
Retention and detention pond	Government	47.94	14.44	0.010
	Consultant	61.39		
	Academic	75.73		
Limitation to urbanization	Government	61.35	31.386	0.000
	Consultant	47.99		
	Academic	89.12		
Frequently maintenance of drainage system	Government	54.03	2.445	0.294
	Consultant	64.03		
	Academic	64.07		
Deepen the river and drainage	Government	47.82	8.737	0.013
	Consultant	63.16		
	Academic	72.27		
Construction of flood diversion	Government	56.26	3.029	0.220
	Consultant	59.91		
	Academic	70.15		
Controlling land use activities	Government	70.18	3.378	0.185
	Consultant	59.75		
	Academic	56.08		
Enhance and strengthen the Act	Government	60.76	0.027	0.986
	Consultant	61.95		
	Academic	61.35		
Installing pump house	Government	77.85	10.465	0.005
	Consultant	55.77		
	Academic	56.25		
Installing gross pollutant trap	Government	56.08	6.176	0.046
	Consultant	58.20		
	Academic	73.82		
Installing flood bypass	Government	55.18	2.729	0.256
	Consultant	61.36		
	Academic	68.32		
Limitation of deforestation	Government	70.10	3.907	0.142
	Consultant	60.80		
	Academic	53.95		
Awareness campaign and education program	Government	75.42	15.202	0.001
	Consultant	63.48		
	Academic	43.10		
Stringent laws and policies	Government	76.92	31.049	0.000
	Consultant	67.89		
	Academic	32.58		
Increase the flood response and awareness	Government	84.39	28.513	0.000
	Consultant	59.66		
	Academic	41.60		

(continued)

Table 4 (continued)

Flash flood mitigation option	Type of organization	Mean rank	Chi-square	Sig. (p-value)
Planting vegetation activities	Government	68.21	15.965	0.000
	Consultant	68.41		
	Academic	40.52		

According to previous study [17], flood bypass must be included in structural flood mitigation works. Installing flood bypass can reduce flood level at certain spots, but it will increase the flood level at other locations [33].

Lastly, mitigation options to reduce flash flood in Kuala Lumpur that received similar perception from the three groups are limitations of deforestation ($p > 0.05$). Based on previous study [34] stated that deforestation activities affect the environmental changes that can result in future climate changes and flooding. Moreover, if the controlling and limiting deforestation activities are conducted, it can reduce the probability of flooding and climate change that may occur in the future [34].

The findings show that there exist a difference of opinion in answering the questionnaire due to the respondent's background of knowledge and experience. This is also due to different roles and responsibilities each organization has in flood mitigation projects.

5 Conclusions

From the perception of these three groups, the suitable mitigation options were identified in reducing flash flood in Kuala Lumpur. These findings may assist the government and local authorities in selecting the mitigation plans to reduce flash flood in Kuala Lumpur to ensure the government in achieving Kuala Lumpur Structure Plan 2020 and National Physical Plan (RFN-2) goals. These findings also may assist local authorities and society for the preparedness of flash flood.

References

1. M.A. Mohit, G.M. Sellu, Mitigation of climate change effects through non-structural flood disaster management in Pekan Town, Malaysia. Procedia Soc. Behav. Sci. **85**, 564–573 (2013)
2. A. Shafie, A case study on floods of 2006 and 2007 in Johor, Malaysia, Fort Collins, CO, 2009
3. Berita Harian, Ibu negara dilanda banjir kilat, April 2011
4. Utusan, KL, Selangor dilanda banjir kilat, October 2013
5. New Straits Times, Flash flood in Kuala Lumpur—New Straits Times, 2013

6. Department of Irrigation and Drainage, Flood Management—Programme and Activities, December 2013
7. N.W. Chan, N.A. Zakaria, A.A. Ghani, T.Y. Lian, Integrating official and traditional flood hazard management in Malaysia, in *1st International Conference on Managing Rivers to the 21st Century: Issues and Challenges*, 2004
8. E. Opolot, Application of remote sensing and geographical information systems in flood management: A review. Res. J. Appl. Sci. Eng. Technol. **5**(10), 1884–1894 (2013)
9. Union of Concerned Scientists, After the storm: The hidden health risks of flooding in a warming world, Climate Change and Your Health, March 2012
10. S.N. Jonkman, Global perspectives of loss of human life caused by floods. Nat. Hazards **34**, 151–175 (2005)
11. S. Fan, Uncertainties in flood disaster management. Watershed Manag. 81–89 (2003). doi:10.1061/40706(266)8
12. A. Chelan, I. Alan, A. Uğurlu, Causes and effects of flood hazards in Turkey, in *International Congress on River Basin Management*, 2007
13. C.P. Konrad, Effects of urban development on floods, US. Geol. Surv. Sci. Chang. World. **76** (3) (2003)
14. L. Wei, Analysis for the flood affected by region tidal, 2005
15. D.J. Parker, C.H. Green, P.M. Thompson, *Urban Flood Protection Benefits—A Project Appraisal Guide* (Gower Technical Press, Aldershot, 1987)
16. N.W. Chan, *A Contextual Analysis of Flood Hazard Management in Peninsular Malaysia* (School of Geography and Environmental Management, Middlesex University, 1995)
17. P.F. Gyriech, *Sustainable Management of Flood Disasters in the Upper East Region* (Department of Planning, College of Architecture and Planning, Ghana, 2011)
18. D.W. Pearce, *Environmental Economics* (Longman, London, 1976)
19. K.C. Yee, Study of flood wall system for a river rehabilitation and flood control method, Universiti Teknologi Malaysia, 2006
20. A.K. Etuonovbe, The devastating effects of environmental degradation—A case study of the Niger delta region of Nigeria, TS06J – Hydrography and the Environment, 2011
21. A. Azwani, Studies of flash floods in Johor Bahru: Air Molek case study area, Universiti Teknologi Malaysia, 2007
22. M.N. Ezemonye, C.N. Emeribe, Flood characteristics and management adaptations in parts of the Imo river system. Ethiop. J. Environ. Stud. Manage. **4**(3), 56–64 (2011)
23. M. Adetunji, O. Oyeleye, Evaluation of the causes and effects of flood in Apete, Ido local government area, Oyo State, Nigeria. Civil Environ. Res. **3**(7), 19–27 (2013)
24. M.A. Tariq, N.V. Giesen, Floods and flood management in Pakistan. Phys. Chem. Earth (2012). doi:10.1016/j.pce.2011.08.014
25. A.N. Khan, B. Khan, S. Qasim, S.N. Khan, Causes, effects and remedies: A case study of rural flooding in District Charsadda, Pakistan. J. Manage. Sci. **7**(1) (2013)
26. R.V. Krejcie, D.W. Morgan, *Determining Sample Size for Research Activities* (National Emergency Training Center, Emmitsburg, MD, 1970)
27. M. Sambasivan, Y.W. Soon, Causes and effects of delays in Malaysian construction industry. Int. J. Project Manage. **25**(5), 517–526 (2007)
28. C.C. Wing, Managing flood problems in Malaysia. Buletin Ingeniur, 1971
29. S.A. Khan, *Flood Management in Bangladesh* (Environmental Engineering, Tampere Polytechnic University of Applied Sciences, 2007)
30. M. Faiz, Hydrological characteristic and parameters for rainfall event at Northern, Central and Peninsular Malaysia, 2010
31. S. Sarantakos, *A toolkit for quantitative data analysis using SPSS [covers versions 10 through to 15]* (Palgrave Macmillan, Basingstoke, 2007)

32. B.E. Montz, E. Gruntfest, Flash flood mitigation: Recommendations for research and applications. Global Environ. Chang. B Environ. Hazards **4**, 15–22 (2002)
33. Department of Irrigation and Drainage, *Flood Management*, vol. 1 (2009)
34. R. Few, R.S. Kovats, M. Ahern, F. Matthies, *Floods, health and climate change: a strategic review* (Tyndall Centre for Climate Change Research, 2004)

Part II
Flood Modelling

GIS Application in Surface Runoff Estimation for Upper Klang River Basin, Malaysia

R. Suzana, T. Wardah, and A.B. Sahol Hamid

Abstract Estimation of surface runoff depth is a vital part in any rainfall-runoff modeling. It leads to stream flow calculation and later predicts flood occurrences. Geographic information system (GIS) is an advanced and opposite tool used in simulating hydrological model due to its realistic application on topography. This chapter discusses on calculation of surface runoff depth for two selected events by using GIS with curve number method for Upper Klang River basin. GIS enables map intersection between soil type and land use that later produces curve number map. The results show good correlation between simulated and observed values when R^2 is more than 0.7.

Keywords Surface runoff • Geographic information system • Curve number method

1 Introduction

Malaysia is a tropical country that experiences rain almost one whole year around, and for some regions, it is heavier. The West Coast of Peninsular is subjected to localized and convective storms generated by the inter-monsoon seasons. Convective storms are extremely variable in time and space and can produce very intense rainfall rates that lead to flooding. With annual rainfall volume ranging from 200 to 500 billion cubic meters (bcm), Malaysia is considered to have an equatorial climate with high humidity.

With heavy rainfall received, flood is a norm in Malaysia. To date, 15 major flood events have been recorded since 1926. Those floods were triggered by the rain in monsoon season which are divided into four main types of monsoon rain: Southwest monsoon, Northeast monsoon, and two inter-monsoons that carry intense rain especially during late evening.

R. Suzana (✉) • T. Wardah • A.B. Sahol Hamid
Flood Control Research Center, Faculty of Civil Engineering, Universiti Teknologi MARA, 40450 Shah Alam, Selangor, Malaysia
e-mail: suzana_ramli@yahoo.com

S.H. Abu Bakar et al. (eds.), *ISFRAM 2014*, DOI 10.1007/978-981-287-365-1_6

Rain is one of the most influential subjects in hydrology. The prediction or the estimation of the rain has substantial contribution to water resources engineering and management. Problems such as water supply and flood can be analyzed by studying the rainfall behavior as well as the rainfall amount calculation. The use of weather radar for rainfall estimation for Malaysian tropical condition is getting more attention especially for flood forecasting purposes [1–6].

There are many purposes of hydrological modeling. Most importantly, hydrological model can aid the engineers and hydrologist in water management plans. The plans include construction of dams, reservoirs, river mitigation works that need estimation, and prediction of water quantities and also flow rate. From the model, we can also determine the increase in stream flow due to new development with current geographic information system (GIS) knowledge being integrated in the model. After all the calculations, predictions, and simulation work, hydrological model can help to assess the best management practice to solve many engineering problems.

GIS simplified the analysis of digital elevation models (DEMs), soils, land use, topography, and others. Therefore, the application of GIS in hydrological modeling is getting more crucial and delivers better workmanship by integrating GIS and rainfall-runoff models. According to [7], GIS technology can be used for data management, storage together with visualization, besides model creation, execution, and calibration. Meanwhile, [8] added that GIS can also be used as a medium to store, analyze, and integrate spatial and attribute data concerning runoff, slope, drainage, and infrastructure. [9] stated that the application of GIS in hydraulic modeling can fasten the processing work in a river system with huge amount of data. He used GIS as a tool to develop flood risk map for Selangor River in Malaysia and produced the automatic flood map delineation for a big river within short time interval.

2 Case Study and Data Collection

The Klang River Basin in Selangor has experienced flooding for more than a decade. Since it is located in the midst of one of the busiest areas in Malaysia—between Selangor and Kuala Lumpur—it suffers from urbanization and a high population. The size of the Klang River Basin is 1,288 km^2 with a total stream length of approximately 120 km. Located at 3°17′N, 101°E to 2°40′N, 101°17′E, it covers areas in Sepang, Kuala Langat, Petaling Jaya, Klang, Gombak, and Kuala Lumpur. For the purpose of rainfall-runoff estimation, only the upper part of Klang river basin is included in this study with area of 468 km^2. The upper section is under jurisdiction of three local authorities which are Ampang Jaya, Ulu Langat, and Gombak. 50 % of the upper land had been developed for residential and commercial areas as well as for industrial zone. The remaining land is reserved for forestry. According to [10], upper part of Klang river basin comprises of some hills with vertical slopes and rising to 1,430 m with maximum elevation of 15–20 m above sea level. The area receives annual mean rainfall around 1,900–2,600 mm, especially at the foothills. Little agricultural activity can be found in the upper section of Klang river basin with most of the land covered with forest and urban areas. Meanwhile, the average pervious percentage of the upper catchment is reported to be around 66 % with forest cover comprising 33.6 % of the total basin area [11].

The hydrological data such as rainfall, river discharge, and water levels were obtained from the Hydrology Division of the Drainage and Irrigation Department (DID) of Malaysia. Nine rain gauge stations have been selected in this study, which are located near the catchment area of Upper Klang River Basin, Selangor. In this study, hourly rainfall data from the DID covering from January 2009 until December 2009 were used. Table 1 shows the nine stations included in the study. Malaysian Remote Sensing Agency (MACRES) supplies the latest satellite image for the land use mapping using SPOT-5 (270/343/8) Pansharp All Bands. The satellite image is converted to digital image using ERDAS IMAGINE by GIS to produce land use map. For the purpose of obtaining curve number, CN for this model, land use map will be used together with soil map at scale 1:25,000 (Figs. 1 and 2)

Table 1 Sub-catchments in KRB

Subbasin no.	Station's name
1	JPS KL
2	JPS Ampang
3	Lading Edinburgh
4	Ibu Pejabat km 11, Gombak
5	Empangan Genting
6	Kg. Sg. Tua
7	Km 16, Gombak
8	Kg. Sg. Sleh
9	Air Terjun Sg. Batu

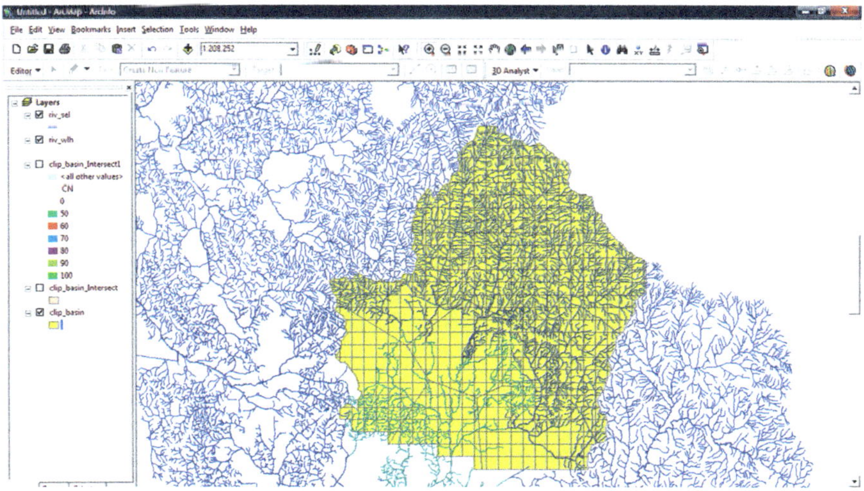

Fig. 1 Klang River Basin (KRB)

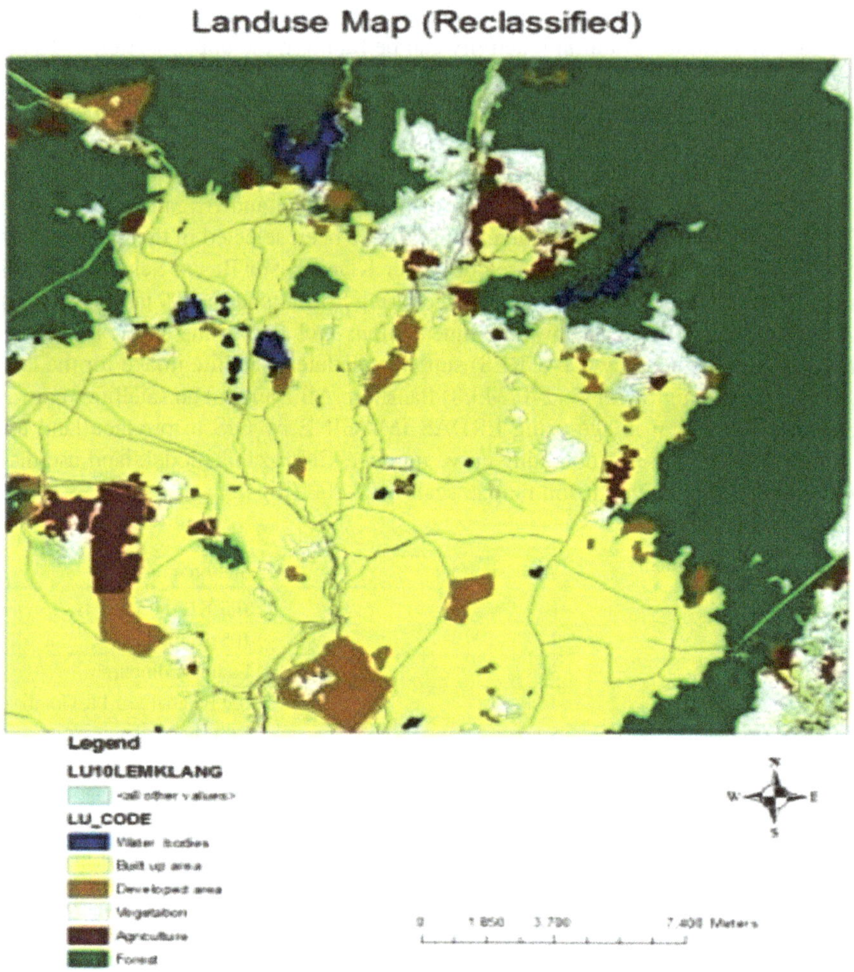

Fig. 2 Land use map for KRB

3 Methodology

Using GIS tool to determine surface runoff is becoming popular among researchers. Due to its simplicity and practicality as a tool to simulate the rainfall-runoff model, GIS generates better analysis for hydrological modeling. In this research, ArcView 9.3 and ERDAS IMAGINE have been deployed to calculate the surface runoff for Upper Klang River Basin by using grid-based curve number method. Work from ArcView 9.3 of GIS includes basin delineation, gridding by 1 km × 1 km, layer intersection between soil map, land use map, and gridded map. After integrating both maps from Department of Agriculture (DOA) of Malaysia with curve number table from Natural Resources Conservation Services (NRCS) (1986), this model

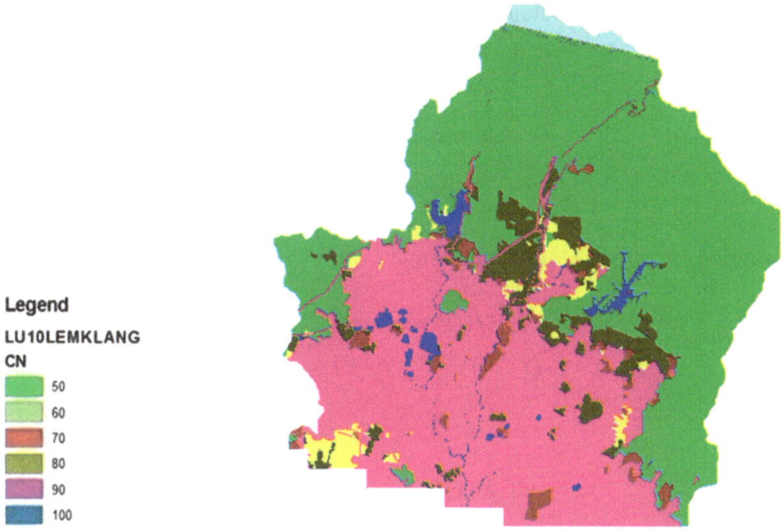

Fig. 3 Curve number map

will produce a new curve number map and finally generates map of surface runoff. Formula from SCS curve number method is applied to produce the estimated amount of excess runoff for the investigated area. ERDAS IMAGINE on the other hand generates raster image from satellite image in order to produce latest land use map. Later, this map is utilized in ArcView 9.3 to calculate the surface runoff (Fig. 3).

Curve number for all sub-catchments will be determined by using GIS with soil type and land use maps. By using physical grid-based method, each cell/grid will be treated individually to evaluate the CN for each unit. Important information for curve number includes soil cover type, soil treatment, soil hydrological condition, and antecedent moisture condition where those parameters can be obtained from maps, tables, or diagrams. Below is the curve number equation that was used in the rainfall-runoff calculation.

$$\frac{F}{S} = \frac{Q}{P - I_a} \tag{1}$$

where

F = actual retention after runoff begins (mm)
S = watershed storage (mm)
Q = actual direct runoff (mm)
P = total rainfall (mm)
I_a = initial abstraction (mm)

The amount of actual retention F can be expressed as

$$F = (P - I_a) - Q \tag{2}$$

Initial abstraction (all losses before runoff begins) is given as

$$I_a = 0.2S \tag{3}$$

Substituting equation (3) into equation (1) will yield to

$$Q = (P - I_a)^2/((P - I_a) + S) \quad \text{for} \quad P \geq 0.2S$$
$$Q = (P - 0.2S)^2/(P + 0.8S) \tag{4}$$

Where

$$S = (1{,}000/\text{CN}) - 10\,(\text{in inch}) \quad \text{or}$$
$$S = \frac{25{,}400}{\text{CN}} - 254\,(\text{in mm}) \tag{5}$$

4 Results and Discussion

The Upper Klang River basin has been delineated into nine sub-catchments using Thiessen Polygon in GIS. For the purpose of surface runoff depth estimation, two events dated 13th March 2009 and 14th April 2009 have been selected. Results can be observed from Figs. 4 and 5. For event dated 13 March 2009, the runoff depth between simulated and observed values gives correlation $R^2 = 0.7$, while for 14th April 2009, the result shows better performance when R^2 is 0.953 (as shown in Figs. 6 and 7).

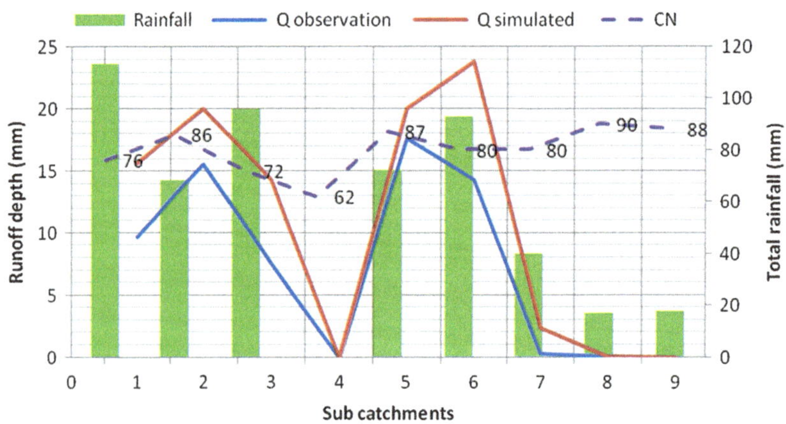

Fig. 4 Runoff depth for storm event 13032009

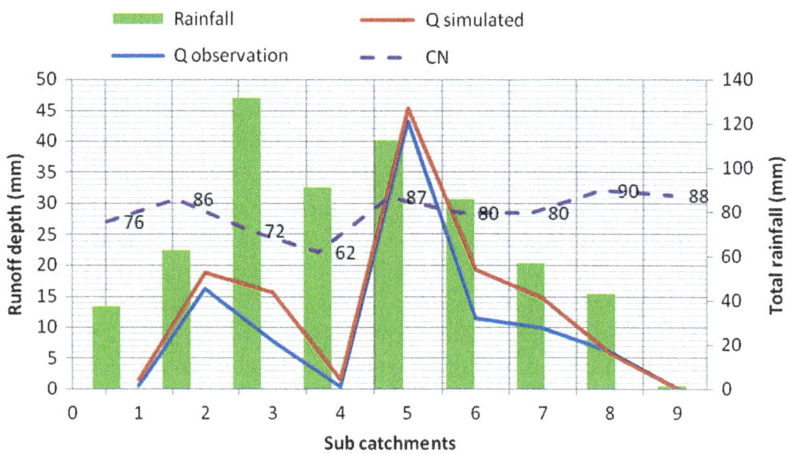

Fig. 5 Runoff depth for storm event 14052009

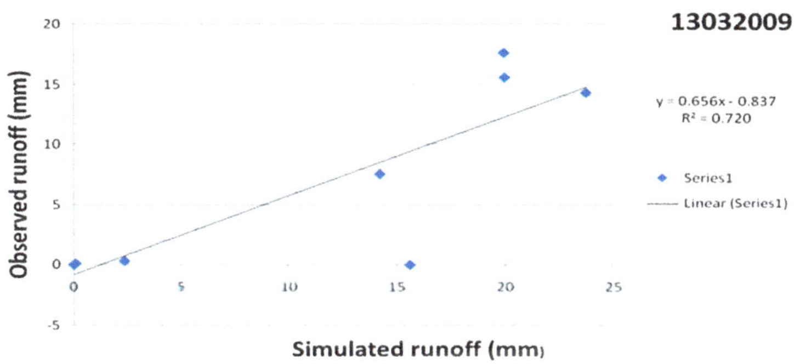

Fig. 6 Correlation assessment for event dated 13032009

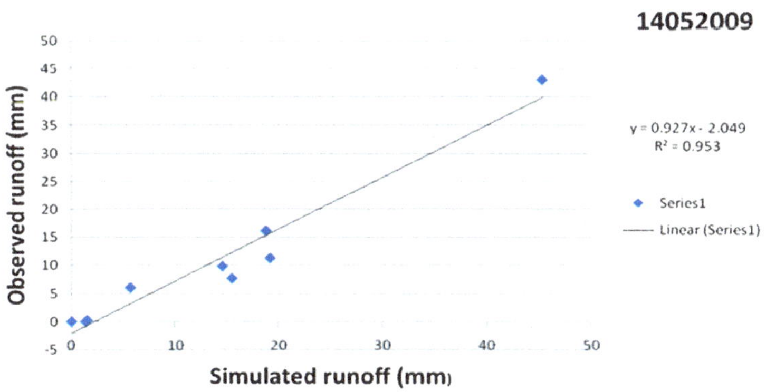

Fig. 7 Correlation assessment for event dated 14052009

The estimation depends hugely on two factors which are precipitation and curve number index. The curve number based on land use and soil type as well as antecedent moisture content in this chapter ranges from 62 to 90.

An excessive amount of rain and higher curve number index lead to greater surface runoff depth. For example, in subbasin 6, with total rainfall amount of 90 mm and curve number of 80, the estimated runoff depth is 24 mm for the first event and for subbasin 4, no runoff depth since there is no precipitation recorded on 13th March 2009. Meanwhile, on 14th April 2009, subbasin 5 shows maximum runoff depth of more than 40 mm with 110 mm in total rainfall and 87 for curve number index. The lowest runoff goes to subbasin 4 which has 62 curve number index with 90 mm of total rain. The results explain that the surface runoff depth varies with land use. Lower runoff depth can be expected from lower value of curve number, for example, 62 that represents vegetation area. Residential area with parking lots and streets with curve number value of 90 enhance the chances of higher runoff depth if accompanied with heavy rain.

For event on March 13th 2009, the highest surface runoff estimation was recorded at sub-catchment 6 with 23.6 mm (estimated data) compared to 14.29 mm (observed data), which show 67 % difference. The total rainfall for sub-catchment 6 is 93 mm with curve number index of 80. Although sub-catchment 1 recorded the highest number of rainfall received with 113.5 mm with CN index of 76, the surface runoff calculation is slightly smaller than sub-catchment 6 with 15.6 mm and 9.65 mm for both simulated and observed data, respectively. It indicates the important factor of CN index which directly refers to land use and vegetation cover that affect the calculation of surface runoff depth. For Upper Klang River Basin, the minimum value for surface runoff on 13th March 2009 was recorded at sub-catchment 8, whereas in subareas 4 and 9, no rainfall data is captured. For other areas, the average rainfall received was 69 mm with average CN value of 81. 38.5 % difference is also calculated between average simulated and average observed data for sub-catchments 2, 3, 5, and 7.

For event dated May 14th 2009, the highest surface runoff estimation was recorded at sub-catchment 5 with 45.32 mm (estimated data) compared to 43.18 mm (observed data), which shows only 5 % difference. The total rainfall for sub-catchment 5 is 112.5 mm with curve number index of 87. Although sub-catchment 3 recorded the highest number of rainfall received with 131.5 mm with CN index of 72, the surface runoff calculation is slightly smaller than sub-catchment 5 with 15.6 mm and 7.75 mm for both simulated and observed data respectively. For sub-catchment 4, where total rainfall received was 91 mm with CN index 62 produced a very small runoff depth of 1.5 mm. While for subarea 7, the rainfall data was recorded as 56.7 mm, which was less compared to subarea 4, and the CN for sub 7 is 80. Those data give higher value of surface runoff estimation even though the rain amount was smaller than other areas with value of simulated data of 14.6 mm and observed value was 10 mm (Tables 2 and 3).

Table 2 Statistical measurement for event 14052009

Event	13032009	
Subarea	Q observation	Q simulated
1	9.645	15.627667
2	15.5456	19.97
3	7.548	14.2354
4	0	0
5	17.6168	19.93
6	14.289	23.755
7	0.318	2.354
8	0.108	0.073549
9	0	0
Total		
Mean error	4.5022462	
Mean absolute error	4.509902	
Root mean square error (RMSE)	6.729893	
Bias	1.7310767	

Table 3 Statistical measurement for event 13032009

Event	13032009	
Subarea	Q observation	Q simulated
1	9.645	15.627667
2	15.5456	19.97
3	7.548	14.2354
4	0	0
5	17.6168	19.93
6	14.289	23.755
7	0.318	2.354
8	0.108	0.073549
9	0	0
Total		
Mean error	4.5022462	
Mean absolute error	4.509902	
Root mean square error (RMSE)	6.729893	
Bias	1.7310767	

5 Conclusion

The application of GIS in surface runoff estimation delivers good performance based on the above results. Curve number method is very appropriate for GIS simulation works due to its realistic application on topography such as land use category and hydrological soil type. In hydrological models, curve number method has been widely deployed especially with GIS technologies to produce direct runoff due to its low input database and its simplicity.

Acknowledgment The authors thank RMI of UiTM for providing RIF Fund (07/2012) for this project, Malaysia Meteorological Department (MMD) and Drainage and Irrigation Department (DID) for providing data radar, rainfall, and last but not least Faculty of Civil Engineering, Universiti Teknologi MARA, Malaysia.

References

1. T. Wardah, S.Y. Sharifah Nurul Huda, S.M. Deni, B. Azwa, Radar rainfall estimates comparison with kriging interpolation of gauged rain, in *Colloquium on Humanities, Science and Engineering, CHUSER 2011*, Penang, 2011, art. no. 6163877, pp. 93–97
2. R. Suzana, T. Wardah, A.S. Hamid, Radar hydrology: New Z/R relationships for Klang River Basin Malaysia based on rainfall classification. World. Acad. Sci. Eng. Technol. **59** (2011)
3. R. Suzana, T. Wardah, Radar hydrology: New Z/R relationships for Klang River Basin, Malaysia, in *International Conference on Environment Science and Engineering. IPCBEE*, vol. 8 (IACSIT, Singapore, 2011)
4. R. Suzana, T. Wardah, Radar hydrology: New Z/R relationships for quantitative precipitation estimation in Klang River Basin, Malaysia. Int. J. Environ. Sci. Dev. **2**(3), (2011)
5. S.Y. Sharifah Nurul Huda, T. Wardah, R. Suzana, A. Hamzah, M.S. Muhammad Faiz, Improved estimation of radar rainfall bias over Klang River Basin using a Kalman filtering approach, in *2012 I.E. Symposium on Business, Engineering and Industrial Applications (ISBEIA)*, 2012, art. no. 13285358, pp. 368–373
6. T. Wardah, R. Suzana, S.Y. Sharifah Nurul Huda, Multi-sensor data inputs rainfall estimation for flood simulation and forecasting, in *2012 I.E. Colloquium on Humanities, Science and Engineering (CHUSER)* (IEEE, December 2012), pp. 374–379
7. E.R. Vivoni, D.D. Sheehan, Using NEXRAD rainfall data in as Arc view-based hydrology model as an educational tool (2012), retrieved from http://www.proceeding.esri.rom/library/userconf/pro
8. D. Ramakrishnan, A. Bandyopadyay, K.N. Kusuma, SCS-CN and GIS-based approach for identifying potential water harvesting sites in the Kalim Watershed, Mahi River Basin, India. J. Earth Syst. Sci. **118**(4), 335–368 (2009)
9. A.J. Hassan, A. Ab Ghani, R. Abdullah, Development of flood risk map using GIS for Sg. Selangor basin (2006), retrieved from redac.eng.usm.my/html/publish/2006_11.pdf
10. K. Abdullah, Kuala Lumpur—Re-engineering a flooded confluence, in *14th Professor Chin Fung Kee Memorial Lecture*, Kuala Lumpur, 2004
11. Perunding Zaaba, *Interim Report of Integrated Flood Forecasting Modeling of Sungai Klang* (Department of Irrigation and Drainage Malaysia, Malaysia, 2009)
12. Department of Agriculture Malaysia, 2013

The Development and Application of Malaysian Soil Taxonomy in SWAT Watershed Model

K. Khalid, M.F. Ali, and N.F. Abd Rahman

Abstract The concept of watershed modeling is embedded in the interrelationships of geo-spatial data and hydro-meteorological data and represented through mathematical abstractions. The behavior of each process is controlled by its own characteristics as well as by its interaction with other processes active in the catchment. The hydrological models vary from empirical models to stochastic models of various kinds and finally to the more recent distributed models. In recent years, distributed watershed models have been increasingly used to implement alternative management strategies in the areas of water resource allocation, flood control, impact assessments for land use and climate change, and pollution control. Many of these models share a common base in their endeavor to incorporate the heterogeneity of the watershed and the spatial distribution of topography, vegetation, land use, soil characteristics, and rainfall. This study attempts to provide a framework of hydrological assessment using a Malaysian soil data as a soil characteristics input in Soil Water Assessment Tool (SWAT) model in its place of using the USDA Soil Taxonomy database. This chapter reports a SWAT simulation output using a Malaysian soil data from the SWAT input file. It was found that the SWAT model can be successfully applied for hydrological evaluation of the Langat River basin, Malaysia. To the Langat River basin, SCS runoff curve number (CN), base flow alpha factor (ALPHA_BF), and groundwater delay (GW_Delay) were found to be the most sensitive input parameters. The works attempting a hydrological simulation using a local soil data directly from Malaysian soil database are still in the programming stages. A same methodology is suggested to be practiced in other open source hydrological models.

Keywords Malaysian soil data • Soil Water Assessment Tool • Watershed modeling • Langat River basin

K. Khalid (✉) • M.F. Ali • N.F. Abd Rahman
Faculty of Civil Engineering, Universiti Teknologi MARA, Shah Alam, Selangor, Malaysia
e-mail: khairikh@pahang.uitm.edu.my

S.H. Abu Bakar et al. (eds.), *ISFRAM 2014*, DOI 10.1007/978-981-287-365-1_7

1 Introduction

In Malaysia, the Info Works Collection System (CS) and Stormwater Management Model (SWMM) are among the most widely used software to model drainage systems [1]. However, many more hydrological models that were found in the literature have been utilized for the watershed modeling study in the country. These included a Hydrologic Engineering Centre—The Hydrologic Modeling System (HEC-HMS) software [2–4]; followed was the MIKE SHE models [5], and finally the most current model is the Soil Water Assessment Tools (SWAT) software [6, 7]. Among these models, the physically based distributed model SWAT is well established for analyzing the impacts of land management practices on water in large complex watersheds [8].

SWAT, which is a public domain model, has been successfully used by researchers around the world for distributed hydrologic modeling and management of water resources in watersheds with various climates and terrain characteristics. Comprehensive review of SWAT model applications, calibrations, and validations is given by [9, 10]. SWAT has many parameters to be calibrated on the streamflow, sediment, and for other environmental purposes. In order to calibrate a streamflow alone, SWAT needs to consider about 26 related input parameters [11]. The SWAT model is chosen in the study since it is available with a source code that allows any modification on the features and processes to suite the present study. The model provides the continuous-time simulation to facilitate the real watershed responses in long simulation periods. The semi-distributed model is sufficient enough in accessing the critical subbasin towards the surface runoff.

Most of the hydrological models use the USDA database when dealing with the soil data. In common practice, there are two options of the modelers' utilized soil data in the model, either straight away using the parents name of the USDA soil series which is comparable with local soil series family or including a new local soil data in the model input file. Both of the options have their own advantages and disadvantages. The first option is more towards the judgment of soil properties which is based on the soil parent materials and not on the local soil series itself. The second one is more preferable to be used due to the accuracy of the results and genuineness in describing the soil property in the watershed. But, the weaknesses of the second method are that the input soil data are only applicable on that specific hydrological model. The same procedure needs to be repeated by the modeler for other hydrological model. The best option to accommodate the soil data to be used in many different hydrological models is to compile and maintain the local soil databases on the website. These data then can easily be linked to the specific model for any simulation purposes.

The study attempts to provide a framework for assessing hydrological model using a Malaysian soil data in SWAT semi-distributed watershed model. This chapter reports a SWAT simulation output using Malaysian local soil data, and a current work progresses on using a third option as a new alternative in hydrological model. The first section gives some basic explanation on the classification and properties of Malaysian soil data. This is important for researcher to embed the soil properties in the range of hydrological modeling. The second section discusses the importance of Langat River basin in terms of water resources-related study with explanations on the geo-spatial conditions of the river basin. Then, the third section explained on the method and materials being utilized in the study and followed by the SWAT model input and setup. Next, the fourth section discusses on the SWAT simulation output of the basin together with the sensitivity analysis of the input parameters. Finally, the chapter concludes by identifying the key issues and giving some directions for future research.

2 Development of Malaysian Soil Data

The development of soil science in Malaysia has mainly been spearheaded by soil surveys, and other branches of soil science have mainly played a fortifying role. Soil surveys in each of the three regions of Malaysia initially developed separately and to a large extent still remain separate. The development of soil surveys in Malaysia can be divided into three main periods [12]. The soils are generally studied with close reference to the parent material. Soil parent materials range from various forms of alluvium, sandstones, and mudstones to basic and ultrabasic igneous rocks, giving rise to a wide spectrum of soil types.

A basic unit of soil mapping in Malaysia is a soil series. The system is based on the USDA Soil Taxonomy, 1992, and to date, more than 240 soil series had been established and rearranged to 11 groups [13]. These groups are based on parent material characteristics of the dominant soils, landscape, and methods of soil formation, and this is summarized in Fig. 1 [14]. Basically, soils are classified under two main categories: mineral soils and organic soils. Soils under mineral soils group then can be classified based on the soil taxonomy, but the criteria and definition were modified to suit the local conditions. While the subclassifications of the organic soils are worked out based on the level of organic matters and its depth using modified criteria of soil taxonomy.

The soil properties are strongly affected by three forces, such as hydraulic conductivity, diffusivity, and water holding capacity. The hydraulic conductivity expresses how easily water flows through soil and is a measure of the soil's resistance to flow. A saturated hydraulic conductivity refers to the hydraulic conductivity at full saturation. Many of the infiltration equations in the hydrological models were using the saturated hydraulic conductivity as a parameter since it is easier to determine compared to the unsaturated hydraulic conductivity or the diffusivity. Diffusivity is the proportion of the hydraulic conductivity to the differential water capacity or the flux of water per unit gradient of water content in the absence of other force fields. Since diffusivity is directly proportional to hydraulic conductivity, usually only the saturated hydraulic conductivity is applied in the approximate infiltration equations.

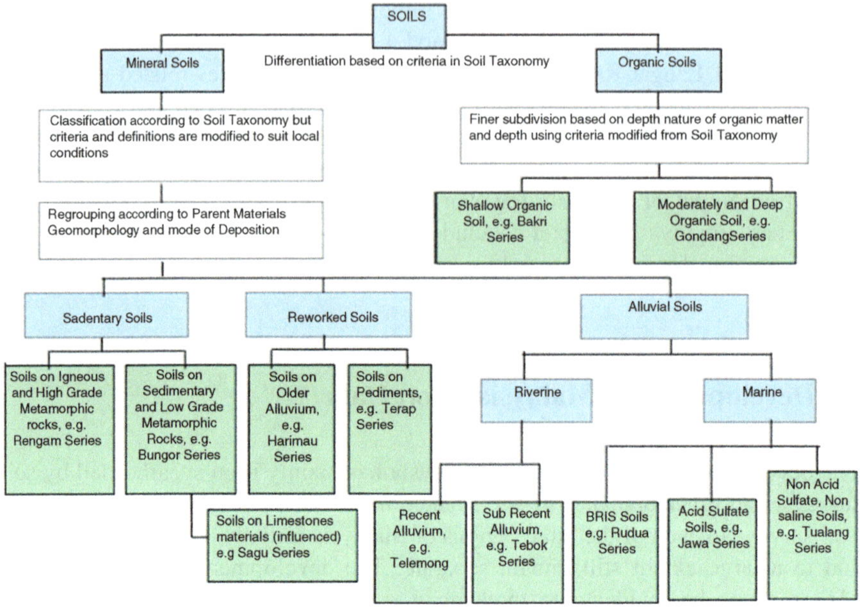

Fig. 1 Subdivision of soil in Peninsular Malaysia

Water holding capacity is the amount of water that can hold due to the pore size distribution, texture, structure, percentage of organic matter, chemical composition, and current water content.

Antecedent or initial water content affects the moisture gradient of the soil at the wetting front, the available pore space to store water, and the hydraulic conductivity of the soil. The drier the soil initially, the steeper the hydraulic gradient and the greater the available storage capacity and increase in the infiltration rate. The wetting front proceeds more slowly in drier soils because of the greater storage capacity, which fills as the wetting front proceeds.

3 Study Area

Langat River basin occupies the south parts of Selangor and small portion of Negeri Sembilan and Wilayah Persekutuan, Malaysia. The main stream of the Langat River, which stretches for 141 km, has a total catchment area of 2,271 km^2 and lies within latitudes of 2°40′152″N to 3°16′15″N and longitudes of 101°19′20″E to 102°1′10″E. The main tributary, Langat River flows from the main range (Titiwangsa Range) at the Northeast of Hulu Langat District in south-southwest direction and draining into the Straits of Malacca as in Fig. 2. Topographically, this basin can be divided into three geographic regions, i.e., the mountainous area of the north, the undulating land in the center of the basin, and the flat flood plain at the

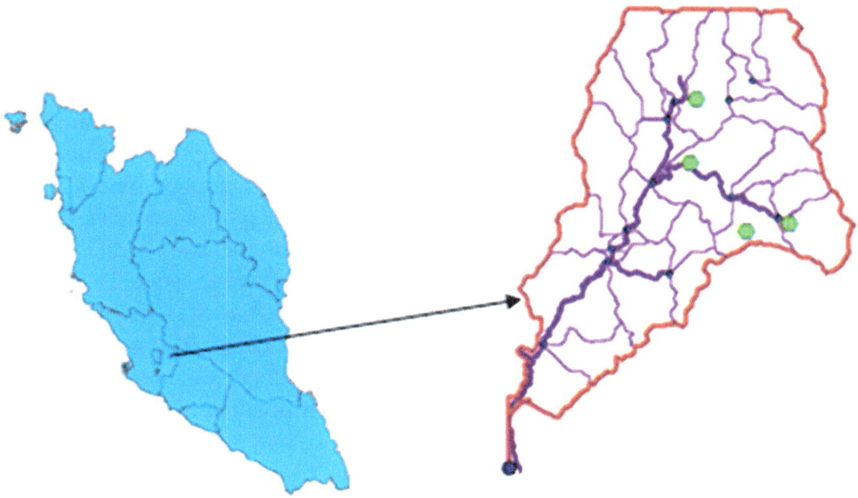

Fig. 2 Location of Langat River basin and its subbasins with hydro-meteorological stations

downstream of Langat River. The upper part of the basin is selected as a study area. The industrial sector is also minimal in the study area. The average rainfall is about 2,400 mm, and the highest months (April and November) show rainfall amount above 250 mm, while the lowest is in June, about the average of 100 mm.

Numerous studies have been conducted on the basin related to water resources and hydrologic behavior of the basin. The basin which became a first watershed in the country is initiated towards implementation of Integrated River Basin Management (IRBM) [15]. Many researchers had observed and studied on the hydrological processes of the basin including a historical water discharges study, the impact of land used change on discharge and direct runoff, sustainable groundwater resources and environmental management, the flood hazard mapping, the water supply, water quality, and a river bed properties study of the river basin. The Langat watershed has experienced many flooding disasters, mostly in Kajang and Hulu Langat areas in the year 2011 and 2012. All these studies and information show the importance and need of widespread sustainable water resources management in the Langat River basin.

4 Method and Materials

Soil Water Assessment Tool (SWAT) is continuous time, spatially distributed model designed to simulate water, sediment, nutrient, and pesticide transport at a catchment scale. The hydrologic cycle simulated by SWAT is based on the water balance equation (1).

$$\mathrm{SW}_t = \mathrm{SW}_0 + \sum_{i=1}^{t} \left(R_{\mathrm{day}} - Q_{\mathrm{surf}} - E_{\mathrm{a}} - w_{\mathrm{seep}} - Q_{\mathrm{qw}}\right)_i \tag{1}$$

In which, SW_t is the final soil water content (mm water), SW_0 is the initial soil water content in day i (mm water), t is the time (days), R_{day} is the amount of precipitation in day i (mm water), Q_{surf} is the amount of surface runoff in day i (mm water), E_{a} is the amount of evapotranspiration in day i (mm water), w_{seep} is the amount of water entering the vadose zone from the soil profile in day i (mm water), and Q_{qw} is the amount of return flow in day i (mm water). To estimate surface runoff, two methods are available. These are the SCS curve number procedure USDA Soil Conservation Service [16] and the Green & Ampt infiltration method [17]. In this study, the SCS curve number method was used to estimate surface runoff. Hargreaves method was used for the estimation of potential evapotranspiration (PET) [18].

5 SWAT Model Input and Setup

The spatially distributed data (GIS input) needed for the ArcSWAT interface include the Digital Elevation Model (DEM), soil data, land use, and stream network layers. Data on weather and observed streamflow were also used for the prediction of streamflow and calibration purposes. DEM was derived mainly from a contour map of 20 m interval and a digital river network, which were provided by Department of Survey and Mapping Malaysia (JUPEM). The land use map of a study area was obtained from Department of Agriculture, Malaysia. The land use map needs to be reclassified according to the specific land cover types such as type of crop, pasture, and forest. The dominant land used in the study area is primary forest reserve (64.80 %), followed by rubber (18.04 %), urban area (7.58 %), and orchard agriculture (3.69 %).

The SWAT model requires different soil textural, physical and chemical properties such as soil texture, available water content, hydraulic conductivity, bulk density, and organic carbon content for different layers of each soil type. These data were provided by the Department of Agriculture, Malaysia. The majority of the study area is covered by a steepland (64.8 %) and followed by a Renggam-Jerangau soil series (23.20 %), Telemong-Akob-Local Alluvium (8.00 %), and Munchong-Seremban (3.24 %). Figure 3 shows the land used and the soil types in the study area.

SWAT requires daily meteorological data that can either be read from a measured data set or be generated by a weather generator model. The weather variables used in this study are daily precipitation and minimum and maximum air temperature for the period 1999–2010. These data were obtained from the Department of Irrigation and Drainage (DID) and Department of Environmental (DOE) Malaysia for stations located within and around the watershed (Fig. 1). A weather generator

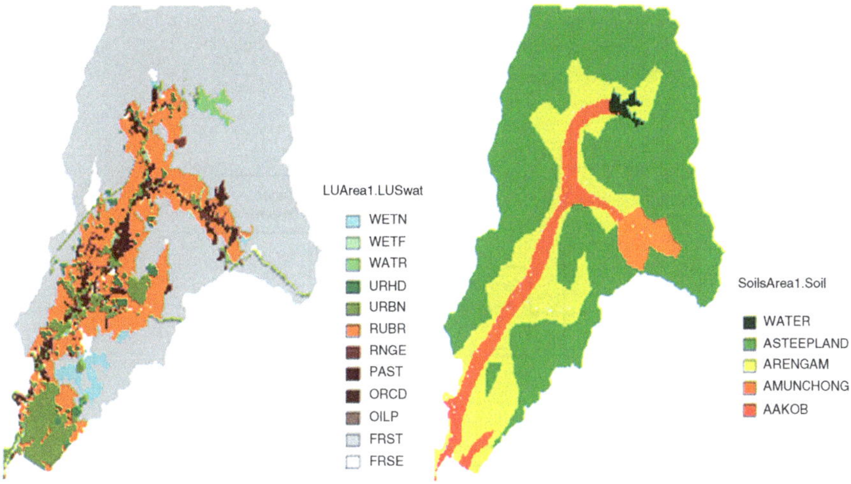

Fig. 3 Land use and soil type distribution maps in study area

developed by [19] was used to fill the gaps due to missing data. Daily river discharge values for the Kajang streamflow station were acquired from Department of Irrigation and Drainage (DID) Malaysia.

The model setup involved five steps: (1) data preparation; (2) subbasin discretization: (3) HRU definition; (4) parameter sensitivity analysis; (5) calibration and uncertainty as shown in Fig. 4. The third option of assessment of Malaysia soil series from a database into the SWAT model can only take place after all the five of the above procedures are successfully completed. The insertion of Malaysia soil series in the SWAT input table as a second option of the study is prepared in the SWAT.2009mdb database input file as shown in Fig. 5. The subbasin discretization processes only focused on the 305.3 km^2 upper part of the Langat River basin.

After setting up the model, the default simulations of stream flow using the default parameter values were conducted in the Langat River basin for the calibration period. The default simulation outputs were compared with the observed streamflow. In this study, the automatic calibration was conducted after the model was manually calibrated and reached a stage where the differences between observed and simulated flows were minimized and shown improved objective function values. The simulation used 12 years of historical daily rainfall data where, the first 2 years, starting from 1st January 1999 to 31st December 2000 were utilized for warming-up the model, followed by next 5 years for model calibration and used for validation processes for the last 5 years data. The parameter sensitivity analysis was done using the ArcSWAT interface for the whole catchment area [20]. The calibration and uncertainty analysis were done using a Sequential Uncertainty Fitting (SUFI-2) algorithm [21, 22].

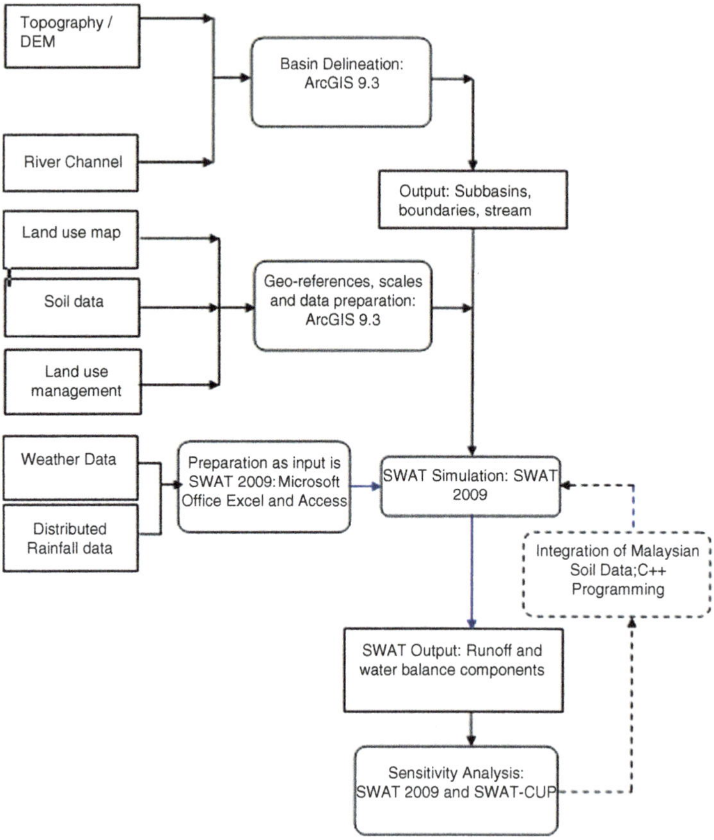

Fig. 4 Steps in model design shows the insertion of Malaysia soil series in SWAT model

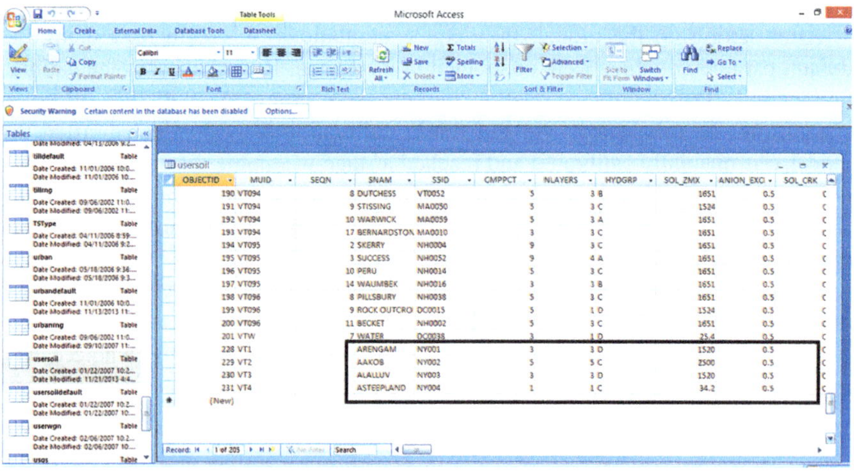

Fig. 5 Malaysia soil series in the SWAT.2009mdb database input file

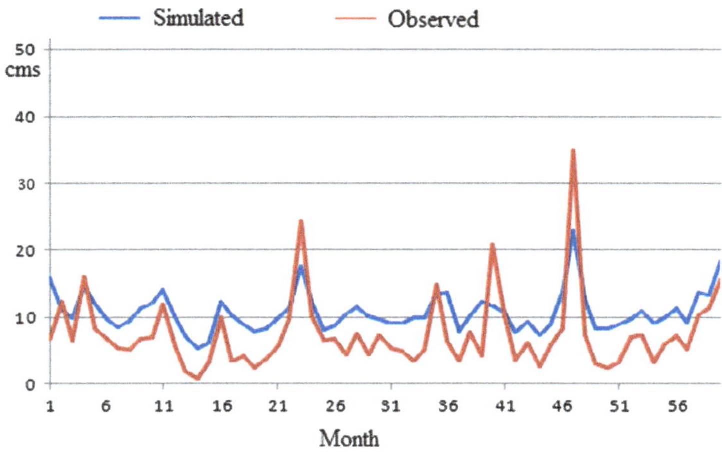

Fig. 6 SWAT 5 years streamflow simulation output

6 SWAT Simulation and Calibration

The comparison of default simulation output with the observed streamflow data of Kajang streamflow station showed an agreement between the observed and simulated flow results. Parameters manually adjusted were evaporation compensation factor (ESCO), curve number (CN), available water holding capacity of the soil layer (Sol_AWC, mm/mm), saturated hydraulic conductivity (Sol_K, mm/h), and surface runoff lag time. The manual calibration was time intensive, but it helped to get better automatic calibration results.

A calibration output obtained as in Fig. 6 shows simulated streamflow is lying slightly above the observed value. A value of coefficient of determination, R^2 of 0.69 was gained on a simulation and the value can be considered as good achievement of the manual calibration processes and minimal enough to access the SWAT performance using a Malaysian soil data as the third option of the study.

7 Sensitivity Analysis

There are a few methods available in assessing the sensitivity of input parameters in hydrological models. In SWAT model, input parameters can be either manually adjusted in the SWAT model or can be accessed in the SWAT-CUP. SWAT-CUP is a computer program for calibration of SWAT models and the programs link SUFI-2 algorithms to SWAT. It enables sensitivity analysis, calibration, validation, and uncertainty analysis of SWAT models. In SUFI-2, parameter uncertainty accounts for all sources of uncertainties such as uncertainty in driving variables, conceptual model, parameters, and measured data.

Twelve hydrological parameters were tested for one-at-a-time sensitivity analysis for the simulation of the stream flow in the study area. The parameters include (1) CN, SCS runoff curve number; (2) OV_N, Manning's "*n*" value for overland flow; (3) SOL_AWC, available water capacity of the soil layer (mm H_2O/mm soil); (4) ESCO, soil evaporation compensation factor; (5) EPCO, plant uptake compensation factor; (6) SURLAG, surface runoff lag time (days); (7) Alpha_BF, baseflow alpha factor (days); (8) GW-REVAP, groundwater "revap" coefficient; (9) GW_DELAY, groundwater delay (days); (10) GW_QMN, threshold depth of water in the shallow aquifer required for return flow to occur (mm); (11) REVAP_MN, threshold depth of water in the shallow aquifer for "revap" to occur (mm); and (12) RCHARG_DP, deep aquifer percolation fraction. These parameters are reported as the most frequent input parameters that were used in the calibration process of surface runoff and baseflow as been reported in previous 64 selected SWAT watershed studies [10].

One-at-a-time sensitivity shows the sensitivity of a variable to the changes in a parameter if all other parameters are kept constant at some value. Ten iterations were conducted for each input parameter in order to gain the best fitted input value and overall coefficient of determination. SCS runoff curve number (CN) was found to be the most sensitive parameter for a Langat River basin. The value became a first input parameter need to be adjusted in SWAT calibration processes and the same finding were experienced by many previous researchers [10]. Other two most sensitive input parameters were groundwater delay (GW_Delay) and base flow alpha factor (ALPHA_BF). A comparison of input parameters on a one-at-a-time sensitivity analysis is shown in Fig. 7.

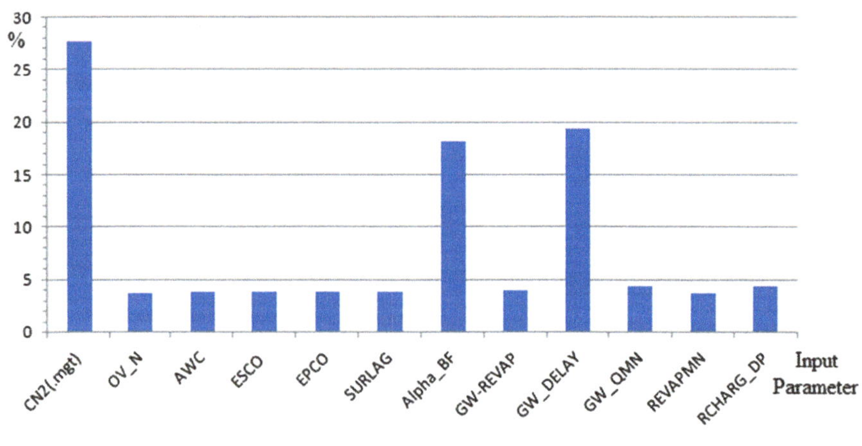

Fig. 7 Comparison of input parameters on a one-at-a-time sensitivity analysis

8 Conclusion

A streamflow of the upper part of the Langat River basin was successfully modeled by SWAT2009 model using a Malaysian soil data. The SCS runoff curve number (CN), base flow alpha factor (ALPHA_BF), and groundwater delay (GW_Delay) were found to be the most sensitive input parameters by using the SUFI-2 algorithm. This model needed further adjustment on input data by including other soil parameters that comprise of a saturated hydraulic conductivity, moist bulk density USLE equation soil erodibility (K) factor, and moist bulk density into one-at-a-time sensitivity analysis. The works attempting a hydrological simulation using a local soil data directly from Malaysian soil database are still in the programming stages. The next step should be assessed on the performance of other open source hydrological model using the same methodology of local soil data.

Acknowledgment The project was funded by Ministry of Education under the Exploratory Research Grant Scheme (ERGS) 600-RMI/ERGS 5/3 (26/2012) and Research Management Institute (RMI) of Universiti Teknologi MARA (UiTM), Malaysia.

References

1. Y.S. Liew, Z. Selamat, Review of urban stormwater drainage system management. Malaysia Water Res. J. **1**, 22–32 (2011)
2. S. Alaghmand, R. Abdullah, I. Abustan, M.A.M. Said, B. Vosoogh, GIS-Based river basin flood modelling using HEC-HMS and MIKE 11 – Kayu Ara river basin, Malaysia. J. Environ. Hydrol. **20**, 1–16 (2012)
3. Y.M. Mustafa, M.S.M. Amin, T.S. Lee, A.R.M. Shariff, Evaluation of land development impact on the tropical watershed hydrology using remote sensing and GIS. J. Spatial Hydrol. **5**(2), 6–30 (2011)
4. P.Y. Julien, A. Ab. Ghani, N.A. Zakaria, R. Abdullah, C.K. Chang, Case study: Flood mitigation of the Muda River, Malaysia. J. Hydraul. Eng. **136**(4), 251–261 (2010)
5. A.B.E. Rahim, I. Yusoff, A.M. Jafri, Z. Othman, A. Ghani, Application of MIKE SHE modelling system to set up a detailed water balance computation. Water Environ. J. **26**(4), 490–503 (2012)
6. A.H. Lai, F. Arniza, Application of SWAT hydrological model to Upper Bernam River Basin, Malaysia. IUP J. Environ. Sci. **5**(2), 7–19 (2011)
7. M.F. Ali, N.F. Abd Rahman, K. Khalid, Discharge assessment by using integrated hydrologic model for environmental technology development, in *4th International Conference on Key Engineering Materials (ICKEM 2014)*, Bali, 22–23 Mar 2014
8. J.G. Arnold, N. Fohrer, SWAT2000: Current capabilities and research opportunity in applied modeling. Hydrol. Proc. **19**(3), 563–572 (2005)
9. P.W. Gassman, M.R. Reyes, C.H. Green, J.G. Arnold, The soil and water assessment tool: Historical development, applications, and future directions. Trans. ASABE **50**(4), 1211–1250 (2007)
10. J.G. Arnold, D.N. Moraisi, P.W. Gassman, K.C. Abbaspour, M.J. White, R. Srinivasan, C. Santhi, R.D. Harmel, A. Van Griensven, M.W. Van Liew, N. Kannan, M.K. Jha, SWAT: Model use, calibration, and validation. Trans. ASABE **55**(4), 1491–1508 (2012)

11. P. Shi, Y. Hou, T. Xie, C. Chen, X. Chen, Q. Li, S. Qu, X. Fang, R. Srinivasan, Application of a SWAT model for hydrological modeling in the Xixian watershed, China. J. Hydrol. Eng. **18**, 1522–1529 (2013)
12. S. Paramananthan, L.J. Uyo, L. Melling, Historical development of soil sciences in Malaysia, in *18th World Congress of Soil Science*, Philadelphia, PA, 2006. http://crops.confex.com/crops/wc2006. Accessed 14 Feb 2014
13. Department of Agriculture (DOA), *Panduan Mengenali Siri-Siri Tanah Utama Di Semenanjung Malaysia* (Jabatan Pertanian Semenanjung Malaysia, Kuala Lumpur, 2008). 118 pp. ISBN 987-983-047-144-0
14. J.S. Lim, Y.K. Chan, Major soil mapping units in Peninsular Malaysia, in *Proceedings on Workshop on Recent Developments in Soil Genesis and Classification*, Kuala Lumpur, 1991, pp. 113–133
15. M.B. Mokhtar, R. Elfithi, A.H.H. Shah, Collaborative decision making as a best practice in integrated water resources management; A case study on Langat basin, Malaysia, in *Proceeding of the 1st Southeast Asia Water Forum*, Chiangmai, Thailand, 2004, pp. 333–354
16. USDA Soil Conservation Service, *National Engineering Handbook; Section 4 Hydrology* (USDA Soil Conservation Service, Washington, DC, 1972). Chapters 4–10
17. W.H. Green, G.A. Ampt, Studies on soil physics, 1. The flow of air and water through soils. J. Agric. Sci. **4**, 11–24 (1911)
18. G.L. Hargreaves, J.P. Riley, Agricultural benefits for Senegal River basin. J. Irrigat. Drain. Eng. **111**(2), 113–124 (1985)
19. J. Schuol, K.C. Abbaspour, Using monthly weather statistics to generate daily data in a SWAT model application to West Africa. Ecol. Model. **201**, 301–311 (2007)
20. A. Van Griensven, T. Meixner, S. Grunwald, T. Bishop, M. Diluzio, R. Srinivasan, A global sensitivity analysis tool for the parameters of multi-variable catchment models. J. Hydrol. **324** (1–4), 10–23 (2006)
21. K.C. Abbaspour, C.A. Johnson, M.T. van Genuchten, Estimating uncertain flow and transport parameters using a sequential uncertainty fitting procedure. Vadose Zone J. **3**(4), 1340–1352 (2004)
22. K.C. Abbaspour, *User Manual for SWAT-CUP, SWAT Calibration and Uncertainty Analysis Programs* (Swiss Federal Institute of Aquatic Science and Technology, Eawag, Duebendorf, Switzerland, 2012). 93 pp

Contribution of Climate Forecast System Meteorological Data for Flow Prediction

Milad Jajarmizadeh, Sobri Harun, Kuok King Kuok, and Norman Shah Sabari

Abstract One of the difficulties in hydrological models is the collection of meteorological data especially in large catchments. This issue is more obvious in the case of semidistributed hydrological models, which need long periods of meteorological data such as precipitation and temperature to perform watershed practices such as hydrological cycle and sustainable development. Moreover, this issue is significant for arid regions that generally have sparse data and lack of weather stations, especially in developing countries. In this study, Climate Forecast System Reanalysis model had been applied for Soil and Water Assessment Tool hydrologic simulator generate daily flow prediction for a catchment including dry climate. Required data for development of hydrologic simulator have been prepared in Geographic Information System database. Then, model has been calibrated via semiautomatic method namely Sequential Uncertainty Fitting 2. Results of study show that application of renewed meteorological data is promising for flow prediction. Also, accuracy of model according to Nash and Sutcliffe obtained efficiency of 0.54 for calibration and 0.45 for validation, respectively. In summary, it can be concluded that results' quality classified as good for calibration and fair for validation according to Nash and Sutcliffe efficiency.

Keywords Watershed modeling • Meteorological data • CFSR database • SWAT

1 Introduction

One of the main problems in the simulation of catchment hydrology is accessibility of hydro-meteorological data to built database of model [1]. Traditional methods include investigation and the collection of data in the field involving the installation and maintenance of a network. It is clear that it is costly and time consuming. Direct measurement of meteorological data has been frequently cited in various literatures

---

M. Jajarmizadeh (✉) • S. Harun • N.S. Sabari
Universiti Teknologi Malaysia, Skudai, Johor, Malaysia
e-mail: milad_jajarmi@yahoo.com

K.K. Kuok
Faculty of Engineering, Swinburne University of Technology, Sarawak, Kuching, Sarawak, Malaysia

S.H. Abu Bakar et al. (eds.), *ISFRAM 2014*, DOI 10.1007/978-981-287-365-1_8

as the best way to recognize atmospheric hydrology catchment, but it takes a long investment in time and cost [2]. Basically, conventional methods of meteorological measurement are costly, time consuming, and difficult because of inaccessible terrain in many of the watersheds. Al-Qurashi et al. [3] stated that one of the obstacles for applying computerized hydrological modeling in arid and semiarid regions is due to sparse data and high spatial variability. Arid and semiarid regions are particular problems for hydrological modelers since meteorological observation is usually hampered by sparse rainfall and runoff gauge networks [4]. Recently, globally renewed weather data have been introduced for application in hydrological modeling. The motives include evaluation of data accuracy and applicability in critical regions. Therefore, a variety of atmospheric data for watershed management and decision-making have been used, contributing to model accuracy. For instance, Modern-Era Retrospective Analysis for Research Applications (MERRA), the European Centre for Medium-Range Weather Forecasts (ECMWF) interim Reanalysis (ERA-Interim), and the National Centers for Environmental Prediction's Climate Forecast System Reanalysis (CFSR) for hydrological models as observed data have been applied [5].

Recently, the Climate Forecast System (CFS) was prepared at National Center for Environmental Prediction (NCEP), and it is fully coupled with ocean–land–atmosphere dynamical seasonal prediction system. CFS model includes two main structures. The atmospheric infrastructure of CFS is a modified sample in resolution on the Global Forecast System (GFS). Indeed, GFS has been used to prepare global weather prediction model at NCEP. In addition, the ocean infrastructure includes the Geophysical Fluid Dynamics Laboratory (GFDL) Modular Ocean Model version 3 (MOM3). Summary of advantages for CFS system data can be (1) the atmosphere–ocean combination spans almost across the globe and (2) CFS is an entirely joined method with no flux modification.

Climate Forecast System and Reanalysis (CFSR) dataset has been developed as part of the Climate Forecast System at the National Center for Environmental Prediction (NCEP) for soil and water assessment tool (SWAT) [6]. CFSR has been available for 30 years from 1979 to 2010 and includes features of atmospheric-ocean-land-sea ice system. Also, it covers changing CO_2 and other trace gasses and solar discrepancies. CFSR has been developed with lower spatial resolution (38 km) and temporal data stream (hourly and daily). Details on CFSR model and more literature can be found in [1].

CFSR data have been used in semidistributed hydrological model by [1]. The result of using CFSR data was promising for stream flow simulation and model performance. Then, CFSR data can provide a new challenge for ungauged watershed modeling across the globe. Weiland et al. [7] used CFSR data for finding the optimal method for deriving potential evaporation on a global scale. In other research, Dile and Srinivasan [4] used CFSR data for hydrological predictions in Nile River basin. They found that conventional weather stations have better performance in hydrologic simulations. However, they stated that using CFSR data required further assessment in arid zone. The use of conventional and climate forecast systems offer promising results. Therefore, it requires more study across the globe in relation to hydrological phenomena. Also, contribution of climate forecast systems can play an important role in ungauged watersheds.

In this study, Climate Forecast System and Reanalysis (CFSR) database has been assigned for SWAT hydrologic simulator for a catchment inclusion arid climate in southern Iran. Basically, collection of meteorological data is a serious obstacle in the implementation of hydrological simulation such as flow prediction. Therefore, this research seeks to understand the general impact of using CFSR database for daily discharge prediction in Roodan Plain (case study). Indeed, scope of this research involves application of remote sensing meteorological data, namely CFSR and feeding this data to SWAT as hydrologic simulator for daily flow simulation. Also, calibration and validation of developed model have been considered with a semiautomatic program namely SWAT-CUP (Calibration and Uncertainty Program) for Roodan Plain.

2 Methodology

2.1 Case Study

Roodan Plain is located in southern Iran. The area of catchment is 10,570 km^2. Roodan Plain is located between maximum elevation of 2,720 m in the north and as low as 92 m in outlet (Fig. 1). Roodan Plain includes two features namely mountainous in the north and east direction and plain in the center and south part. For the period of 1978–2008, the average annual precipitation was 215 mm [8]. The heaviest rainfall is from October to March. For the same period, the mean daily temperature was 25 °C. Soil types are a mix of clay, silt, and sand. Generally, the climate of Roodan is arid to semiarid with seasonal rainfall mostly in autumn and winter. The dominant land cover of Roodan Plain is range brush and mixed grassland and shrubs. Roodan Plain is classified as low to medium developing catchment in southern Iran with the potential for agricultural planning and collecting of surface water resources. Agricultural planning and sustainable development are key programs in this basin.

2.2 SWAT Hydrologic Simulator and CFSR Database

Soil and water assessment tool is a hydrologic simulator that was developed from several earlier models by Arnold for the United States Department of Agriculture in the early 1990s. It is favorable for evaluation of land management practices on water, sediment, agricultural, and chemical yields in various scale catchments. Also, SWAT classified as watershed scaled, continuous time model have been considered for the study of long-term results on large and complex basins. This

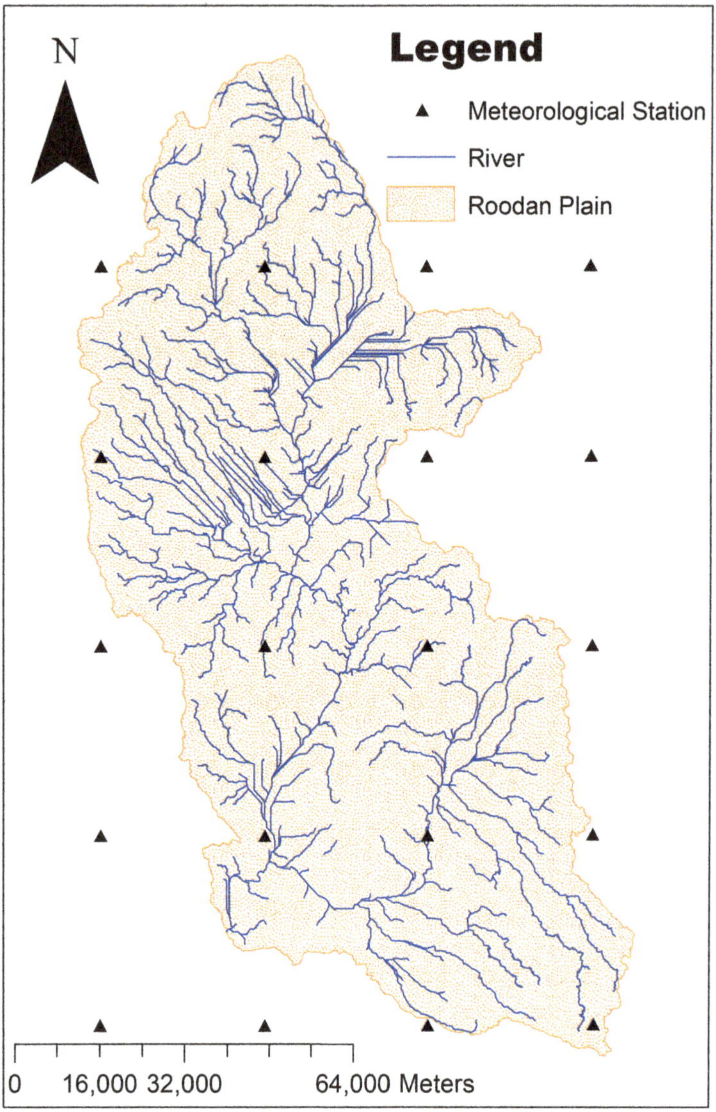

Fig. 1 Roodan Plain with presentation of CFSR stations across the watershed located in southern Iran

hydrologic simulator includes semiempirical and semiphysical features. SWAT as semidistributed simulator requires Geographic Information System (GIS) database and meteorological information. Basic database of SWAT includes digital elevation modeling (DEM), soil and land cover maps, and hydro-meteorological data.

Also, cognition of ground water and subsurface flow might be helpful in the development of model for hydrologic cycle of river basin. SWAT model has evolved from various progressions and updates. Three official versions of SWAT for application are SWAT 2005, SWAT 2009, and SWAT 2012. In this study, SWAT version 2009 has been used for preparing Roodan modeling. A comprehensive explanation of SWAT model and related information are available in recent studies [9, 10].

In this study, general database for SWAT simulator has been developed with DEM and land use and soil maps. In regard to meteorological stations, Climate Forecast System and Reanalysis database has been used for precipitation and temperature in daily time frame over 20 stations across the watershed. Soil map provides information such as soil hydrologic group, moist bulk density, available soil water capacity, saturated hydraulic conductivity, and percentage of clay, silt, sand, and rock availability in soil sample. Also, land use database includes two basic types of information, hydrological and crop type parameters. In total, DEM is based on the delineation of stream river networks and geometric features derived from topographic map. For this study, soil map of Food and Agriculture Organization of the United Nations (FAO) was used which linked 5,000 soil types in SWAT model.

Land use map of Roodan was prepared based on the satellite image of Landsat7, observation of various parts of case study (2007–2008), available land use map (1:25,000), and obtained statistics of agricultural area from agriculture organization of Hormozgan. Then, the daily rainfall and temperature were used from 20 meteorological stations across Roodan Plain on 3/1/2014 8:38 AM. The stations cover the entire watershed area between south Latitude 26.78 to North Latitude 29.14, and West Longitude 56.71 to East Longitude 57.98 (Fig. 1). Finally, Roodan Plain is divided into 69 subbasins and 689 hydrologic response units (HRUs) as they have more individual combined features of land use, soil type, and slop classes for watershed in routing phase. Calibration and validation of model were performed with SWAT-Calibration and Uncertainty analysis (SWAT-CUP) [10]. This program included Sequential Uncertainty Fitting Program 2 (SUFI-2) algorithm to calibrate the developed SWAT model. Additional details of SWAT-CUP and SUFI-2 can be found in a new study by [10].

3 Results and Discussion

In this study, two factors derived from SUFI2 algorithm are considered for evaluation of the quality of model results [11]. The *P*-factor criteria prepare a measure of the model's ability to capture uncertainties. Ideally, the *P*-factor is required to obtain value of unity. The *R*-factor provides the quality of the calibration. Ideally, the value should be approximately around zero. 95 percent prediction uncertainty (95PPU) as indicated in the Figs. 2, 3, and 4 is overall uncertainty in the output. The interpretation of *P*-factor and *R*-factor, both present the capability of the model calibration and interpretation on uncertainty [10]. Moreover, it has been discussed on Nash–Sutcliffe coefficient (NS) for both calibration and validation periods

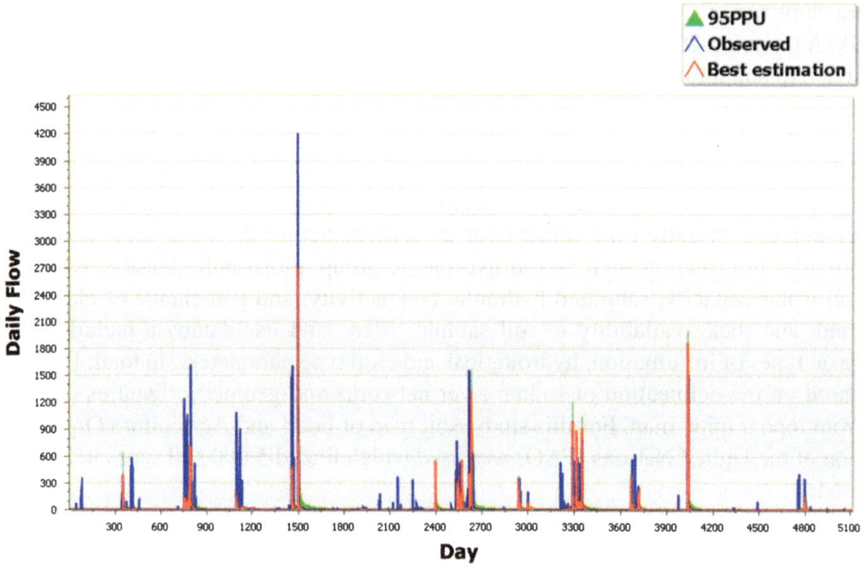

Fig. 2 First iteration of calibration including 500 simulations during 1989–2002

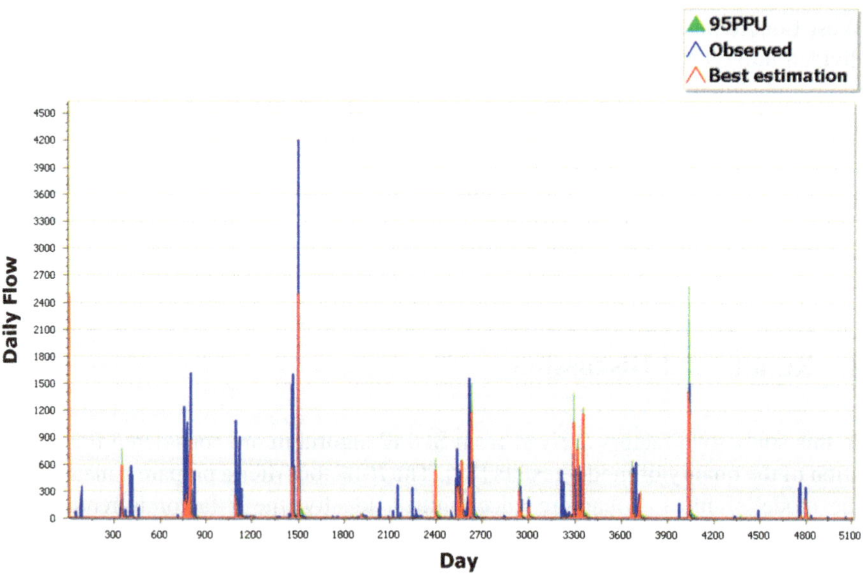

Fig. 3 Second iteration of calibration including 350 simulations during 1989–2002

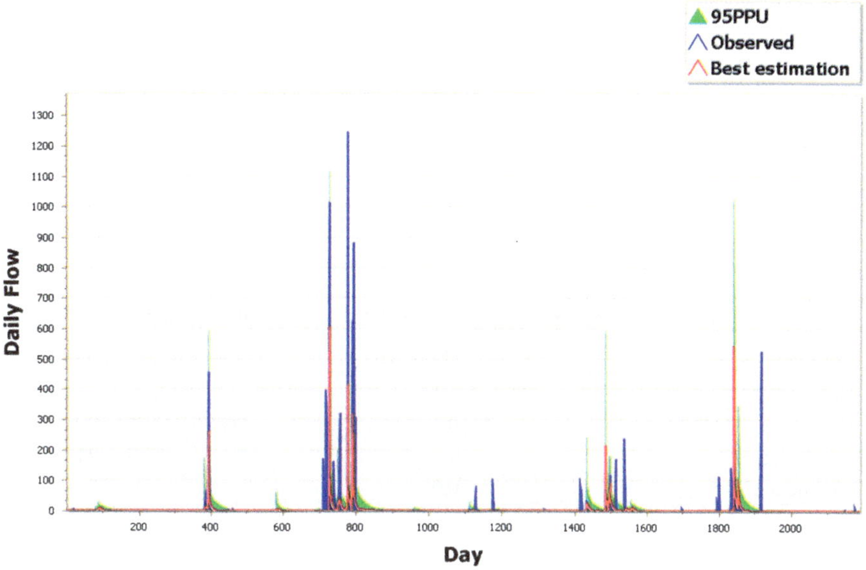

Fig. 4 Trend analysis for validation period during 2003–2008

[12]. Calibration period has been performed for 1989–2002 and validation involved 2003–2008 for recent years. Indeed, calibration period includes a long period to consider a variety of flow distribution in case study from base flow to peak flow. Concerning results, in the first iteration, *P*-factor and *R*-factor obtained 0.87 and 0.21. In the second iteration, *P*-factor and *R*-factor were calculated as 0.44 and 0.18, respectively (first iteration includes 500 simulations and second iteration includes 350 simulations) for calibration period. Indeed, first and second iterations are every repetition of a mathematical or computational procedure applied to the result of a previous application, typically as a means of obtaining successively closer approximations to the solution of a problem. The goal of increasing iterations was to increase NS value. In calibration period, NS obtained 0.52 and 0.54 in first and second iteration, respectively between observed and simulated data. In regard to quality of model accuracy, range of NS coefficient shows calibration has been classified in good manner [13]. In general, calibration of model by evaluation of mentioned criteria was satisfactory. On the other hand, for validation period, *P*-factor and *R*-factor obtained are 0.80 and 0.24, respectively. Value of NS is 0.45 for validation period and it has been classified as fair accuracy [13]. Moreover, *P*-factor and *R*-factor are good for model accuracy in validation period according to [11].

Also, trend analysis as graphical presentation has been drawn for calibration and validation periods. Figures 2 and 3 show trend analysis for calibration period for first and second iterations.

Figures 2 and 3 show increasing iterations have no obvious effect on improving the result, so iteration 2 (Fig. 3) has been chosen for evaluation of final results, and

Table 1 General statistic analysis for observed and predicted flows in calibration period (1989–2002)

Index	Observed	Simulated
Mean	13.6	6.6
Median	2.44	0.32
Std. deviation	101	62.3
Minimum	0	0
Maximum	4,210	2,503

Table 2 General statistics for observed and predicted flows in validation period 2003–2008

Index	Observed	Simulated
Mean	5.17	4
Median	0.4	0.14
Std. deviation	49	24.5
Minimum	0.02	0
Maximum	1,248	606.4

calibration has not continued in further iteration. Based on Fig. 3, the model includes underestimation for flow prediction in highest event. Highest event is 4,209 m^3/s recorded in 1993, and the corresponding predicted event is 2,503 m^3/s during calibration period. However, it should be noted that SWAT simulator is not suitable for single event evaluation, and this model is continuous based simulation [14]. For validation period, trend analysis shows an underestimation for rainy seasons, mostly February and March. Also, highest flow has been recorded in 2005 with 1,248 m^3/s, and corresponding predicted flow obtained is 416 m^3/s. In general, SWAT has a fair simulation involving CFSR data in a large catchment. Also, trend analysis indicated that CFSR data are promising for modeling and cognition of high flow events. The underestimation of flows might be related to a lack of information on groundwater behavior and subsurface flow in catchment. Moreover, Dile and Srinivasan [4] stated that they found obvious underestimation in flow prediction in monthly simulation via CFSR database for SWAT simulator in a fresh study. Moreover, they obtained unsatisfactory results for some selected basins located in North Africa. However, the result of this study indicated good (calibration) to fair (validation) including CFSR database for southern Iran.

In regard to further assessment, statistical analysis has been applied for both calibration and validation periods. Table 1 shows general statistic for Roodan Plain. Mean values for observed and simulated flows of 13.6 and 6.6, respectively, have been obtained. In addition, median is 2.44 and 0.32 for observed and simulated flows. Generally, SWAT including CFSR data underestimated flow prediction as shown in Table 1. Also, under prediction can be seen in Table 2 in validation period. However, the values of general statistics have slightly better discrepancies between observed and simulated data in comparison with calibration period.

4 Conclusion

In this research, Climate Forecast System Reanalysis has been applied for a large arid catchment in southern Iran. Daily flow prediction has been simulated with Soil and Water Assessment Tool (SWAT) as quasi-distributed model. Evaluation of model has involved P-factor and R-factor that are related with Sequential Uncertainty Fitting 2 procedure as quasi-automatic calibration method. A long period from 1989 to 2002 including a variety of flow distribution was considered for adjustment of model and validation of model was for a more recent period from 2003 to 2008. Results of model with contribution of CFSR data in a large-scale basin show that model has fair flow prediction. However, according to Nash and Sutcliffe, efficiency calibration has been classified as good and validation as fair quality. Moreover, P-factor and R-factor were satisfactory for calibration and validation. Also, general statistical analysis reveals that predicted flows have lower values for average, median, and standard deviation for both calibration and validation against observed data.

In general, SWAT for high flow events were underestimated, but cognition of high events via simulation was satisfactory using CFSR database. This issue indicated that CFSR data are promising for hydrological simulators for flow prediction and decision-making in critical subjects in a watershed. Also, they can be used in catchment with lack of information for meteorological data in conservative way for studies on watershed development and management.

Acknowledgment We deeply appreciate the Research Management Center (RMC) of Universiti Teknologi Malaysia for funding this research under postdoctoral fellowship scheme. We are thankful to all members of consultant engineers of Ab Rah Saz Shargh Corporation in Iran and the Regional Water Organization, Agricultural Organization, and Natural Resources Organization of Hormozgan province, Iran. We also wish to acknowledge Dr. Philip W. Gassman at the Center for Agricultural and Rural development, Iowa State University for introduction of SWAT model and related new contributions.

References

1. D.R. Fuka, M.T. Walter, C. MacAlister, A.T. Degaetano, T.S. Steenhuis, Z.M. Easton, Using the Climate Forecast System Reanalysis as weather input data for watershed models. Hydrol. Process. **28**(22), 5613–5623 (2013). doi:10.1002/hyp.10073
2. F. Konukcue, A. Istanbulluoglu, I. Kocaman, Determination of the water yields for small basins in semi-arid area: Application of the modified Turk method to the Turkey's condition. J. Cent. Eur. Agr. **6**, 263–268 (2005)
3. A.A.H. Al-Qurashi, N. McIntyre, H. Wheater, C. Unkrich, Application of the Kineros2 rainfall-runoff model to an arid catchment in Oman. J. Hydrol. **355**, 91–105 (2008)
4. Y.T. Dile, R. Srinivasan, Evaluation of CFSR climate data for hydrologic prediction in data-scarce watersheds: An application in the Blue Nile River Basin, J. Am. Water Resour. Assoc. **50**(5), 1226–1241 (2014)

5. R.A. Smith, C.D. Kummerow, A comparison of in situ, reanalysis, and satellite water budgets over the Upper Colorado River Basin. J. Hydrometeorol. **14**, 888–905 (2013)
6. S. Saha, S. Nadiga, C. Thiaw, J. Wangw, W. Wang, Q. Zhang, H.M. Van Den Dool, H.L. Pan, S. Moorthi, D. Behringer, D. Stokes, M. Peña, S. Lord, G. White, W. Ebisuzaki, P. Peng, P. Xie, The NCEP climate forecast system. Am. Meteorol. Soc. **19**, 3483–3517 (2006)
7. F.C. Sperna Weiland, C. Tisseuil, H.H. Durr, M. Vrac, L.P.H. van beek, Selecting the optimal method to calculate daily global reference potential evaporation from CFSR reanalysis data for application in a hydrological model study. Hydrol. Earth Syst. Sci. **16**, 983–1000 (2012)
8. A.R.S. Shargh, Comprehensive studies of water resource management for Roodan watershed, Synthesis report of Roodan. Consulting Water Resource Engineering Corporation, register code 14800 Mashhad, Iran, 2009
9. P.W. Gassman, A.M. Sadeghi, R. Srinivasan, Applications of the SWAT model special section: Overview and insights. J. Environ. Qual. **43**(1), 1–8 (2014)
10. J.G. Arnold, D.N. Moriasi, P.W. Gassman, K.C. Abbaspour, M.J. White, R. Srinivasan, C. Santhi, R.D. Harmel, A. van Griensven, M.W. Van Liew, N. Kannan, M. Jha, SWAT: Model use, calibration, and validation. Am. Soc. Agr. Biol. Eng. **55**(4), 1491–1508 (2012)
11. K.C, Abbaspour, M. Vejdani, S. Haghighat, SWATCUP calibration and uncertainty programs for SWAT, in *Proceedings of the International Congress on Modelling and Simulation (MODSIM'07)*, Modelling and Simulation Society of Australia and New Zealand, Australia, 2007
12. P. Krause, D.P. Boyle, F. Base, Comparison of different efficiency criteria for hydrological model assessment. Adv. Geosci. **5**, 89–97 (2005)
13. P.B. Parajuli, N.O. Nelson, L.D. Frees, K.R. Mankin, Comparison of AnnAGNPS and SWAT model simulation results in USDA-CEAP agricultural watersheds in south-central Kansas. Hydrol. Process. **23**, 748–763 (2009)
14. S.L. Neitsch, J.G. Arnold, J.R. Kiniry, J.R Williams, Soil & Water Assessment Tool Theoretical Documentation Version 2009. Grassland, Soil and Water Research Laboratory—Agricultural Research Service Blackland Research Center—Texas AgriLife Research, Texas Water Resources Institute Technical Report No. 406 Texas A&M University System College Station, TX, 2011

Hydrological Modeling in Malaysia

Jazuri Abdullah, Nur Shazwani Muhammad, and Pierre Y. Julien

Abstract Hydrological modeling in Malaysia using two-dimensional fully distributed and physically based model is relatively new. Basic guidelines in choosing model dimensions (i.e., 1D, 1D–2D, 2D, and 3D) are discussed. The application of 3D model will give more accurate results if the groundwater movement is considered but requires more time to simulate and prepare input data. However, the 1D model is less accurate in representing the real topography of the study area, even though requires short time to simulate. This chapter also gives an overview about the hydrological modeling in Malaysia which ranges from 1D to 3D. Several researches on hydrology using different model were reviewed. The success of each model will be highlighted.

Keywords Hydrological modeling • Lumped model • Distributed model

1 Introduction

For the past 5 years, the frequency and magnitude of floods in Malaysia have been relatively high. Generally, floods happen between November and February each year due to the Monsoon climate. These floods have caused lots of damages but also provide lots of valuable information. In recent years, rapid development within a river watershed has resulted in higher runoff and decreasing river capacity. These, in turn, resulted in an increase in flood frequency and magnitude [1]. The government has been spending a lot of money on flood mitigation projects in urban and rural areas. Improper methods for predicting peak discharge, time to peak, and volume of water lead to inappropriate channel design [2].

J. Abdullah (✉)
Flood Control Research Center, Universiti Teknologi MARA, Shah Alam, Selangor, Malaysia
e-mail: jazuri9170@salam.uitm.edu.my

N.S. Muhammad
Department of Civil Engineering, Universiti Kebangsaan Malaysia, Bangi, Selangor, Malaysia

P.Y. Julien
Department of Civil and Environmental Engineering, Colorado State University, Fort Collins, CO, USA

S.H. Abu Bakar et al. (eds.), *ISFRAM 2014*, DOI 10.1007/978-981-287-365-1_9

In Malaysia, the prediction of flood frequency using stochastic models is common. The statistical concept [3–6] and artificial neural network (ANN) [7–9] are the preferred methods as compared to other stochastic models. This method is well developed and widely tested. However, this method is reliable if the recurrence intervals of the peak discharge do not exceed the lengths of recorded peak discharges data [10, 11]. In addition to this, the maximum flood hydrographs are also important for the solution of many hydrological and environmental problems.

Increasing demands on computer models that can estimate more precisely the peak discharge, time to peak, and volume of water is a need. This is important to allow more adequate warning time to be issued, which leads to a better response from relevant agencies in protecting public, property, and infrastructure. However, computer models are still relatively new in Malaysia even though it has been widely used in many other countries [12].

Selection of one-dimensional model is the most common approach to carry out the hydrological modeling [13]. The engineers are aware of the importance of using two-dimensional model in simulating the hydrological events. However, in Malaysia, particularly, the lack of reliable basic data such as DEM, land use, and soil type are main obstacles.

The main purpose of this chapter is to review the use of hydrological modeling in Malaysia. The criteria for selection of the model also will be discussed and summarized.

2 For the Selection of Hydrological Model

There are several well-known hydrological models currently in use. The availability of source code is one of the main criteria for model selection. The model must also have the ability to support the fully distributed parameters and the two-dimensional overland routing approach. Some models use either a semidistributed or lumped (Fig. 1a) approach, which does not consider the spatial variability of processes, boundary condition, and watershed geometric characteristics. A fully distributed model (Fig. 1b) is expected to give better results than a semidistributed model [15]. The two-dimensional overland (Fig. 2b) routing is important as compared to one-dimensional overland (Fig. 2a) routing because it is very helpful to analyze outputs, which provides more information. An extra-added value to the model is the ability to work with raster GIS database. The availability of rainfall and flow data is also considered.

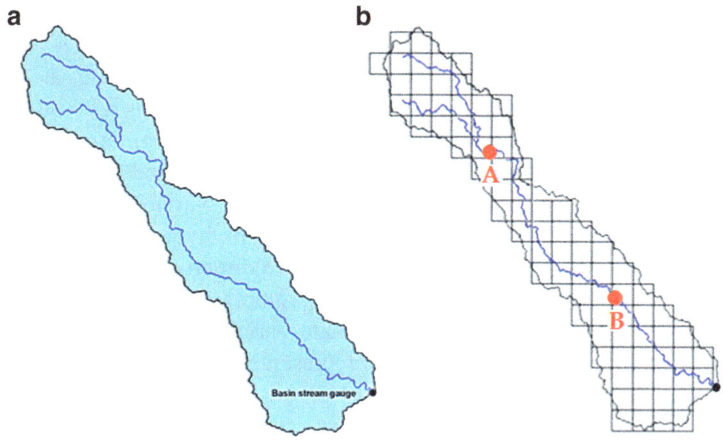

Fig. 1 Lumped and distributed [14]. (**a**) Lumped model. (**b**) Distributed model

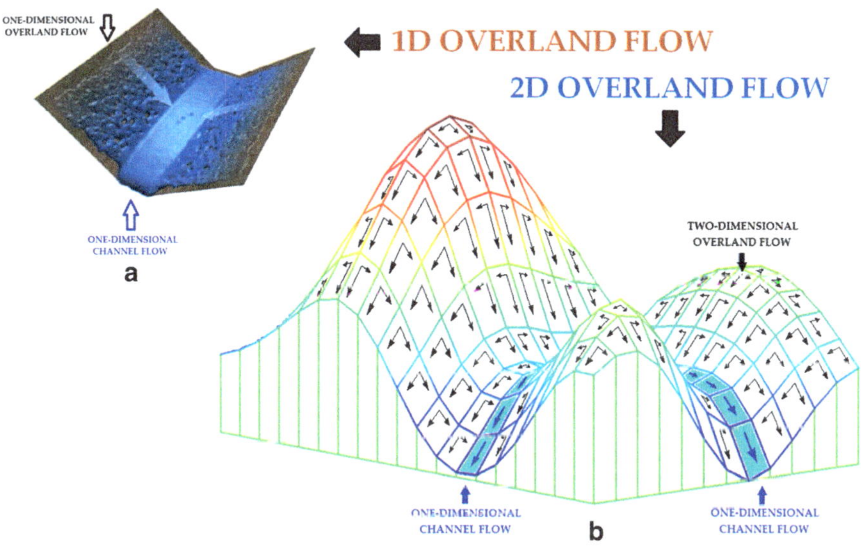

Fig. 2 Comparison of overland flow. (**a**) 1D overland flow (modified from [14]) and (**b**) 2D overland flow

3 Lumped Versus Distributed Models

Lumped models (Fig. 1a) have been used for more than 50 years to estimate flow at the watershed outlet. The simplification of many watershed characteristics may affect the simulation results. The parameters used in this model are spatially averaged and uniform across the watersheds [16, 17], and the number of parameters is less [18]. However, in reality, these input data vary.

A number of questions remain as to how the variability of rainfall and watershed characteristics impact runoff to generate the streamflow at the watershed outlet [19–22]. Nowadays, instead of lumped modeling, distributed modeling, as shown in Fig. 1b, is becoming a more favorable approach in research. This is because most of the models are compatible to work with GIS, the emergence of large data sets and the increased efficiency of powerful computer to simulate and display the results [21]. Distributed models better represent the spatial variability of factors that control runoff, thus enhancing the predictability of hydrologic processes [15, 23]. These models usually use parameters that are directly related to the physical characteristics of the watershed, including topography (i.e., elevation), soil type, channel properties, and land use. The climate variability can also be taken into account as reported by [24]. Results are presented in the form of spatial and temporal characteristics [25–27].

4 Selection of the Complexity of the Model

The risks of not being able to represent the topography of the watersheds, the difficulty in getting solution, and the application of the hydrological models at different size of watersheds are the main concerns in selecting the complexity of the hydrological model (CHM). Figure 3 shows the "trade-off diagram" for the CHM

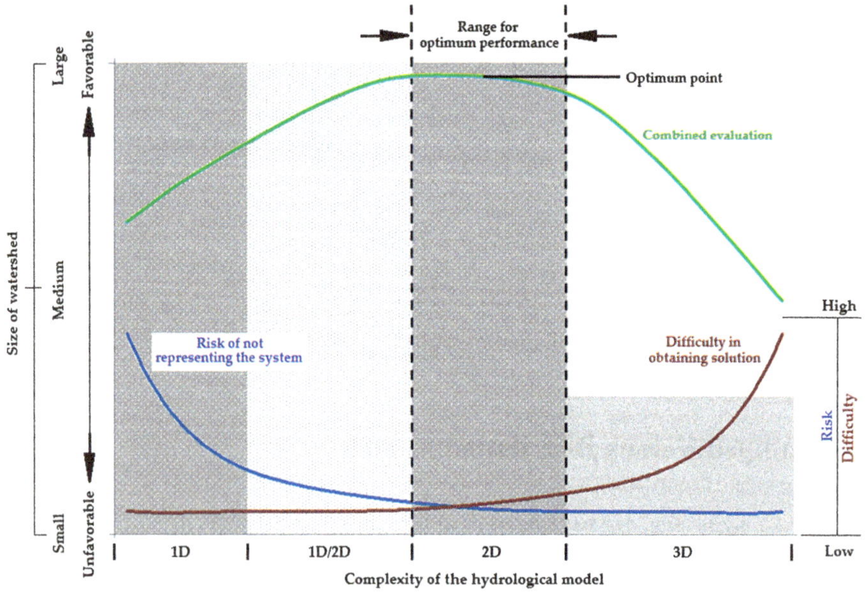

Fig. 3 "Trade-off diagram" in selecting dimensions of hydrological modeling (modified after [28])

(i.e., one-, integrated one-two, two-, and three-dimensional hydrological modeling) and size of watershed.

Generally, the choice of CHM depends on the project objectives [29, 30] and scopes, the knowledge and skills of the modeler, resources constrain [28], and time and length scale [31, 32]. In addition to these, the optimization and presentation of the final output should be considered, as described by [33]. Choosing a complex hydrological model will represent the characteristics of the watershed better, but it will be difficult to obtain the solution. Another factor that should also be considered is the size of watershed. A simpler model was usually selected when a large size of watershed is to be modeled. From Fig. 3, the 1D and 2D models are more favorable to simulate hydrological model for any size of watershed. Conversely, the application of 3D model in hydrological model to variety of size of watershed is scarce [31, 34]. According to them, a 1D or 2D model is sufficient to simulate this distribution as compared to 3D, which may not be realistic because it currently is very costly.

4.1 Risk of Not Presenting the System

In hydrological modeling, the representation of the system should be as accurate as it supposed to be. The representation of the system can be extracted directly from the digital elevation model (DEM). This is the most important data because topography controls runoff and watershed boundaries [35]. The shape and timing of the hydrograph have shown to be a function of size, slope, shape, soil types, storage capacity, land use, and climatic variables. When a model is able to reflect the principle of how a watershed functions hydrologically, then the possibility to extrapolate beyond current situation with reliable prediction may be possible [36]. Rainfall intensity and duration are the major driving forces of the rainfall-runoff process, followed by watershed characteristics that translate the rainfall input into an output hydrograph at any point of the watershed.

4.2 Difficulty in Obtaining Solution

The difficulty in obtaining solution covers: (1) ease to use and to prepare the input data, (2) model accuracy, (3) hydrologic parameters consistency, (4) sensitivity of the output when parameters change, (5) storage (in computer hard drive) require for the output, (6) data limitations, (7) computer time simulation, and (8) availability of data. The availability of data is the most important in selecting the CHM [37]. In general, the 1D model can predict flow and produce hydrograph when it has been calibrated and validated.

According to [38, 39], the basic idea in the selection of models is to adopt the simplest model (i.e., easy to use and apply) that will provide acceptable results. However, the ease of application will also depend upon the individual experience of the modeler, both in the use of the model and the knowledge of the watershed. Generally, the complexity of the model is strongly related to the ease of the application. That means, the simpler models normally require the least effort to apply and least effort in calibration and validation as compare to more complex model [40–42].

Study conducted by [43] concluded that the accuracy of the model may vary and mostly inconclusive, and therefore controversial. However, studies show that most rainfall-runoff models will predict runoff and streamflow with similar accuracy [40–42, 44–48]. The accuracy of the model is determined by availability of the input data and observed input and output time series at various locations in a watershed [37]. The accuracy of the model can be measured using model performance evaluation techniques as suggested by [49, 50], and [51]. The sensitivity analyses of a model will reveal information on the relative importance of many input parameters as well as uncertainty in the model output [52].

5 Hydrological Modeling in Malaysia

In Malaysia, models from the United Kingdom (UK), United States of America (USA), and Australia are widely used for rainfall-runoff simulations. References [53–57] used the commercial software InfoWorks River Simulation (IWRS), while [58] used InfoWorks Collection System (IWCS) from the UK to simulate rainfall runoff. Hydrological model from USA such as HEC [59–61], L-THIA program [62], MIKE [63–65], and MAYA 3D have been used to simulate flood events [66]. The 2DSWAMP and XP-SWMM models from Australia are used by [67] and [68] to simulate runoff. These models except L-THIA are not publicly available. Most hydrological modeling studies in Malaysia were carried out using a one-dimensional approach except [65] and [67], which are two-dimensional approaches.

Commercial softwares from the UK, namely IWRS and IWCS have been widely used in simulating hydrological processes. Reference [58] used the IWCS model in their case study at Tanjong Malim, Perak, to draft a comprehensive stormwater management and flood mitigation plan for local authority. They found that this model has the ability to model the interaction between rivers and urban drainage. These results were very useful to design the flood mitigation plan based on the impact of various design storm events to the study area. Additionally, the study also provides the local authority valuable information to plan for existing and future land use changes. References [53–56] used the IWRS model to simulate the impact of runoff on the floodplains and the water quality of the river before and after the floods. They successfully simulated these events, and the information is very useful to the city council for flood mitigation design and water quality management. The IWRS software simulated the flood events at the Damansara Catchment (Kg. Melayu Subang—upstream, Taman TTDI Jaya, Batu 3, and Taman Sri Muda) in 2006, 2007, and 2008 [57]. The model has the ability to simulate and produce hydrograph that is very useful in designing structures such as retention ponds and flood walls especially in the low-lying area (Taman TTDI Jaya and Batu 3 in Shah Alam, Selangor).

Reference [59] used the commercial software HEC-HMS to determine the runoff and hydrograph characteristics modeling for an oil palm plantation at the Skudai River watershed. From the high index of the model performance (calibrated and validated models efficiency index of 0.81 and 0.82, respectively), they suggested that the model can be used for filling the missing runoff from rainfall data. The HEC-HMS software estimated the flood at Johor River [60]. Good agreement was shown in the evaluation of peak discharge, and the model performance is close to unity. The HEC-2 model has been adopted by [61] to predict water surface profiles for Langat River at Selangor and Linggi River at Negeri Sembilan. The HEC-2 model was developed by the US Army Corps of Engineers especially to compute water surface profile. This model has successfully predicted the water level at Linggi River, Negeri Sembilan with a small error. However, the application of this model to the Langat River, Selangor, does not reach good agreement. Therefore, they concluded the model can be applied at tropical rivers with reasonable error if the input data is good.

Modeling the effects of mangroves on tsunamis has been applied using commercial software from Australia, namely 2DSWAMP by [67]. This model has been used to investigate the pattern of mangrove trees distribution and the diameter that can affect the attenuation of tsunamis at Merbok Estuary, Kedah. A one-dimensional hydrodynamic model, namely XP-SWMM, has been used by [68] to simulate flood water of the Damansara River at TTDI, Selangor. They studied the time of water filling and volume of flood discharge (m^3/s) over the flood plain. They were successful in producing a Flood Hazard Mapping for Urban Area (FHMUA). A free commercial program, L-THIA (developed by Purdue University), simulated the runoff at Pinang River, Pulau Pinang [62]. Reference [65] analyzed the flood event at Damansara River, Selangor using MIKE-FLOOD. The two-dimensional simulation provides crucial information with regard to the direction and rate of flood propagation, the flood inundation extents as well as flood depths and flood durations which cannot be achieved using one-dimensional simulation.

References

1. C.K. Chang, A. Ab-Ghani, R. Abdullah, N.A. Zakaria, Sediment transport modeling for Kulim River—A case study. J. Hydro-Environ. Res. **2**(1), 47–59 (2008)
2. MSMA, Urban Stormwater Management Manual (MSMA Manual). Jabatan Pengairan dan Saliran Malaysia (DID—Department of Irrigation and Drainage, Malaysia) (2000), 13-3
3. J. Suhaila, A.A. Jemain, Fitting the statistical distributions to the daily rainfall amount in Peninsular Malaysia. Jurnal Teknologi, Universiti Teknologi Malaysia **46**(C), 33–48 (2007)
4. J. Suhaila, A.A. Jemain, Fitting the statistical distribution for daily rainfall in Peninsular Malaysia based on AIC criterion. J. Appl. Sci. Res. **4**(12), 1846–1857 (2008)

5. W.Z. Wan-Zin, A.A. Jemain, K. Ibrahim, J. Suhaila, M.D. Sayang, A comparative study of extreme rainfall in Peninsular Malaysia: With reference to partial duration and annual extreme series. Sains Malaysiana **38**(5), 751–760 (2009)
6. W.Z. Wan-Zin, A.A. Jemain, K. Ibrahim, The best fitting distribution of annual maximum rainfall in Peninsular Malaysia based on methods of L-moment and LQ-moment. Theor. Appl. Climatol. **96**, 337–344 (2009)
7. N.I.A. Nor, S. Harun, A.H.M. Kassim, Radial basis function modeling of hourly streamflow hydrograph. J. Hydrol. Eng. **12**(1), 113–123 (2007)
8. T. Wardah, S.H.A. Bakar, A. Bardossy, M. Maznorizan, Use of geostationary meteorological satellite images in convective rain estimation for flash-flood forecasting. J. Hydrol. **356**, 283–298 (2008)
9. M. Sulaiman, A. El-Shafie, O. Karim, H. Basri, Improved water level forecasting performance by using optimal steepness coefficients in an artificial neural network. Water Res. Manag. **25**, 2525–2541 (2011)
10. L.S. Kuchment, A.N. Gelfan, Estimation of extreme flood characteristics using physically based models of runoff generation and stochastic meterological inputs. Water Int. **27**(1), 77–86 (2002). International Water Resources Association
11. M. Eberle, H. Buiteveld, J. Beersma, P. Krahe, K. Wilke, Estimation of extreme floods in the river Rhine basin by combining precipitation-runoff modelling and a rainfall generator, in *International Conference on Flood Estimation*, March 6–8, Berne, Switzerland, ed. by M. Spreafico et al. CHR Report II-17, International Commission for the Hydrology of the Rhine basin (CHR), Lelystad, The Netherlands, 2002
12. A. Ab. Ghani, N.A. Zakaria, R.A. Falconer, River modeling and flood mitigation; Malaysian perspective. Proc. Inst. Civ. Eng. Water Manag. **162**, 1–2 (2009)
13. DNA, *Klang River Basin Environmental Improvement and Flood Mitigation Project*, Final Report (2002)
14. COMET, *Distributed Hydrologic Models for Flow Forecasts* (Part 1) (2012), http://www.meted.ucar.edu/hydro/DHM/dhm2/part1/print.htm#page_4.0.0. Accessed 1 Jan 2012
15. A.A. El-Nasr, G.A. Jeffrey, J. Feyen, J. Berlamont, Modelling the hydrology of a catchment using a distributed and a semi-distributed model. Hydrol. Process. **19**(3), 573–587 (2005)
16. D.L. Johnson, A.C. Miller, A spatially distributed hydrologic model utilizing raster data structures. Comput. Geosci. **23**(3), 267–272 (1997)
17. S.M.S. Shah, P.E. O'Connell, J.R.M. Hosking, Modelling the effects of spatial variability in rainfall on catchment response. 1. Formulation and calibration of a stochastic rainfall field model. J. Hydrol. **175**, 67–88 (1996)
18. J.C. Refsgaard, Paramerization, calibration and validation of distributed hydrological models. J. Hydrol. **198**, 69–97 (1997)
19. D.A. Woolhiser, Search for physically based runoff model—A hydrologic El Dorado? J. Hydraul. Eng. **122**(3), 122–129 (1996)
20. M.B. Smith, P.G. Konstantine, The distributed model intercomparison project (DMIP)—Preface. J. Hydrol. **298**, 1–3 (2004)
21. M.B. Smith, S. Dong-Jun, I.K. Victor, M.R. Seann, Z. Ziya, D. Qingyun, M. Fekadu, C. Shuzheng, The distributed model intercomparison project (DMIP): Motivation and experiment design. J. Hydrol. **298**, 4–26 (2004)
22. T.M. Carpenter, P.G. Konstantine, Continuous streamflow simulation with the HRCDHM distributed hydrologic model. J. Hydrol. **298**, 61–79 (2004)
23. B.E. Vieux, J.E. Vieux, Vflo™: A real-time distributed hydrologic model, in *Proceedings of the 2nd Federal Interagency Hydrologic Modeling Conference*, Las Vegas, NV, 2002
24. M.J. Shultz, J.C. Robert, Development, calibration, and implementation of a distributed model for use in real-time river forecasting, in *Proceedings of the 3rd Federal Interagency Hydrologic Modeling Conference*, Reno, NV, 2006
25. B.E. Vieux, G.M. Fekadu, Ordered physics-based parameter adjustment of a distributed model, in *Calibration of Watershed Models*, ed. by Q. Duan, H.V. Gupta, S. Sorooshian,

A.N. Rousseau, R. Turcotte, vol. 6 (American Geophysical Union, Washington, DC, 2003), pp. 267–281
26. M. Velleux, Spatially distributed model to assess watershed contaminant transport and fate, Ph.D. thesis, Department of Civil and Environmental Engineering, Colorado State University, 2005
27. M. Velleux, J. England, P.Y. Julien, TREX: Spatially distributed model to assess watershed contaminant transport and fate. Sci. Total Environ. **404**(1), 113–128 (2008)
28. D.E. Overton, M.E. Meadows, *Stormwater Modeling* (Academic, New York, 1976)
29. S.A. Dooge, Problems and methods of runoff-rainfall modeling, in *Mathematical Models for Surface Water Hydrology*, ed. by T.A. Cirinai, U. Maione, J.R. Wallis (Wiley, New York, 1977)
30. M.B. McPherson, *Urban Runoff Control Planning*, U. S. Environmental Protection Agency Report EPA-600/9-78-035, 1978
31. M. Church, Scales of process, modes of analysis: Multiples scales in rivers, in *Proceeding of 6th International Gravel-Bed Rivers Conference*, 2006
32. R.B. Grayson, G. Blöschl, Chapter 14: Summary of pattern comparison and concluding remarks, in *Spatial Patterns in Catchment Hydrology: Observations and Modelling*, ed. by R.B. Grayson, G. Blöschl (Cambridge University Press, Cambridge, 2000), pp. 355–367
33. H. Scoging, A.J. Parsons, A.D. Abrahams, Application of a dynamic overland-flow hydraulic model to a semi-arid hillslope, Walnut Gulch, Arizona, in *Overland Flow: Hydraulics and Erosion Mechanics*, ed. by A.J. Parsons, A.D. Abrahams, 1st edn. (Chapman & Hall, New York, 1993)
34. CWCB (Colorado Water Conservation Board), Chapter 12: Unique hydraulic conditions. *Floodplain Stormwater and Criteria Manual* (CWCB, WRC Engineering Inc., Denver, CO, 2008)
35. B.E. Vieux, Chapter 2: Data sources and structure, in *Distributed Hydrologic Modeling Using GIS*, ed. by B.E. Vieux. Water Science and Technology Library, 2nd edn. (Springer, New York, 2004)
36. M. Sivapalan, P.C. Young, 134: Downward approach to hydrological model development, in *Encyclopedia of Hydrological Sciences*, ed. by M.G. Anderson (Wiley, Chichester, 2004)
37. P.B. Bedient, W.C. Huber, Chapter 6: Urban hydrology, in *Hydrology and Floodplain Analysis*, ed. by P.B. Bedient, W.C. Huber, 3rd edn. (Prentice Hall, Upper Saddle River, NJ, 2002)
38. H.V. Knapp, A. Durgunoglu, T.W. Ortel, A review of rainfall-runoff modeling for stormwater management. *SWS Contract Report 516*, U. S. Geological Survey, Illinois District, 1991
39. A. Leopardi, E. Oliveri, M. Greco, Two-dimensional modeling of floods to map risk-prone area. J. Water Resour. Plann. Manag. **128**(3), 168–178 (2002)
40. WMO (World Meteorological Organization), Intercomparison of conceptual models used in operational hydrological forecasting. *WMO Operational Hydrology Report No. 7*, WMO—No. 429, Geneva, Switzerland, 1975
41. J. Abbott, Testing of several runoff models on an urban watershed. *U. S. Army Corps of Engineers*, Hydrologic Engineering Center Technical Paper No. 59, 1978
42. M. Franchini, M. Pacciani, Comparative analysis of several conceptual rainfall-runoff models. J. Hydrol. **122**, 161–219 (1991)
43. M.B. McPherson, *Urban Runoff Control Planning*, U. S. Environmental Protection Agency EPA-600/9-78-035, 1978
44. C.N. Papadakis, H.C. Preul, Testing of methods for determination of urban runoff. ASCE J. Hydraul. Div. **99**(HY9), 1319–1335 (1973)
45. D.P. Heeps, R.G. Mein, Independent comparison of the three urban runoff models. ASCE J. Hydraul. Div. **100**(HY7), 995–1009 (1974)
46. J. Marsalek, T.M. Dick, P.E. Wisner, W.G. Clarke, Comparative evaluation of three urban runoff models. Water Resour. Bull. **11**(2), 306–328 (1975)

47. K.M. Loague, R.A. Freeze, A comparison of rainfall-runoff modeling techniques on small upland catchments. Water Resour. Res. **21**(2), 229–248 (1985)
48. C.S. Melching, B.C. Yen, H.G. Wenzel, Output reliability as guide for selection of rainfall-runoff models. ASCE J. Water Resour. Plann. Manag. **117**(3), 383–398 (1991)
49. D.R. Legates, G.J. McCabe, Evaluating the use of "goodness-of-fit" measures in hydrologic and hydroclimatic model validation. Water Resour. Res. **35**(1), 233–241 (1999)
50. P. Krause, D.P. Boyle, F. Base, Comparison of different efficiency criteria for hydrological model assessment. Adv. Geosci. **5**, 89–97 (2005)
51. D.N. Moriasi, J.G. Arnold, M.W. Van-Liew, R.L. Bingner, R.D. Harmel, T.L. Veith, Model evaluation guidelines for systematic quantification of accuracy in watershed simulations. Am. Soc. Agric. Biol. Eng. **50**(3), 885–900 (2007)
52. W. James, A.W. Kuch, Chapter 9: Sensitivity-calibration decision-support tools for continuous SWMM modeling: A fuzzy-logic approach, in *Modeling the Management of Stormwater Impacts*, ed. by W. James, Monograph 6, Proceeding of Conference on the Stormwater and Water Quality Management Modeling Conference, Toronto (Computational Hydraulics International, Guelph, ON, 1997)
53. D.Y. Mah, F.J. Outuhena, S. Said, Use of Infoworks river simulation (RS) in Sungai Sarawak Kanan modeling. J. Inst. Eng. Malaysia **68**(1), 1–9 (2007)
54. D.Y. Mah, S.H. Lai, R.B. Chan, F.J. Putuhena, Investigative modeling of the flood bypass channel in Kuching, Sarawak, by assessing its impact on the inundations of Kuching-Batu Kawa-Bau Expressway. Struct. Infrastruct. Eng. 1–10 (2010)
55. D.Y.S. Mah, C.P. Hii, F.J. Putuhena, S. H. Lai, River modeling to infer flood management framework. *Technical Note*, 2011
56. S. Said, D.Y.S. Mah, P. Sumok, S.H. Lai, Water quality monitoring of Maong River, Malaysia, in *Proceedings of the Institute Civil Engineers, Water management*, vol. 162, 2009, pp. 35–40
57. A.N.A. Ali, J. Ariffin, Model reliability assessment: A hydrodynamic modeling approach for flood simulation in Damansara catchment using InfoWorks RS. Adv. Mater. Res. **250–253**, 3769–3775 (2011)
58. L.C. Siang, R. Abdullah, N.A. Zakaria, A.A. Ghani, C.C. Kiat, Modelling urban river catchment: A case study of Berop River, Tanjong Malim, Perak, in *2nd International Conference on Managing Rivers in the 21st Century: Solution Towards Sustainable Rivers Basin*, 2007, pp. 165–171
59. Z. Yusop, C.H. Chan, A. Katimon, Runoff characteristics and application of HEC-HMS for modeling stormflow hydrograph in oil palm catchment. Water Sci. Technol **56**(8), 41–48 (2007)
60. M.A.M. Razi, J. Ariffin, W. Tahir, N.A.M. Arish, Flood estimation studies using hydrologic system (HEC-HMS) for Johor River, Malaysia. J. Appl. Sci. **10**(11), 930–939 (2010)
61. T.A. Mohammed, S. Said, M.Z. Bardaie, S.N. Basri, Numerical simulation of flood levels for tropical rivers. IOP Conf. Ser. Mater. Sci. Eng. **17**, 1–10 (2011)
62. M.Y. Izham, U. Md. Uznir, A.R. Alias, K. Ayob, Georeference, rainfall-runoff modeling and 3D dynamic simulation: Physical influence, integration and approaches, COM. Geo, in *First International Conference on Computing for Geospatial Research and Application*, Washington, DC, 2010, pp. 1–8
63. L. Billa, S. Mansor, A.R. Mahmud, Spatial information technology in flood early warning systems: An overview of theory, application and latest developments in Malaysia. Disast. Prev. Manag. **13**(5), 356–363 (2004)
64. L. Billa, S. Mansor, A.R. Mahmud, A. Ghazali, Hydro-meteorological modeling and GIS for operational flood forecasting and mapping, in *The 2nd International Conference on Water Resources and Arid Environment*, 2006, pp. 1–16
65. S.P. Lim, H.S. Cheok, Two-dimensional flood modeling of the Damansara River. Proc. Inst. Civ. Eng. Water Manag. **162**, 13–24 (2009)
66. J.N. Ghazali, A. Kamsin, A real time simulation of flood hazard, in *Fifth International Conference on Computer Graphics, Imaging and Visualization*, 2008, pp. 393–397

67. F.Y. Teo, R.A. Falconer, B. Lin, Modelling effects of mangroves on tsunamis. Proc. Inst. Civ. Eng. Water Manag. **162**, 3–12 (2009)
68. M.E. Toriman, A.J. Hassan, M.B. Gazim, M. Mokhtar, S.A. Sharifah-Mastura, O. Jaafar, O. Karim, N.A. Abdul-Aziz, Integration of 1-d hydrodynamic model and GIS approach in flood management study in Malaysia. Res. J. Earth Sci. **1**(1), 22–27 (2009)

Multiday Rainfall Simulations for Malaysian Monsoons

Nur S. Muhammad and Pierre Y. Julien

Abstract This study examines the suitability of the discrete autoregressive and moving average [DARMA(1,1)] model to simulate the sequences of daily rainfall data in Malaysia. The daily monsoon rainfall data recorded at Subang Airport are used to test this modeling approach. The autocorrelation function and probability distributions of wet and dry run lengths estimated from the DARMA(1,1) model matched the sample values quite well. Both theoretical and sample autocorrelation functions slowly decay to zero at day 15. The theoretical probability distribution for two consecutive wet days estimated for the DARMA(1,1) is 0.1966, while the observed rainfall data give a probability of 0.2066. Additionally, the sum of squared errors for the DARMA(1,1) model were very small, i.e., 0.0015. Furthermore, two simulations were done, i.e., 100 samples of 9,600 days (simulation A) and a very long sequence of 1,000,000 days (simulation B). This was done to test the capability of DARMA(1,1) to model a long sequence of daily rainfall. The statistics examined in this study include the lag-1 autocorrelation coefficient (lag-1 ACF) and the maximum wet and dry run lengths. Generally, the statistics of generated rainfall for simulation A fall within two standard deviations from the sample. The diversions of these statistics were reasonable, considering the sample size used in this study. The estimated lag-1 ACF for simulation B was slightly lower than the sample. The maximum wet and dry run lengths were much higher than the observed data because of the different sample sizes. It is concluded that the DARMA(1,1) model is able to simulate the long sequences wet and dry days and preserving the statistics within reasonable accuracy.

Keywords Multiday rainfall • Monsoon rainfall precipitation • Stochastic modeling • DARMA(1,1)

N.S. Muhammad (✉)
Department of Civil and Structural Engineering, Faculty of Engineering and Built Environment, Universiti Kebangsaan Malaysia, Bangi 43600, Selangor, Malaysia
e-mail: shazwani.muhammad@ukm.edu.my

P.Y. Julien
Department of Civil and Environmental Engineering, Colorado State University, Fort Collins, CO 80523, USA

S.H. Abu Bakar et al. (eds.), *ISFRAM 2014*, DOI 10.1007/978-981-287-365-1_10

1 Introduction

Malaysia is located near the equator and experiences hot and humid climate throughout the year. The country is influenced by two major seasons, namely the North East (NE) and South West (SW) monsoons. The NE monsoon typically occurs from November to March; while the SW monsoon is from May to September. April and October are known as inter-monsoons. These monsoons bring lots of moisture, and the country receives a total rainfall of between 2,000 and 4,000 mm with 150–200 rainy days annually [1].

Therefore, multiday rainfall events are common in Malaysia and cause particularly devastating floods, especially on large watersheds [2]. Historical events include the two extreme Kota Tinggi floods in December 2006 and January 2007, which resulted from more than 350 and 450 mm of cumulative rainfall in less than a week. The estimated economic loss reached half a billion US dollars and more than 100,000 local residents had to be evacuated [3].

Considering the nature of climate in Malaysia and the devastating consequences of multiday rainfall events as discussed above, generating the sequences of daily rainfall using stochastic model should be given serious consideration. Additionally, the planning and designing of water resources projects require the analysis of reliable and long-term hydrological data such as rainfall and streamflow, which can be scarce in developing countries such as Malaysia. Therefore, engineers should consider generating synthetic hydrological data using the parameters estimated from the records available in their analysis.

The low order discrete autoregressive family models, i.e., discrete autoregressive [DAR(1)] and discrete autoregressive and moving average [DARMA(1,1)], are frequently used in simulating the sequence of daily rainfall. DAR(1) is also known as the first-order Markov Chain. Reference [4] introduces the concept of Markov Chain model to simulate the occurrence of daily rainfall at Tel Aviv. This model assumes that the probability of rain depends only on the current state (wet or dry) and will not be influenced by its past behavior. References [5–13] are among the studies that were successful in modeling the sequence of rainy and dry days using first-order Markov Chains.

Reference [7] applied the first-order Markov Chain to produce ten synthetic sequences of daily rainfall at Universiti Pertanian Malaysia (UPM), Serdang, Selangor, Malaysia. The authors gathered the daily rainfall data from 1968 to 1978 and divided the data into 11 states according to the amount. The simulations were done for the different monsoon seasons in Malaysia: the Northeast (from November to March), two transitional periods (April and October), and the Southwest (from May to September). They found that the first-order Markov Chain was able to reproduce the daily rainfall of any length in the area. However, the synthetic daily rainfall was generated for a period of 1 year only. Therefore, this research did not indicate if the first-order Markov Chain is able to simulate long daily rainfall sequences.

Reference [14] discusses the optimum order of Markov Chain for daily rainfall during North East (NE) and South West (SW) monsoons using two different thresholds, i.e., 0.1 and 10.0 mm. Eighteen rainfall stations in Peninsular Malaysia were used in this study. They found that the optimum order of a Markov Chain varies with the location, monsoon seasons, and the level of threshold. For stations located in the northwestern and eastern regions of Peninsular Malaysia, the occurrence of rainfall (threshold level 10.0 mm) for both monsoons can be represented using a first-order Markov Chain model. Other than that, Markov Chain models of higher order are suitable to represent rainfall occurrence, especially during the NE monsoon, for both levels of threshold. This study shows that the rainfall events in Peninsular Malaysia requires a longer memory model than first-order Markov Chain to simulate the sequence of daily rainfall. However, high order Markov Chain increased the model uncertainty because more parameters have to be used [15]. It also makes the calculations more complex. These disadvantages can be overcome by DARMA(1,1) model.

The DARMA model is a simple tool to model stationary sequences of dependent discrete random variables with specified marginal distribution and correlation structure [16]. The model is stationary; therefore, the rainfall data should be divided into their respective seasons in order to consider the seasonal variations. References [17] and [18] use this model to simulate the sequence of daily rainfall using data collected from several stations in the Netherlands, Suriname, India, and Indonesia. They concluded that that DARMA(1,1) is successful in simulating the daily rainfall in tropical and monsoon areas, where prolonged dry and wet seasons may occur. The DARMA(1,1) model provides longer persistence than the first-order Markov Chain. Other studies that use this model to simulate the sequence of daily rainfalls include [19–23].

This chapter discusses the simulation of daily rainfall using the DARMA(1,1) model. The daily rainfall measurements from the Subang Airport were used in this study. This station is chosen because it provides a long and reliable record of 52 years, i.e., from 1960 to 2011. Separate analyses are conducted for NE and SW monsoons. However, only the results for NE monsoons will be presented here.

2 DARMA(1,1) Model

The DARMA(1,1) model is represented as [16]

$$X_t = U_t Y_t + (1 - U_t) A_{t-1} \tag{1}$$

with

$$X_t = \begin{cases} Y_t & \text{with probability } \beta \\ A_{t-1} & \text{with probability } (1-\beta) \end{cases}$$

where U_t is an independent random variable taking value of 0 or 1 only such that

$$P(U_t = 1) = \beta = 1 - P(U_t = 0) \tag{2}$$

Y_t is independent and identically distributed (i.i.d) random variable having a common probability of $\pi_k = P(Y_t = k)$ and $k = 0, 1$, and A_t is an autoregressive component given by

$$A_t = \begin{cases} A_{t-1} & \text{with probability } \lambda \\ Y & \text{with probability } (1-\lambda) \end{cases}$$

It should be noted that A_t is a first-order Markov Chain, and the process of simulation is assumed to start at A_{-1} [18]. This variable has the same probability distribution as Y_t but is independent of Y_t. The X_t is not Markovian, but (X_t, A_t) forms a first-order bivariate Markov Chain.

The theoretical autocorrelation function of the DARMA(1,1) model is [18]

$$\operatorname{corr}(X_t, X_{t-k}) = r_k(X) = c\lambda^{k-1}, \quad k \geq 1 \tag{3}$$

where r_k is the lag-k (days) autocorrelation function and

$$c = (1-\beta)(\beta + \lambda - 2\lambda\beta) \tag{4}$$

The sample autocorrelation function (r_k) for the time series is estimated based on the sequences of dry and rainy days and not the rainfall amounts [22].

$$r_k = \left[\sum_{t=1}^{N-k} (x_t - \overline{x})(x_{t+k} - \overline{x})\right] \left[\sum_{t=1}^{N} (x_t - \overline{x})^2\right]^{-1} \tag{5}$$

$$\overline{x} = \frac{1}{N}\sum_{t=1}^{N} x_t \tag{6}$$

where N is the sample size.

Three parameters of the DARMA(1,1) model are π_0 or π_1, λ, and β. The parameter λ may be estimated from the lag-1 autocorrelation coefficient as given in (3) and (4). The parameters π_0 or π_1 may be estimated from (7) and (8).

$$\pi_0 = \frac{\overline{T_0}}{\overline{T_0} + \overline{T_1}} \tag{7}$$

$$\pi_1 = 1 - \pi_0 \tag{8}$$

where $\overline{T_0}$ is the mean run length for dry days, and $\overline{T_1}$ is the mean run length for wet days.

The estimation of λ may be determined by minimizing (9) using and [18] suggested using the ratio of the second to the first autocorrelation coefficients as an initial estimator for λ, as shown in (10).

$$\phi(\lambda) = \sum_{k=1}^{M} \left[r_k - c\lambda^{k-1}\right]^2; \quad k \geq 1 \tag{9}$$

$$\hat{\lambda} = \frac{r_2}{r_1} \tag{10}$$

in which M is the total number of lags considered, and c can be determined from the lag-1 autocorrelation coefficient of the DARMA(1,1) model. Then β was estimated from

$$\hat{\beta} = \frac{(3\hat{\lambda} - 1) \pm \sqrt{(3\hat{\lambda} - 1)^2 - 4(2\hat{\lambda} - 1)(\hat{\lambda} - \hat{c})}}{2(2\hat{\lambda} - 1)} \tag{11}$$

The probability distributions of wet and dry run lengths for the DARMA(1,1) model are given in [16].

3 Results and Discussion

3.1 *Simulating the Sequences of Daily Rainfall Using DARMA(1,1)*

The occurrence of rainfall event in this study is treated as a discrete variable. Threshold value is determined using the Von Neumann ratio [24]. The definition of wet is any day with rainfall of more than 0.1 mm, and a dry day received less than or equal to the said amount. This value is chosen because it ensures homogeneity of the time series.

The first step in simulating the sequences of daily rainfall using DARMA(1,1) is to estimate the model parameters, i.e., π_0 or π, λ, and β. The average wet and dry run lengths are calculated from the observed daily rainfall dataset, and the values are $\overline{T_1} = 3.00$ days and $\overline{T_0} = 2.19$ days. Following this, the estimated probabilities of a wet and dry day are $\hat{\pi}_1 = 0.58$ and $\hat{\pi}_0 = 0.42$, respectively. The parameter λ is calculated using (9), based on the Newton–Raphson iteration techniques and initial

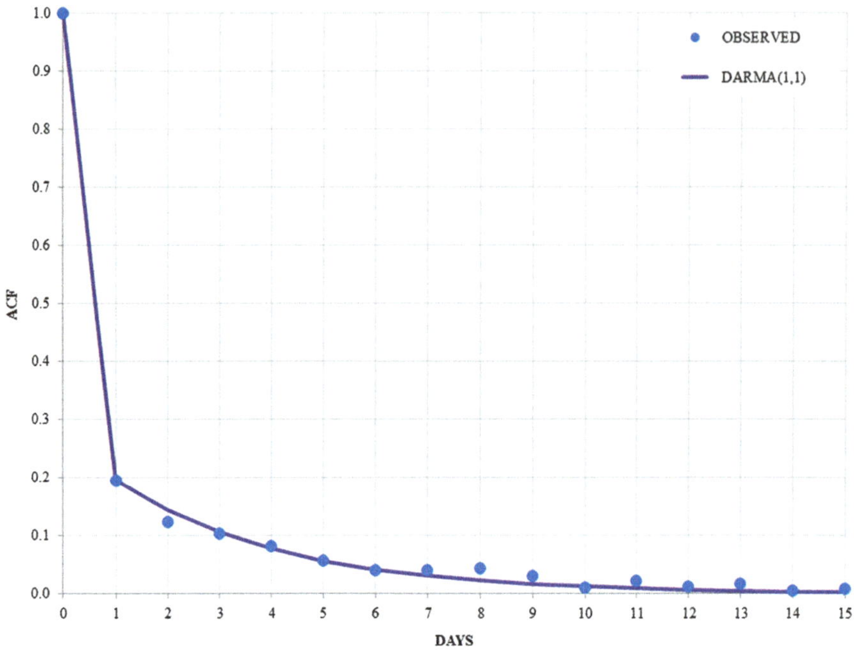

Fig. 1 Sample and theoretical ACF

estimation using (10). Then (11) is applied to estimate the parameter β. These give estimated model parameters as $\hat{\lambda} = 0.7339$ and $\hat{\beta} = 0.5775$.

After all parameters were determined, the theoretical and sample autocorrelation functions (ACFs) are estimated using (3) and (5), respectively. These values are compared using graphical method, as shown in Fig. 1. Excellent agreements between the sample and theoretical ACFs for the estimated DARMA(1,1) model are shown. The ACFs matched even though the number of lag (day) increased. Both sample and theoretical ACFs decay slowly and eventually reach to nearly zero at day 15.

Other than the ACFs, the theoretical and sample probability distributions of wet and dry run lengths were also estimated and compared to further examine the suitability of DARMA(1,1) model to simulate the sequence of daily rainfall.

Excellent agreements were observed between the theoretical and sample probability distributions of wet and dry run lengths. For example, the theoretical probability distribution for two consecutive wet days estimated for the DARMA (1,1) was 0.1966, while the observed rainfall data give a probability of 0.2066. Additionally, the sum of squared errors for the DARMA(1,1) model was very small, i.e., 0.0015. The probability distributions of wet run lengths are illustrated in Fig. 2. Based on the analyses performed on the ACFs and probability distributions of wet and dry run lengths, the authors concluded that the DARMA(1,1) model is suitable to represent the occurrence of daily rainfall at Subang Airport.

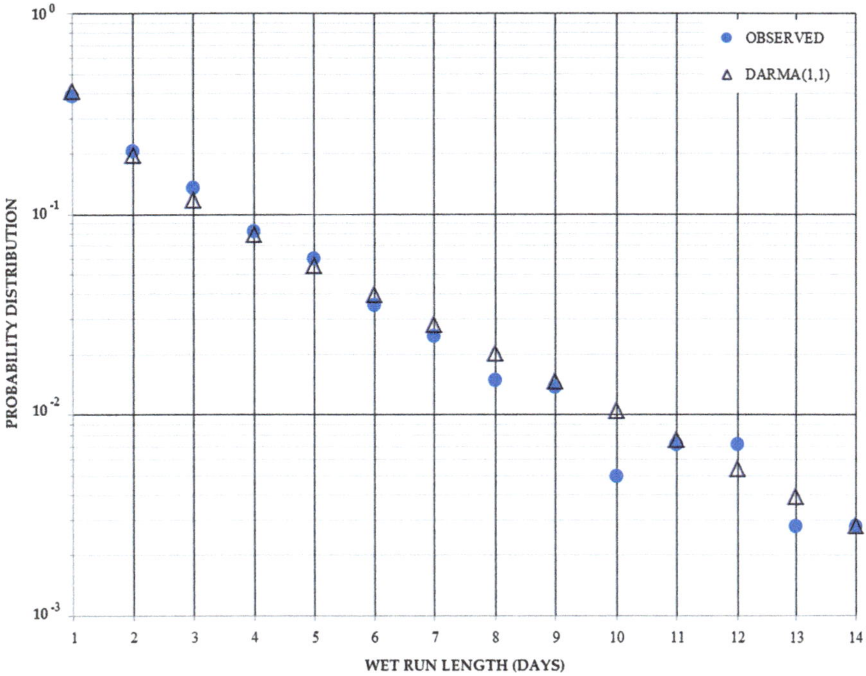

Fig. 2 Probability distribution of wet run lengths

We further analyzed the simulations of daily rainfall sequences using DARMA (1,1) model. A total of two simulations are performed, i.e., simulation A and simulation B. Simulation A was performed to compare it with the statistics of measured rainfall at Subang Airport, in order to ensure that the model applications were persistent and examine the variability of the simulated samples. This simulation consists of 100 samples of 9,600 days. A sample size of 9,600 days is chosen because this is about the same as the sample data for NE monsoon. On the other hand, simulation B consists of a sample with the size of 1,000,000 days. This was done to test the capability of DARMA(1,1) to model a long sequence of daily rainfall.

The statistics examined in this study include the lag-1 autocorrelation coefficient (lag-1 ACF) and the maximum wet and dry run lengths, which are given in Table 1.

Generally, the statistics of generated rainfall for simulation A fall within two standard deviations from the sample. The diversions of these statistics were reasonable, considering the sample size used in this study.

The estimated lag-1 ACF for simulation B was slightly lower than the sample. The maximum wet and dry run lengths were much higher than the observed data because of the difference in sample sizes. These values are significant in estimating the highest possible consecutive wet and dry days over a long period of time. This information is valuable for water resources engineers to plan their strategy. For

Table 1 Statistics for observed and simulated daily rainfall during NE monsoon

Statistics	Sample	Simulation A		Simulation B
		Mean	Standard deviation	
Lag-1 ACF	0.196	0.179	0.012	0.181
Maximum wet run length (days)	31	24	4	34
Maximum dry run length (days)	21	16	3	25

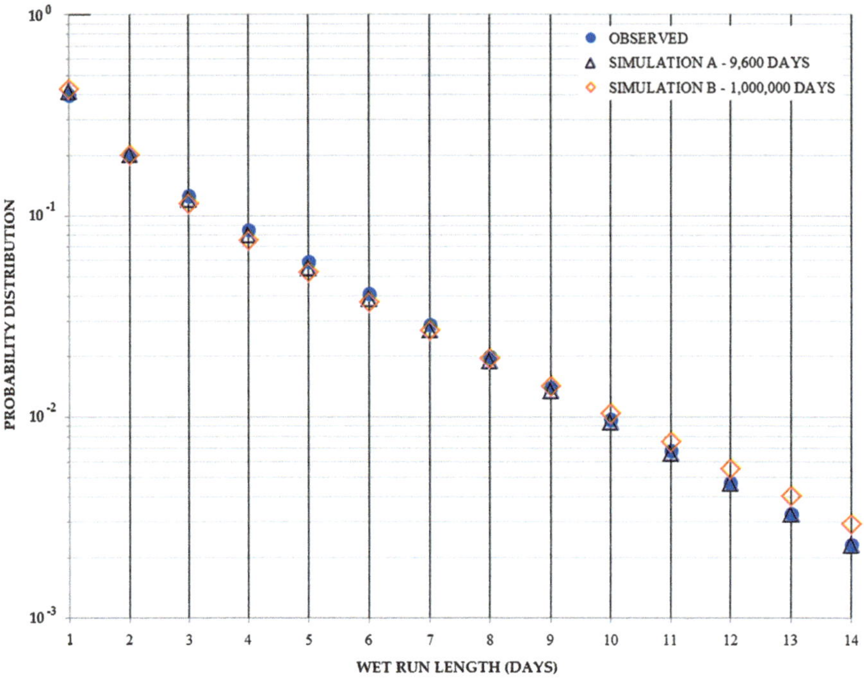

Fig. 3 Probability distributions of wet run lengths generated from simulations A and B

example, what is the best way to ensure constant water supply if it does not rain for 25 days or longer.

In terms of probability distribution functions of wet and dry run lengths, both simulations A and B show good agreement with the sample. As an example, for wet run lengths of 10 days, the estimated probability distribution of the sample and simulations A and B were 0.0097, 0.0094, and 0.0103, respectively. Similarly, the estimated probability distributions for dry run lengths of 7 days were 0.0130, 0.0134, and 0.0135 for the sample and simulations A and B, respectively. Figure 3 detailed the wet run lengths of the sample and simulations A and B. These results indicate that the DARMA(1,1) model is able to simulate the long sequences wet and dry days. These characteristics are important because Malaysia is affected by extreme floods that occur as a result of multiday rainfall events.

4 Conclusions

The DARMA(1,1) model was applied to the daily rainfall data at Subang Airport from 1960 to 2011. Model parameters were estimated from the sample. The autocorrelation functions of the sample are similar to the theoretical values. Additionally, the probability distributions of wet and dry run lengths also show good agreements between the sample and theoretical estimations. These results indicate that DARMA(1,1) was suitable to model the sequences of daily rainfall at Subang station.

Two simulations were performed, including a very long sequence of daily rainfall (1,000,000 days). The statistics estimated for this exercise include lag-1 autocorrelation functions and maximum wet and dry run lengths. The generated sequences' statistics fall within two standard deviations of sample. Additionally, the probability distributions of wet and dry run lengths estimated for the simulated rainfall sequences using DARMA(1,1) are also comparable with the sample. It is concluded that the DARMA(1,1) model is able to simulate the long sequences wet and dry days and preserving the statistics within reasonable accuracy.

Acknowledgment This study was supported by the Universiti Kebangsaan Malaysia (grant number GGPM-2014-046). Also, we would like to thank the Department of Meteorology, Malaysia, for providing the rainfall data. Their assistance is highly appreciated. The authors also appreciate the comments from three anonymous reviewers that led to the improvement of this article.

References

1. J. Suhaila, A.A. Jemain, Fitting daily rainfall amount in Malaysia using the normal transform distribution. J. Appl. Sci. **7**(14), 1880–1886 (2007)
2. J. Abdullah, Distributed runoff simulation of extreme monsoon rainstorms in Malaysia using TREX, Ph.D. thesis, Department of Civil and Environmental Engineering, Colorado State University, Fort Collins, CO, 2013
3. S. Abu Bakar, M.F. Yusuf, W.S. Amly, Johor hampir lumpuh (…), Utusan Melayu, 15 Jan 2007, http://www.utusan.com.my/utusan/info.asp?y=2007&dt=0115&pub=Utusan_Malaysia&sec=Muka_Hadapan&pg=mh_01.htm. Accessed 11 Feb 2013
4. K.R. Gabriel, J. Neuman, A Markov chain model for daily rainfall occurrence at Tel Aviv. Q. J. Roy. Meteorol. Soc. **88**(375), 90–95 (1962)
5. C.T. Haan, D.M. Allen, J.O. Street, A Markov chain model of daily rainfall. Water Resour. Res. **12**(3), 443–449 (1976)
6. R.W. Katz, Precipitation as a chain-dependent process. J. Appl. Meteorol. **16**(7), 671–676 (1977)
7. M.Z. Bardaie, A.C.A. Salam, A stochastic model of daily rainfall for Universiti Pertanian Malaysia, Serdang. Pertanika **4**(1), 1–9 (1981)
8. J. Roldan, D.A. Woolhiser, Stochastic daily precipitation models 1: A comparison of occurrence process. Water Resour. Res. **18**(5), 1451–1459 (1982)

9. M.J. Small, D.J. Morgan, The relationship between a continuous-time renewal model and a discrete Markov chain model of precipitation occurrence. Water Resour. Res. **22**(10), 1422–1430 (1986)
10. O.D. Jimoh, P. Webster, The optimum order of a Markov chain model for a daily rainfall in Nigeria. J. Hydrol. **185**, 45–69 (1996)
11. T.C. Sharma, Simulation of the Kenyan longest dry and wet spells and the largest rain-sum using a Markov model. J. Hydrol. **178**, 55–67 (1996)
12. S.K. Tan, S.Y. Sia, Synthetic generation of tropical rainfall time series using an event-based method. J. Hydraul. Eng. **2**(2), 83–89 (1997)
13. D.S. Wilks, Multisite generalization of a daily stochastic precipitation generation model. J. Hydrol. **210**, 178–191 (1998)
14. S.M. Deni, A.A. Jemain, K. Ibrahim, Fitting optimum order Markov chain models for daily rainfall occurrences in Peninsular Malaysia. Theor. Appl. Meteorol. **97**, 109–121 (2009)
15. P.A. Jacobs, P.A.W. Lewis, Stationary discrete autoregressive-moving average time series generated by mixtures. J. Time Ser. Anal. **4**(1), 19–36 (1983)
16. P.A. Jacobs, P.A.W. Lewis, Discrete time series generated by mixtures. I: Correlational and runs properties. J. Roy. Stat. Soc. B (Methodol.) **40**(1), 94–105 (1978)
17. T.A. Buishand (ed.), *Stochastic Modelling of Daily Rainfall Sequences* (Veenman & Zonen, Wageningen, 1977)
18. T.A. Buishand, The binary DARMA (1,1) process as a model for wet-dry sequences. Technical Note, Agriculture University, Wageningen, 1978
19. T.J. Chang, M.L. Kavvas, J.W. Delleur, Stochastic daily precipitation modeling and daily streamflow transfer processes. Technical Report No. 146, Purdue University Water Resources Research Center, 1982
20. T.J. Chang, M.L. Kavvas, J.W. Delleur, Modeling of sequences of wet and dry days by binary discrete autoregressive moving average processes. Am. Meteorol. Soc. **23**, 1367–1378 (1984)
21. T.J. Chang, M.L. Kavvas, J.W. Delleur, Daily precipitation modeling by discrete autoregressive moving average processes. Water Resour. Res. **20**(5), 565–580 (1984)
22. J.W. Delleur, T.J. Chang, M.L. Kavvas, Simulation models of sequences of dry and wet days. J. Irrigat. Drain. Eng. **115**(3), 344–357 (1989)
23. K. Cindrić, The statistical analysis of wet and dry spells by binary DARMA (1,1) model in Split, Croatia, in *BALWOIS Conference 2006*, Ohrid, 2006
24. J. Von Neumann, Distribution of the ratio of the mean square successive difference to the variance. Ann. Math. Stat. **12**(4), 367–395 (1941)

Part III
Modeling and Socioeconomic Impact

Estimating Floods from an Ungauged River Basin Using GIUH-Based Nash Model

Bhabagrahi Sahoo and P.G. Saritha

Abstract Among the existing mesoscale rainfall-runoff models, the geomorphological instantaneous unit hydrograph (GIUH) models can predict the streamflow reasonably well in less data availability conditions using the measurable basin geomorphology. In light of this, the rainfall-runoff transformation process of the Baitarani River Basin at Anandapur in eastern India was studied herein using the GIUH-based Nash model. In this model, the gamma function of the Nash model is approximated that does not need the computation of either the complete or incomplete gamma functions. To evaluate the efficacy of this model, four performance evaluation measures of Nash–Sutcliffe efficiency, percentage error in volume, percentage error in peak, and percentage error in time to peak are used. The results reveal that the GIUH-based Nash model performs with the average Nash–Sutcliffe model efficiency of 87.21($\pm$8.05) % with overprediction of the runoff volume by 29.91($\pm$18.76) %. However, this model could estimate the runoff peak more accurately with an average peak error of 3.86 ($\pm$2.50) % having an average time to peak error of 2.08($\pm$4.17) %. Hence, overall, the GIUH-based Nash model has the capability of estimating the basin-scale runoff reasonably well.

Keywords Geomorphology • GIUH • Nash model • Runoff • Ungauged basin

1 Introduction

The extreme events of high flood cause damages, resulting in the loss of human life, property, and crop and other losses due to the disruption of economic activity. Further, estimation of flood peaks is also required for the design of bridges, culverts, waterways, and spillways for dams and the estimation of scour at a hydraulic structure. Hence, estimation of river runoff from any ungauged river basin is very much important for flood management.

Among the various rainfall-runoff models available in the literature, the techniques which need long-term observed time series data are not of much use for the

B. Sahoo (✉) • P.G. Saritha
School of Water Resources, Indian Institute of Technology Kharagpur, Kharagpur, West Bengal 721302, India
e-mail: bsahoo2003@yahoo.com

S.H. Abu Bakar et al. (eds.), *ISFRAM 2014*, DOI 10.1007/978-981-287-365-1_11

flood estimation in most of the small to medium-sized river basins worldwide due to poor streamflow gauging network. In such situations, one may use the technique of regionalization of model parameters using the observed data from the hydro-meteorologically similar river basins. However, determination of the hydro-meteorological similarity of river basins is still a challenging task for river engineers. Hence, the conceptual models, such as, the geomorphologic instantaneous unit hydrograph (GIUH)-based Nash model, which uses the measurable catchment and rainfall characteristics to estimate its model parameters, could be a better option for estimating the runoff from any ungauged river basin.

By linking the IUH with the basin geomorphology, Rodriguez-Iturbe and Valdes [1] advocated the concept of GIUH. These geomorphological parameters are Horton's bifurcation ratio, area ratio, and length ratio [2], in which the channel network is described using the Strahler [3] ordering scheme. Subsequently, to estimate the probability density function of travel times of water, Rinaldo and Rodriguez-Iturbe [4] and Rodriguez-Iturbe and Rinaldo [5] linked this to the stream networks and other landscape features. Subsequently, without using the conventional exponential distributions of stream holding times of a water droplet, Van der tak and Bras [6] introduced the gamma distribution-based GIUHs that better fit the data-driven IUHs. Gupta et al. [7] developed a generalized and simplified form of the GIUH modifying the formulation of Rodriguez-Iturbe and Valdes [1]. However, these GIUH-based approaches mostly account for the catchment geomorphology in a spatially lumped manner to calculate the peak and time to peak of a runoff event, resulting in a triangular hydrograph. Hence, to address this issue of getting the natural shape of the runoff hydrograph, Bhaskar et al. [8], Sahoo et al. [9], and Kumar and Kumar [10] linked the GIUH models with the conventional Clark and Nash IUH models for parameter estimation. Recently, the GIUH models have been advanced for hydrograph synthesis, adding a new extent to hydrologic simulations. All the necessary geomorphologic data can be obtained from the topographic maps or from digital elevation models. Further, the applicability of the Nash model requires the computation of incomplete gamma function. Hence, there is a need to simplify the gamma function in the GIUH-based Nash model framework.

In light of the above discussion, this study has been undertaken to evaluate the suitability of the GIUH-based Nash model to model the rainfall-runoff transformation process of the Baitarani River Basin at Anandapur in eastern India.

2 The GIUH-Based Nash Model

To estimate the lumped runoff from an ungauged river basin, the GIUH formulation by Rodriguez-Iturbe and Valdes [1] provides a triangular runoff hydrograph with its peak and time to peak, respectively, expressed as

$$q_{\rm p} = 1.31\left(V_{\rm p}/L_{\Omega}\right)R_{\rm L}^{0.43} \tag{1}$$

$$t_{\rm p} = 0.44\left(L_{\Omega}/V_{\rm p}\right)\left(R_{\rm B}/R_{\rm A}\right)^{0.55}R_{\rm L}^{-0.38} \tag{2}$$

where $q_{\rm p}$ = peak flow depth per unit time per unit effective rainfall ($\rm h^{-1}$), $V_{\rm p}$ = expected peak velocity (m/s), L_{Ω} = length of the highest order stream (km), $R_{\rm B}$ = Horton's bifurcation ratio, $R_{\rm A}$ = Horton's area ratio, $R_{\rm L}$ = Horton's length ratio, and $t_{\rm p}$ = time to peak (h).

The peak velocity, $V_{\rm p}$, is the climate forcing that depends on the effective rainfall on the river basin. Multiplying (1) with (2) would result in a dimensionless product denoting the river basin characteristics only which is independent of the climate forcing, expressed as

$$q_{\rm p}t_{\rm p} = 0.5764\left(R_{\rm B}/R_{\rm A}\right)^{0.55}R_{\rm L}^{-0.38} \tag{3}$$

Further, the Nash IUH model is given as [11, 12]

$$u(t) = \left[1/(K\Gamma(N))\right](t/K)^{N-1}\exp(-t/K) \tag{4}$$

where N = shape parameter of the Nash IUH model indicating the number of linear reservoirs cascades, K = scale parameter of the Nash IUH model indicating the storage effect of the basin, and $\Gamma(\cdot)$ is the gamma function which can be expressed using the Nemes' approximation as [13]

$$\Gamma(N) = \left(\frac{N}{e}\right)^{N}\sqrt{\frac{2\pi}{N}}\left(1+\frac{1}{12N^2-0.1}\right)^{N} \tag{5}$$

where $e \approx 2.71828$, the Euler's number.

Note that to get better convergence properties of the gamma function, Nemes [13] converted the Stirling–De Moivre asymptotic series approximation [14] to the gamma function into a new one, which is having a better accuracy than the Stirling [15], Laplace, and Ramanujan [16] approximations.

Considering the time to peak of the Nash IUH as $t = t_{\rm p}$, $du/dt = 0$. Consequently, the differentiation of (4) with respect to $t = t_{\rm p}$ will yield the property of the gamma probability density function, expressed as

$$t_{\rm p} = K(N-1) \tag{6}$$

Using (4), one can also express the peak flow (at $t = t_{\rm p}$) as

$$q_{\rm p} = \left[1/(K\Gamma(N))\right]\left(t_{\rm p}/K\right)^{N-1}\exp\left(-t_{\rm p}/K\right) \tag{7}$$

Now, using (5), (6), and (7) into (3), we will get

$$2.71828\left(1-\frac{1}{N}\right)\left(\frac{12N^2-0.1}{12N^2+0.9}\right)^N \exp(1-N)\sqrt{\frac{N}{2\pi}} = 0.5764\,(R_B/R_A)^{0.55}R_L^{-0.38} \tag{8}$$

Note that (8) is a function of the measurable basin characteristics only. By estimating the Nash IUH model parameters, N and K, from the GIUH model given by (1), (2), and (3), the full shape of the runoff hydrograph can be obtained by using (4). This forms the GIUH-based Nash model. The shape parameter, N, can be obtained by solving the nonlinear equation (8) by the Newton–Raphson optimization method. Similarly, the scale parameter, K, can be estimated by using (6) from the known values of N and t_p estimated using (8) and (2), respectively. The Nash IUH ordinates are convoluted with the excess rainfall to estimate the ordinates of the direct surface runoff (DSRO) hydrograph. Addition of baseflow with the ordinates of the DSRO hydrograph yields the total runoff hydrograph.

3 Study Area

The Baitarani River basin at Anandapur (Fig. 1) having an area of about 8,580 km^2 is located in the Eastern India in between the longitudes of 85°10′E and 87°03′E and latitudes of 20°35′N and 22°15′N. This basin receives an average annual rainfall of 1,187 mm. The daily rainfall data in the basin are available at eight gauging stations. Other daily meteorological data, viz., solar radiation, maximum temperature, minimum temperature, dew point temperature, average temperature, relative humidity, and wind speed at $1° \times 1°$ grid size, were downloaded from http://power.larc.nasa.gov. The daily runoff data at Anandapur gauging station was collected from the Central Water Commission (CWC), Bhubaneswar, Odisha, India, for the period of 1996–1999. The digital elevation model (DEM) data of 90×90 m resolution for the study area was downloaded from http://srtm.csi.cgiar.org.

For estimating the geomorphologic parameters of the GIUH-based Nash model, the stream ordering of the basin was carried out using the ArcMAP9.3 software which envisaged that this is a fourth order basin, having the length of the highest order stream, $L_\Omega = 19.39$ km. The values of Horton's bifurcation ratio (R_B), length ratio (R_L), and area ratio (R_A) of this basin are 4.05, 2.15, 4.89, and 19.39 km, respectively.

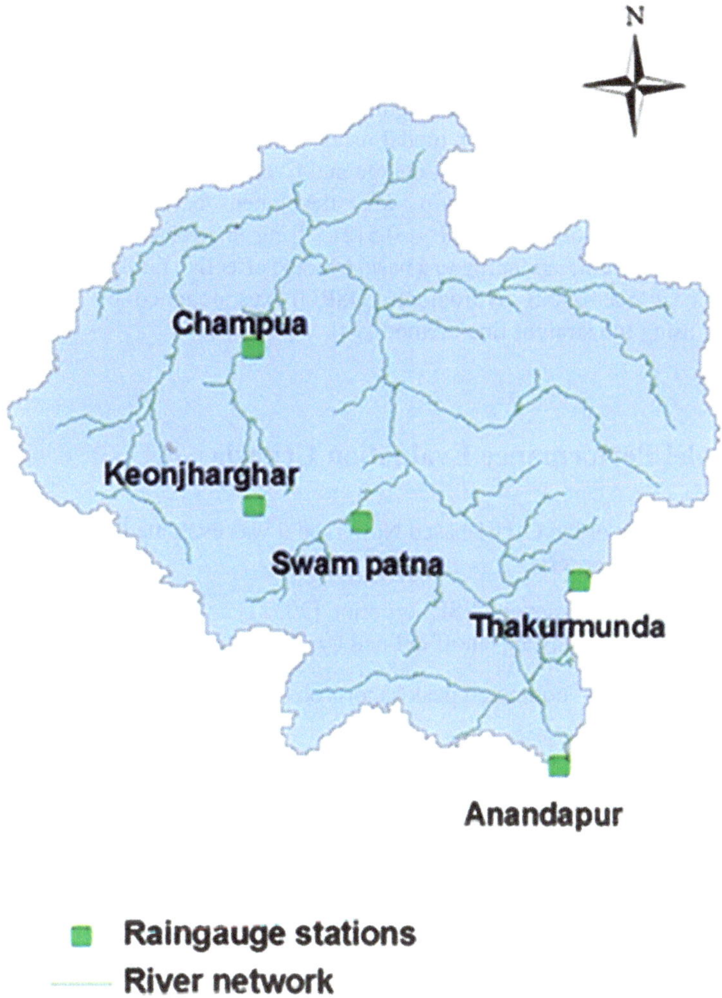

Fig. 1 The Baitarani River basin at Anandapur

4 Development of Effective Rainfall — Peak Velocity Relationship

The effective rainfall, I_r, in the Baitarani River basin for the years 1996–1999 was computed by water balance method as

$$I_r = R - A_b - ET_c - F \tag{9}$$

where R = basin average rainfall estimated using the Thiessen polygon method, A_b = abstraction losses (considered as 4 % in this study), ET_c = crop evapotranspiration estimated as $ET_c = K_c \times ET_o$ in which K_c = average crop coefficient estimated based on the areal weight of the individual crop coefficients of different land use and land cover classes, ET_o = reference evapotranspiration determined using the FAO-24 radiation method [17, 18], and F = infiltration rate calculated by the Horton's infiltration model using the experimental field data.

Using the sparsely measured discharge and river cross-section data at the basin outlet, a regression relationship was developed between I_r and V_p as $V_p = 1.319I_r^2 - 0.096I_r + 0.194$ ($R^2 = 0.512$). Using this relationship, the value of V_p was computed corresponding to a basin average effective rainfall of I_r. Similarly, the direct surface runoff hydrograph (DSRO) was obtained by separating the baseflow using the straight line method [19].

5 Model Performance Evaluation Criteria

The performance of the GIUH-based Nash model was evaluated using four performance evaluation criteria:

1. Nash–Sutcliffe efficiency (NSE) criterion [20, 21];
2. Percentage error in peak runoff defined by

$$\text{PEP} = (1 - \text{computed peak runoff/observed peak runoff}) \times 100;$$

3. Percentage error in time to peak runoff defined by

$$\text{PETP} = (1 - \text{computed time to peak/observed time to peak}) \times 100; \text{ and}$$

4. Percentage error in volume,

$$\text{PEV} = (1 - \text{computed outflow flood volume/observed outflow flood volume}) \times 100.$$

6 Results and Discussion

Table 1 shows the estimated GIUH-based Nash model parameters for four rainfall-runoff events. This reveals that the K values of the GIUH-based Nash model vary from 1.089 to 1.807, whereas N, which is a function of basin geomorphology only, remains constant at 2.992.

Table 1 Parameters of the GIUH-based Nash model for the selected rainfall-runoff events

Event date	N	K (h)
June 20, 1996	2.992	1.089
August 04, 1996	2.992	1.092
June 22, 1997	2.992	1.696
October 13, 1999	2.992	1.807

Figure 2 shows the reproduction of the total runoff hydrographs by the GIUH-based Nash model at the basin outlet for the four selected storm events.

It can be envisaged from this figure that the rising limbs of almost all the events are reproduced well, except the October 1999 event. However, the recession limbs are always overpredicted. This may be attributed to the imprecision involved in estimating the storage coefficient of K, which is a function of the peak velocity, V_p. Further, V_p is a function of the excess rainfall intensity, which is estimated using regression analysis. Hence, the uncertainty involved in the estimation of V_p could be the main reason behind the inaccuracy in reproducing the recession limb of the storm hydrographs. Conversely, there is a good reproduction of the runoff peaks and time to peaks in all the storm events considered herein (see Fig. 3).

Further, Table 2 shows that the Nash–Sutcliffe model efficiency, error in peak flood estimate, error in time to peak estimate, and error in flood volume vary from 76.71 to 94.96 %, −6.71 to 3.80 %, −8.33 to 0.00 %, and −51.91 to −10.91 %, respectively. The performance of the model seems to be very good in computing the high magnitude floods, whereas in case of the low magnitude floods, the model performance decreased. Hence, when the whole basin contributes the runoff of high magnitudes, the errors due to detention storage and other losses are of less importance as compared to the total runoff realized at the outlet indicating a better performance of the linear and lumped GIUH-based Nash model. However, during the low magnitude flood events, these catchment processes are more pronounced, causing more nonlinearity in the rainfall-runoff transformation dynamics. Such a phenomenon might not be captured very well by the linear GIUH-based Nash model presented herein. Moreover, the results in this study reveal that the GIUH model has the tendency of overpredicting the flood volume always for the selected storm events.

The peak value of the GIUH-based Nash model depends upon the Horton's length ratio, peak velocity of flow, and length of the highest order stream. The Horton's length ratio being a small value and, since the peak velocity of flow depends upon the excess rainfall intensity, these parameters do not have much influence on the peak flow. Hence, the factor that is affecting the peak flow is the length of the highest order stream. At the same time, the shape parameter, N, is having a value of 2.992. That means the number of linear reservoirs in the whole basin is 2.992 that is responsible for estimating a slightly higher peak by the GIUH model. In this study, for computing the spatial average of excess rainfall intensity at daily scale, the field-estimated infiltration rate was aggregated into daily scale. The spatial averaging along with temporal aggregation could have added error resulting in a slightly overestimated peak velocity parameter, V_p. Since the value of storage coefficient, K, is inversely proportional to V_p, this would have resulted in a reduced value of K in the GIUH model causing less storage effect in the river basin during the advancement of flood. Subsequently, the runoff in the recession limb, which is generally released from the basin storage representing the basin characteristics, could have resulted in a delayed and overestimated response as computed by the GIUH model.

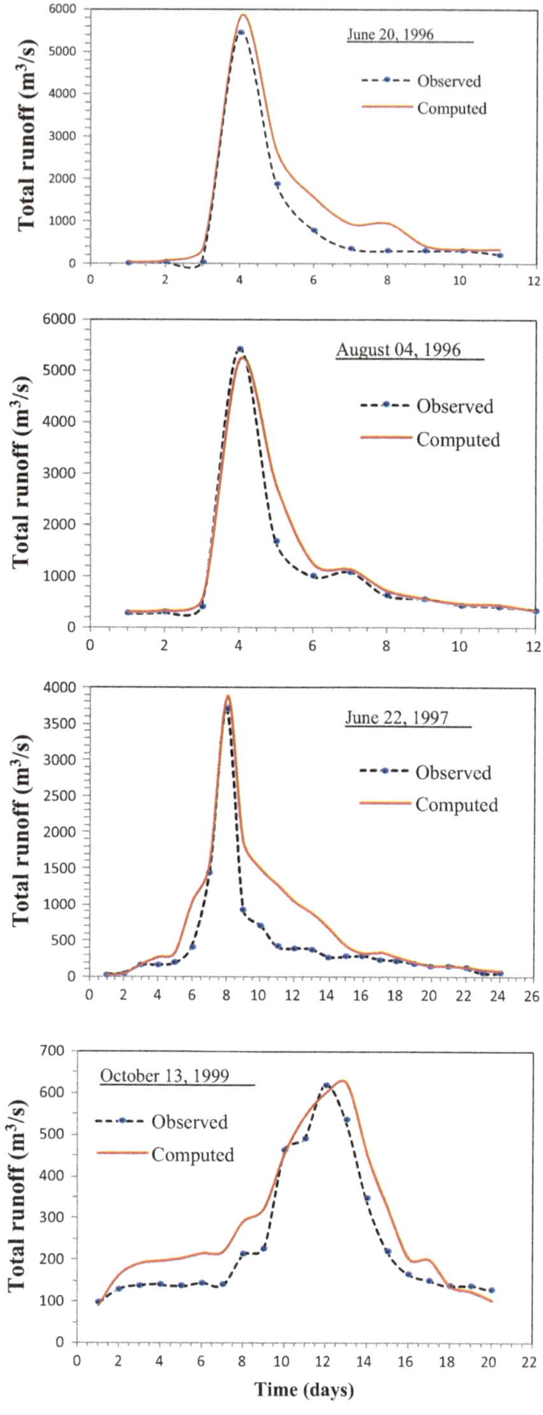

Fig. 2 Reproduction of runoff hydrographs by the GIUH-based Nash model

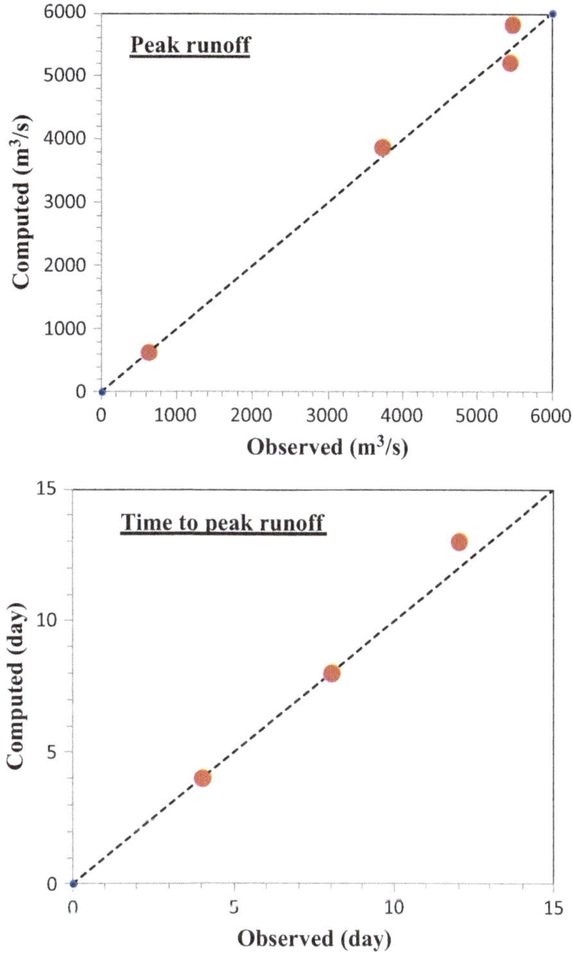

Fig. 3 Scatter plot of observed peaks and time to peaks of the total surface runoffs with that computed by the GIUH-based Nash model

Table 2 Performance evaluation measures of the GIUH-based Nash model

Event date	NSE (%)	PEP (%)	PETP (%)	PEV (%)
June 20, 1996	91.77	−6.71	0.00	−38.57
August 04, 1996	94.96	3.80	0.00	−10.91
June 22, 1997	76.71	−4.28	0.00	−51.91
October 13, 1999	85.40	−0.63	−8.33	−18.27
Absolute average	87.21	3.86	2.08	29.91
Standard deviation	8.05	2.50	4.17	18.76

7 Conclusions

Research on the "predictions in ungauged basins" (PUB) has been given much importance recently by the hydrologists and water resources engineers worldwide. Consequently, the GIUH-based Nash model is applied to model the rainfall-runoff transformation process of the Baitarani River basin. The results reveal that this model has the potential to estimate the basin response reasonably well. However, the model accuracy could be further enhanced by reducing the uncertainty in peak velocity estimation, which is considered as a polynomial function of excess rainfall intensity in this study. Moreover, the results reveal that the GIUH-based Nash model has the capability to be used for the mesoscale rainfall-runoff modeling and real-time flood forecasting.

References

1. I. Rodriguez-Iturbe, J.B. Valdes, The geomorphologic structure of hydrologic response. Water Resour. Res. **15**(6), 1409–1420 (1979)
2. R.E. Horton, Erosional development of streams and their drainage basins: hydrophysical approach to quantitative morphology. Bull. Geol. Soc. Am. **56**(3), 275–370 (1945)
3. A.N. Strahler, Quantitative analysis of watershed geomorphology. Trans. Am. Geophys. Union **38**(6), 913–920 (1957)
4. A. Rinaldo, I. Rodriguez-Iturbe, Geomorphological theory of the hydrological response. Hydrol. Process. **10**(6), 803–829 (1996)
5. I. Rodriguez-Iturbe, A. Rinaldo, *Fractal River Basins: Chance and Self-Organization* (Cambridge University Press, New York, 1997)
6. L.D. Van der Tak, R.L. Bras, Incorporating hillslope effects into the geomorphologic instantaneous unit hydrograph. Water Resour. Res. **26**(10), 2393–2400 (1990)
7. V.K. Gupta, E. Waymire, C.T. Wang, A representation of an instantaneous unit hydrograph from geomorphology. Water Resour. Res. **16**(5), 855–862 (1980)
8. N.R. Bhaskar, B.P. Parida, A.K. Nayak, Flood estimation for ungauged catchments using the GIUH. J. Water Resour. Plan. Manag. **123**, 228–238 (1997)
9. B. Sahoo, C. Chatterjee, N.S. Raghuwanshi, R. Singh, R. Kumar, Flood estimation by GIUH-based Clark and Nash models. J. Hydrol. Eng. **11**, 515–525 (2006)
10. A. Kumar, D. Kumar, Predicting direct runoff from hilly watershed using geomorphology and stream-order-law ratios: case study. J. Hydrol. Eng. **13**, 570–576 (2008)
11. J.E. Nash, The form of the instantaneous unit hydrograph. Int. Assoc. Sci. Hydrol. Publ. **45**(3), 114–121 (1957)
12. J.E. Nash, A unit hydrograph study with particular reference to British catchments. Proc. Inst. Civ. Eng. **17**, 249–282 (1960)
13. G. Nemes, New asymptotic expansion for the gamma function. Arch. Math. **95**(2), 161–169 (2010). doi:10.1007/s00013-010-0146-9. ISSN 0003-889X
14. E. Whittaker, G. Watson, *A Course of Modern Analysis* (Cambridge University Press, Cambridge, 1963), pp. 251–253
15. E.T. Copson, *Asymptotic Expansions* (Cambridge University Press, Cambridge, 1965)
16. S. Raghavan, S.S. Rangachari (eds.), *S. Ramanujan: The Lost Notebook and Other Unpublished Papers* (Narosa Publ. House, Springer, New Delhi, Berlin, 1988)

17. J. Doorenbos, W.O. Pruitt, *Guidelines for Prediction of Crop Water Requirements. FAO Irrigation and Drainage Paper No. 24 (revised)* (Food and Agriculture Organization, Rome, 1977)
18. B. Sahoo, I. Walling, B.C. Deka, B.P. Bhatt, Standardization of reference evapotranspiration models for a sub-humid valley rangeland in the Eastern Himalayas. J. Irrig. Drain. Eng. **138** (10), 880–895 (2012). doi:10.1061/(ASCE)IR.1943-4774.0000476
19. K. Subramanya, *Engineering Hydrology*, 3rd edn. (Tata McGraw-Hill, New Delhi, 2008), pp. 247–250
20. J.E. Nash, J.V. Sutcliffe, River flow forecasting through conceptual models, Part I – A discussion of principles. J. Hydrol. **10**, 282–290 (1970)
21. ASCE Task Committee on Definition of Criteria for Evaluation of Watershed Models of the Watershed Management Committee, Irrigation and Drainage Division, Criteria for evaluation of watershed models. J. Irrig. Drain. Eng. **119**(3), 429–442 (1993)

A GIS and Excel-Based Program to Calculate Flow Accumulation from the Data of Land Use

Jurina Jaafar, Aminuddin Baki Isa, Wardah Tahir, and Fauzilah Ismail

Abstract This chapter presents an approach to calculate flow accumulation by integrating a GIS-based and excel-based calculation with information from real data of land use. The procedure was based on a single-flow algorithm that is extended to consider the soil and land surface roughness. This allows realistic calculations of flow accumulation for any time period. The approach is not restricted to surface water runoff but also can be applied to kinds of mass fluxes since it is time-oriented approach. This chapter presents the principle of flexibility and the functionality of the extensions and gives some applications in the fields of hydrology and hydraulics.

Keywords GIS-based • Single-flow algorithm • Mass fluxes

1 Introduction

Most hydrodynamic and hydraulic models to calculate flow accumulation require as much as possible information about the geometry of the stream or river channel, the cross-section shape, and bottom slope, land use and the relative roughness effects. All these can be expected to affect the river and the surface flow condition. Therefore, data input is very important in models to perform its ultimate performance in describing the real-world system behavior. This requirement is somehow very contradictory when it comes to catchment where data availability is neither adequate nor available to serve the model purposes.

The challenges in hydrodynamic modeling are the quantity and quality of parameters used as an input and later to be used in the calibrating model parameters. This matter of challenges has also been raised up by [1] on his concern about the lack of local runoff data in rainfall-runoff modeling, whereby [1] stressed that in rainfall-runoff modeling, the calibration process is very important as it can be used to adjust for biases in the inputs and when the parameter calibration is performed; dramatically it can enhance model's performance. The media properties such as soil

J. Jaafar (✉) • A.B. Isa • W. Tahir • F. Ismail
Faculty of Civil Engineering, Universiti Teknologi MARA, Shah Alam, Malaysia
e-mail: jurina@salam.uitm.edu.my

S.H. Abu Bakar et al. (eds.), *ISFRAM 2014*, DOI 10.1007/978-981-287-365-1_12

and vegetation are very critical to be analyzed since their information is always unknown and poorly known to the modeler.

From the above reasons, the knowledge of data obtained and the suitability to be used in simulation are very essential to the model performance and calibration. Thus, the usage of further available tools can be directed to a better understanding of the physical features governing the water flow on the surface. The development of hydrodynamic modeling in this study will be further explained in intendance chapter.

2 Methodology

2.1 Digital Elevation Processing (DEM)

Prior to the use of the hydrological model, several processing steps are required to derive the hydrological variables. The processing steps include filling depressions in the DEM to ensure a grid has a flow path to the outlet. The second procedure is to condition the flow direction dataset. The flow direction for a cell is the direction water will flow out of the cell. Table 1 shows the value used to indicate the direction of flow water. Slopes derived from this step are then stored for data input computation of surface flow, followed by the third procedure: conditioning the flow direction data set to create the flow accumulation.

Upon completion of the following procedure (i.e., depression less DEM, flow direction, and flow accumulation), the catchment delineation can be created. At present, buildings and houses are treated as impermeable areas that transfer surface runoff to adjacent cells. This approach is recognized as considerably appropriate in representing flow over urban surfaces [2, 3].

A computational procedure was developed to determine the cell-to-cell connectivity sequence. A Java program was developed for runoff computation that computes the runoff from the highest points in the catchment to the outlet cell. The computation algorithm begins with each start cell following the flow directions of each cell until the outlet cell is reached. Cells are assembled in overland flow paths. Manual extraction of data is carried out during this process. Data preparation is

Table 1 Flow direction of a cell

1	↑
2	↗
4	→
8	↘
16	↓
32	↙
64	←
128	↖

stored manually in Excel program. Detail programming is described in the next section.

2.2 Hydrological Model

Development of the model is time step oriented. The connectivity sequence is designed to compute flow from one cell to another. Thus, within each time step, the surface flow is computed for each cell; the outflow from one cell becomes the inflow to the next downstream cell. Within the hydrological model, infiltration for each cell is derived from the Green and Ampt model. Green and Ampt model is used due to element of physical behaviour described in the model. Approach described by Smith and Goodrich [2] is used in the development of the model. The model development starts by assuming that there is ponding on the surface.

Runoff models predict the temporal distribution of runoff at a catchment outlet based on the temporal distribution of effective rainfall and the catchment characteristic [4]. In hydrology processes, the abstractions refer to infiltration, evaporation, evapotranspiration, and depression storage. Chin et al. [4] referred the most important abstraction are infiltration and depression storage and it is important to emphasize the topography and surface cover of the catchment.

2.3 Hydrological Modelling

Hydrologic cycle shows that the rain which falls into the ground can be possibly divided into two cycles: either infiltrates into the ground or moves on the surface. Referring to the equation, the element included in the lateral flows is the infiltration rate (f). In conjunction with this, the Green and Ampt model is used to calculate the infiltration rate in each cell. In Green and Ampt model [5], the infiltration rate is expressed using Eq. (1).

$$f(t) = K\left(\frac{1}{F_t}\psi \cdot \Delta\theta + 1\right) \tag{1}$$

where $f(t)$ = infiltration rate in cm/h, K = hydraulic conductivity in cm/h, ψ = wetting front suction head in cm, $\Delta\theta$ = available soil moisture content vol./vol. (effective soil porosity minus initial moisture content), and F_t = cumulative infiltration in cm. K is taken as one-half of the saturated hydraulic conductivity.

Since function f is equal to df/dt, Eq. (1) can be integrated to derive an expression based on the cumulative infiltration as a function of time.

$$F_{t+\Delta t} - F_t - \psi \cdot \Delta\theta \cdot \ln\left(\frac{F_{t+\Delta t} + \psi\Delta\theta}{F_t + \psi\Delta\theta}\right) = K \cdot \Delta t \tag{2}$$

where Δt is the time step in seconds and $F_{t+\Delta t}$ and F_t represent cumulative values of infiltration at the end and initial time step.

Equation (2) provides a platform to calculate the incremental amount of infiltration during each time step for the period of surface ponding. This incremental volume is deducted from the incremental rainfall volume added to each cell between time t (initial) and $t+\Delta t$. In this model, a Newton–Raphson method is used to solve for incremental infiltration at $F_{t+\Delta t}$. Let $F2 = F_{t+\Delta t}$ and $F1 = F_t$. Equation (2) can be arranged to form Eq. (3).

$$\mathrm{INF} = F2 - F1 - \psi \cdot \Delta\theta \cdot \ln\left(\frac{F2 + \psi \cdot \Delta\theta}{F1 + \psi \cdot \Delta\theta}\right) - K \cdot \Delta t \tag{3}$$

Differentiating the function with respect to $F2$ gives:

$$\frac{\mathrm{dINF}}{\mathrm{d}F2} = 1 - \frac{\psi \cdot \Delta\theta}{F2 + \psi \cdot \Delta\theta} \tag{4}$$

An initial calculation starts with estimates made for the value of $F2$:

$$F2\mathrm{est} = F1 + \tfrac{1}{2}\Delta\,\mathrm{Rainfall} \tag{5}$$

In Eq. (5), the Δ Rainfall is the incremental rainfall in cm for this time step. Using Newton–Raphson procedure, $F2$ can be computed as follows:

$$F2\mathrm{new} = F2\mathrm{est} - \frac{\mathrm{INP}(F2\mathrm{est})}{\frac{\mathrm{dINF}}{\mathrm{d}F2}(F2\mathrm{est})} \tag{6}$$

The iteration stops when $F2$new and $F2$est are sufficiently close, i.e., when tolerance reaches 0.001.

$$|F2\mathrm{new} - F2\mathrm{est}| < \mathrm{tol} \tag{7}$$

At initial beginning time step, F_t is equal to zero; the potential infiltration rate (f_t) is calculated using Eq. (1). Then the value obtained is compared to the water available in the cell. Water available in the cell is referred as depth of water, the accumulation of inflows from other cells, and the incremental rainfall in the cell. The water held in depression storage is served as infiltration demand. If the available water is insufficient to satisfy the potential infiltration rate (f_t), then the available water will infiltrate, and cumulative infiltration is measured using Eq. (8).

$$F_{t+\Delta t} = F_t + \Delta\text{rain} + \frac{\sum \text{inflows} \cdot \Delta t}{\text{cell area}} + h_1 \tag{8}$$

Inflow is the total flow from adjacent cells. Initial depth h_1 refers to current depth in the cell (previous cell). In other cases where f_t is less than the available surface water, the ponding will occur. In this case $F_{t+\Delta t}$ is computed using Eq. (3) until Eq. (8). This process of computation will produce infiltration (F) that will be used in the next computation to determine surface runoff. The next section will describe the computation of the runoff dynamic.

3 Runoff Dynamic

Runoff is defined as a flow of the rainwater at a small scale (mm), which is not infiltrated by the soil surface (rainfall excess), and is a complex phenomenon with exact flow directions, unit discharges (discharge per unit flow width), and depths varying widely across the surface [6].

The spatial distributed water balance model applies the Mass Conservation Law to illustrate the mass balance within each spatial unit and integrates a momentum equation that defines the water movement between cells.

The most simplified approaches to surface water fluxing of two-dimensional overland flow are based on the kinematics wave approximation of the Saint Venant's equation, and the continuity equation for each cell can be computed as in Eq. (9):

$$h_2 = h_1 + \text{netrain} + \left(\frac{\sum \text{inflows}}{\text{cell area}} - \frac{\text{outflow}}{\text{cell area}}\right) \cdot \Delta t \tag{9}$$

The depth of water in the cell is expressed as h_2 and h_1 in meters. Both are expressed at the beginning and end of the time step. Net-rain is the subtraction of rainfall and infiltration expressed in meters. Σ Inflows is the total of flow from adjacent cells in m^3/s, and outflow is the flow that comes out from the cell. Outflow is calculated using Manning's equation in Eq. (10).

$$\text{Outflow} = (h_2 - d)^{5/3} \cdot \text{slope}^{1/2} \cdot \frac{1}{n} \cdot \text{cell width} \tag{10}$$

Depression storage in meters is expressed as d in Eq. (10). Here, the slope is the cell surface slope in the direction of flow. Cell width is taken perpendicular to the flow and Manning's roughness as (n). Computation begins with h_2 estimation. Equation (10) is rearranged into Eq. (9), to come up with Eq. (11);

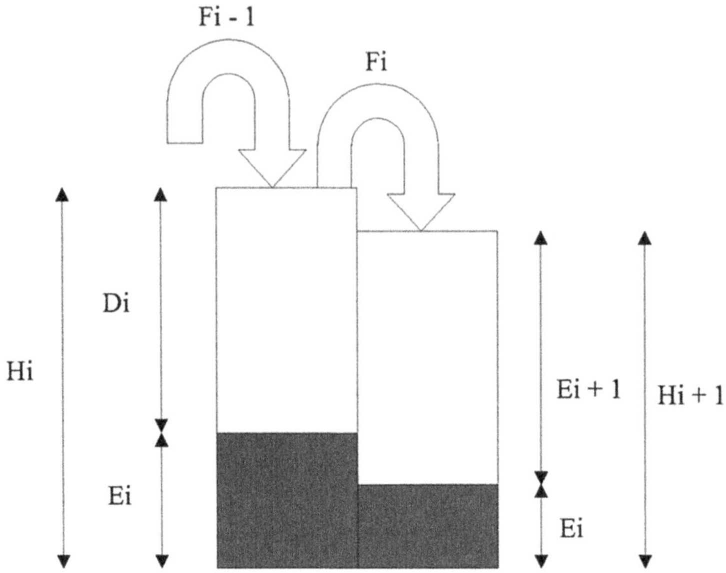

Fig. 1 Geometry of flux between two adjacent cells

$$R = -h_2 + h_1 + \text{netrain} + \frac{\sum \text{inflows}}{\text{cell area}} \cdot \Delta t - b^{5/3} \frac{\text{slope}^{1/2}}{n \cdot \text{cell length}} \Delta t \qquad (11)$$

where R is the average outflow depth. Newton–Raphson method is used to solve the computation.

Figure 1 is the illustration of cell-to-cell connectivity computation.

Figure 2 is the cells' illustration of inflow into the cell, infiltration, and outflow from the cell.

Figure 3 illustrates the water flux flow between the cells [7].

4 Subgrided Method

Subgrided method of mapping land aims to obtain a greater optimization of identifying the referring to knowledge based to substitute to Manning roughness. Thus with these grid GIS tools, the land use changes sensed by the satellite images can be easily updated its characterization into any application. Workflow programming is illustrated accordingly in Fig. 3.

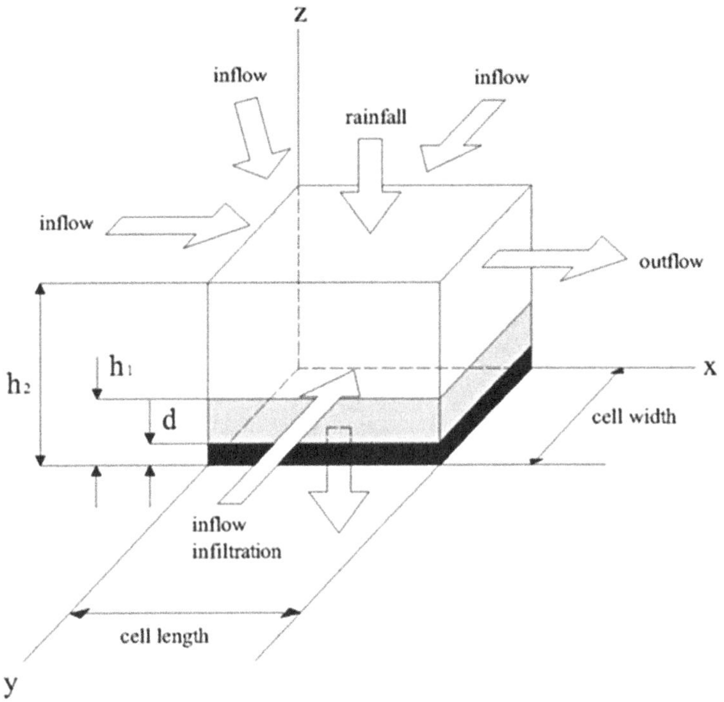

Fig. 2 Idealized cells' illustration for surface flow computation

4.1 Parameters

The data used with the knowledge of the absence of field data test. The soil information used to calculate infiltration is adapted from [8]. Model parameters used are shown in Table 2. Sensitivity analysis carried out by Pretorius [9] had shown that the *K* and Manning's *n* are the most sensitive parameters during the calibration process. Hereby in his study the parameters were adjusted to achieve better agreement between observed and simulated discharges. Since in this study the effects of parameters sensitivity are not being observed, there is no requirement to adjust the values during calibration stage.

Fig. 3 Summary of workflow programming

Table 2 Values of Manning's roughness coefficient, emissivity, depression storage, hydraulic conductivity, effective porosity, and wetting front suction head

Soil	Hydraulic conductivity K (mm/h)	Effective porosity θ (vol./vol.)	Wetting front suction head ψ (cm)
Sandy clay loam	2.0	0.19	220
Landcover	Manning's n	Emissivity	Depression storage d (mm)
Pervious (Agriculture/Forest)	0.25	Sand − 0.76 = 0.24	2.0
Impervious (Urban/ Open area)	0.013	Asphalt − 0.93 = 0.07	1.0

5 Calculation and Programming

It is important here to establish the dummy calculation to compare it as in the programming. The following calculation shows the program calculation from beginning to find infiltration to final outflow within each cell till it reaches the outlet cell (Fig. 4).

6 Conclusion and Future Work

Through this study, it shows that the remote sensing technique and GIS can be fully utilized for extracting as much as possible physical parameters from a catchment. The expected future work of this study is to obtain other data that can be used in runoff simulation, to extract information of the land use classification as an input parameters for the models, and to test the most influenced parameters that affect the water flow by using statistical approach.

Acknowledgment The authors would like to express our deepest gratitude to the Research Management Institute (RMI) UiTM for providing us with the financial support.

References

1. G. Bloschl, Rainfall-runoff modelling of ungauged catchments Collier C.G. (1984), Application of weather radar systems. J. Appl. Meteorol. **3**, 658–665 (2001)
2. R.E. Smith, D.C. Goodrich, Rainfall excess overland flow. Encycl. Hydrol. Sci. **10**, 111 (2006)
3. D. Djokic, D.R. Maidment, *Hydrologic and Hydraulic Modeling Support with Geographic Information Systems* (ESRI, Redlands, 2000)
4. D.A. Chin, A. Mazumdar, P.K. Roy, *Water-Resources Engineering*, vol. 12 (Prentice Hall, Englewood Cliffs, 2000)

Input data:

Rainfall = R = 200 mm = 0.2 m = 20cm
Slope = S = 0.05295
K = 0.1 cm/hr , θ = 0.41 ;φ = 6.13 cm ; θ (0.41) x φ (6.13) = 2.513
N = 1 – 0.91 = 0.09, ε
D = 1 cm = 10 mm = 0.01 m
NR = R –F = 0.2 – 0 = 0.2
Cell 1 (E1)

Step 1 : Infiltration Calculation
Check R and K , K = 0.1 cm/hr : R = 20 cm , F1 initial = 0

Noted : the cell contain information of urban land, infiltration near to zero.
Next : Calculate Outflow
This above values is obtained from simulation.
Result : Direction
-1 -1 -1 32 32
16 64 64 64 -1
16 64 64 64 64
Result / value in each cell:

0.00000	0.00000	0.00000	0.00000	4.64682
0.00000	0.00000	0.00000	6.99029	4.64682
11.41676	11.96498	12.78411	6.99029	0.00000
17.63262	9.09993	8.31317	6.99029	4.64682

End point at A4 from computer generated; 11.41676 + 9.09993 = 17.63262
The result calculated here is a direct calculation without iteration involved.

Fig. 4 Programming algorithm

5. V.T. Chow, D.R. Maidment, L.W. Mays, *Applied Hydrology* (McGraw-Hill, New York, 1988)
6. A.J. Smith, J.V. Turner, Density-dependent surface water–groundwater interaction and nutrient discharge in the Swan–Canning Estuary. Hydrol. Process. **15**(13), 2595–2616 (2001)

7. A.A. Voinov, C.H. Fitz, R. Costanza, Surface water flow in landscape models: 1. Everglades case study. Ecol. Model. **108**, 131–144 (1998)
8. D.A. Chin, *Water-Resources Engineering* (Pearson Education Hall, New Jersey, 2006)
9. M.J. Pretorius, *Flood modelling using data available on the internet*. Doctoral dissertation, University of Cape Town, 2011

Flood Damage Assessment: A Review of Flood Stage–Damage Function Curve

Noor Suraya Romali, Muhammad @ S.A. Khushren Sulaiman, Zulkifli Yusop, and Zulhilmi Ismail

Abstract Nowadays, flood control has been replaced by flood management concept in terms of living with flood, making benefit of it, and minimizing its losses. The success in flood management in any region depends on the evaluation of different types of flood losses. For the assessment of flood damages, this requires the use of stage–damage functions for different categories of land use. A review is presented of the methods used to construct stage–damage function curves for residential, commercial, agricultural, and industrial category. Two main approaches in constructing stage–damage functions are empirical approach, which is based on damage data of past floods, and synthetic approach, which uses damage data collected by interview survey or questionnaire. For a developing country like Malaysia which has limited history and actual flood damage data, the synthetic method is the preferred approach in constructing stage–damage function curve.

Keywords Flood management • Flood damage assessment • Stage–damage function • Land use • Synthetic approach

1 Introduction

Flood damage estimation is an essential element in water resources planning, mainly for the purpose of flood mitigation benefits' evaluation [1]. In conventional practice, the flood management approaches focus on the design standards and structural flood mitigation measures [2]. Normally, flood mitigation structures were designed in order to control up to a certain, predefined design flood, i.e. return period of the design rainfall. In recent years, this structural flood control approach has been changed to a new developed concept which is referred to as "flood risk management" [2]. The degree of protection is determined by broader

N.S. Romali (✉) • M.@S.A.K. Sulaiman
Faculty of Civil Engineering and Earth Resources, Universiti Malaysia Pahang, Gambang, Pahang, Malaysia
e-mail: suraya@ump.edu.my

Z. Yusop • Z. Ismail
Faculty of Civil Engineering, Universiti Teknologi Malaysia, Skudai, Johor, Malaysia
e-mail: zulyusop@utm.my

S.H. Abu Bakar et al. (eds.), *ISFRAM 2014*, DOI 10.1007/978-981-287-365-1_13

considerations than some predefined design flood, while more attention is put on non-structural flood mitigation measures. A significant evolution that can be seen in this context is a dominant transformation from flood hazard to flood risk. In common practice, flood policies are focusing more on the regulation and reduction of flood hazard, i.e. decreasing the probability of occurrence and intensity of flood discharges and inundations [3]. In flood risk management, where flood risk is given more attention, risk is defined as damage that occurs or would be exceeded with a certain probability in a certain time period. Hence, damage aspects are important and need to be well considered on flood risk management [3].

In addition, unpredictable occurrences of disasters, such as flooding, are making people to take protection and prevention. In many countries, people can insure themselves against flooding. In developed countries such as the United States, the United Kingdom, and Australia, flood insurance has been adopted as a part of the tools for residual flood risk management to support and complement non-structural approach. As a result, flood insurance has been incorporated as part of a comprehensive integrated flood risk management. However, in Malaysia, having flood insurance is not a requirement. The awareness of taking flood insurance to protect their properties is still low, and furthermore, there is a lack of incentive from the government to promote flood insurance as an instrument for flood risk management in the country [4, 5]. Although it cannot prevent actual property damages or loss of life as structural measures would do, the advantages of having flood insurance are that it can significantly reduce the economic risk associated with flooding. An insured property damaged by the flood can be replaced quickly without depending on financial aid from the government [6]. As flood insurance has become a part of flood risk management, the flood hazard and the potential flood damage are of great interest to insurance companies [6]. Insurance companies need the information on flood risk data and damage curve to decide the customer prices for the flood insurance [7].

The generation and compilation of an adequate flood damage assessment concern many issues regarding the nature of damage caused by floods, such as

- Proper classification of damage categories considering nature of damage [8]
- Obtaining detailed flood parameters such as flow velocity, depth, and duration at any given location [8]
- Establishment of relationships between flood parameters and damage for different damage categories [8]

2 Flood Damage Assessment

The selection of approaches for flood damage assessment may depend on the types of damages. In undertaking a systematic flood damage assessment, it is important to recognize that flood damage consists of two main components, namely the tangible and intangible types of damage [9].

2.1 Types of Flood Damages

There are two main types of flood damages, i.e. tangible and intangible damage. The damage that can be readily measured in monetary value is the tangible damage, while the damage that cannot be directly measured in monetary terms is known as intangible damage [10]. Moreover, tangible damage is further divisible into two subtypes, i.e. direct and indirect damage. Direct damage is the damage caused to items (e.g. buildings and inventory items) by contact with or submersion in water. In contrast, indirect damage is the damage caused by the interruption of physical and economic networks, such as traffic flow disruption and individual income loss, as well as consequences of business cut-off [11].

In the other view, Merz et al. [3] define direct damages as damages which occur due to the physical contact of flood water with humans, property, or any other object, while indirect damages are the damages that are induced by the direct impacts and occur in space and time, outside the flood event. Both types of damages are further classified into tangible and intangible damages, depending on whether or not they can be assessed in monetary values [12, 13]. Tangible damages are damages that occur to man-made properties or resource flows which can be easily specified in monetary terms, whereas intangible damage is damage to assets which are not traded in a market and are difficult to transfer to monetary values [3].

Several fundamentally different types of flood damage have been discussed by Dutta et al. [8], Penning-Rowsell and Chatterton [14], and Lee and Mohamad [15]. Dutta et al. [8] further categorized the direct and indirect tangible damage into primary and secondary, as shown in Table 1, while a description of examples in Table 2 is given by Merz et al. [3].

2.2 Flood Parameters

The amount of damages resulting from a flood depends on variable flood parameters, such as flood water depth, flood water velocity, year of flooding, duration of flooding, sediment and effluent contents, flooded area covered, and flood warning system [1, 14]. These are also agreed by James and Hall [16] and McBean et al. [17], where they stated that flood damage is actually affected not only by water depth but also by many different factors associated with the local increase of

Table 1 Flood damages category and loss examples [8]

Category			Examples
Tangible	Direct	Primary	Structures, contents, and agriculture
		Secondary	Land and environment recovery
	Indirect	Primary	Business interruption
		Secondary	Impact on regional and national economy
Intangible			Health, psychological damage

Table 2 Examples for different types of damages [3]

	Tangible	Intangible
Direct	Damage to private building and contents; disruption of infrastructure such as roads; railroads; erosion of agricultural soil; destruction of harvest; damage to livestock; evacuation and rescue measures; business interruption inside the flooded area; clean up costs	Loss of life; injuries; loss of memorabilia; psychological distress, damage to cultural heritage; negative effects on ecosystems
Indirect	Disruption of public services outside the flooded area; induced production losses to companies outside the flooded area (e.g. suppliers of flooded companies); cost of traffic disruption; loss of tax revenue due to migration of companies in the aftermath of flood	Trauma; loss of trust in authorities

costs due to the occurrence of flood events. According to Dutta et al. [8], all the factors may be the significant flood parameters that influenced flood damages; however, most previous flood damage assessment studies have chosen water depth as the flood damage variable.

2.3 Approaches of Flood Damages Assessment

Two common flood damage estimation approaches are unit loss models and model applications. The unit loss approach is based on a property by property assessment, either actual or potential [8], while the model applications estimate the linkage effects, or inter-sectoral relationships, of floods within economy [18, 19].

From the literature, it can be summarized that most published information on damage collection and analysis come from the United States, the United Kingdom, Japan, and Australia, which have adopted a unit loss approach. The detailed methodologies for tangible loss estimation had been established by the United Kingdom and Australia [14, 20, 21], while for the United States, Japan, etc., detailed damage estimation methodology is limited to urban damage only [22, 23]. It has been noted that these countries adopt similar approach in damage estimation, i.e. unit loss approach [12, 24–26].

As mentioned earlier by Dutta et al. [8], the establishment of an adequate flood loss estimation model involves many issues due to the nature of the flood damages. Some of the most important issues in flood loss estimation are obtaining detailed flood parameters such as flow velocity, depth, and duration at any given location, proper classification of damage categories, and establishment of relationships between flood parameters and damage for different damage categories [8]. The relationship between flood parameters and flood damage can be represented by stage–damage function, which is developed based on historical flood damage information, questionnaire survey, laboratory experiences, etc. [26, 27].

3 Stage–Damage Function Curves

Stage–damage functions are important components in flood damage estimation model. Normally, stage–damage function curves were developed for estimating flood losses. Stage–damage curves are the first essential stage in flood loss assessment. They are combined with field surveys of property at risk and with hydrological information (probability and extent of flooding, velocity, and the like) to give predictions of event damages from which average annual damages can be calculated [26].

Stage–damage functions may be derived by using these two most commonly used methods, i.e. one is based on damage data of past floods, and the other one is from hypothetical analysis known as synthetic stage–damage functions based on land cover, land use patterns, type of objects, information of questionnaire survey, etc. [8].

3.1 Synthetic Stage–Damage Function Curves

The methodology of synthetic approach was first suggested by White [28]. Synthetic stage–damage curves are based on hypothetical analysis, where it doesn't depend on information from an actual flood event [2]. A detailed procedure of synthetic stage–damage curves for several land use types had been produced by Penning-Rowsell and Chatterton [14]. The procedure had been used to assess flood damage to both residential and commercial properties in the United Kingdom. It also provides an essential input into computer programs that are designed to evaluate the benefits of flood prevention measures.

Synthetic stage–damage curves are of two types, either depending on the existing databases, or by values and loss adjusters surveys, as shown in Fig. 1.

1. *Existing databases*: Estimates of the damage to building fabric were obtained using existing information on the possible effects of flooding on building material and the like. The losses inventory was based upon ownership rates obtained from marketing manuals and consumer research surveys [26].
2. *Valuation surveys*: The alternative approach to the inventory method is to undertake surveys of the different types of dwellings at risk in the flood-prone area. Valuation surveys select a sample of dwellings in each designated dwelling class, and a checklist of possible contents, usually by type of room (kitchen, bedrooms, etc.), is drawn up. For the selected properties, the surveyor (ideally a qualified loss adjuster or valuer) notes all items and their current value based on type, quality, and degree of wear. The survey can include information on the height above the floor of each item or the heights can be taken as standard from house to house. The information for the sample of each dwelling type is then averaged and stage–damage curves constructed [14].

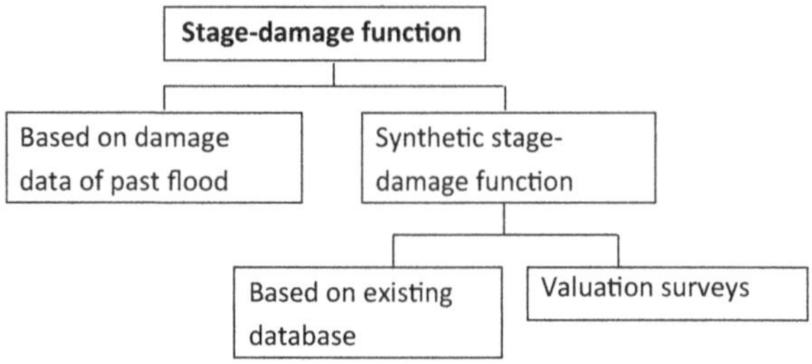

Fig. 1 Types of stage–damage function

3.2 Categories of Stage–Damage Curves

Stage–damage curves may be developed using two types of approaches: either depth–damage or depth–percent damage-based approaches. In depth–damage approach, stage–damage relationships are determined directly from prototype data, and normally, the curves are developed separately for many types of structures. Hence, it is time-consuming and costly [29]. Furthermore, the useful life of the relationship is short. The percent-damage approach defines the flood damage as percentage of the total value of the damaged property, depending on the water depth [29]. With the depth–percent damage approach, the percentages from a depth–percent damage function are multiplied by a replacement value in order to develop a stage–damage relationship [30].

A numbers of scientific articles discussing the development of stage–damage functions are referred to in this review. The articles classified the stage–damage functions into several categories as summarized below:

1. *Residential*: For residential category, building was classified into major categories such as detached, semi-detached, and terrace. These categories were also classified by age and then further subdivided by social class of the occupants [10]. The stage–damage information, following normal practice, was divided into "building fabric" and "inventory".

   A loss estimation model to describe urban flood damage in Japan used stage–damage functions that had been derived from the averaged and normalized data published by the Japanese Ministry of Construction. The respective data used are based on the site survey data accumulated since 1954 [18]. Five depth–percent damage curves had been formulated by Dutta et al. [8]: residential structure (wooden), residential structure (RCC concrete), residential content, non-residential property, and non-residential stocks. Figure 2 shows the stage–damage function curve produced by Dutta et al. [8] for residential structure (wooden) category.

   The determination of depth–percent damage relationship was also carried out by Oliveri and Sontoro [29], considering two building types with two and four storeys, respectively, having different average finishing levels and, consequently, different unit prices. The adopted technique is the same for both cases. The first phase of evaluations excluded the building contents, only analysing the structure. The damage begins when the water reaches a depth of 0.25 m; depth increments by steps of 0.25 m were considered. For each water depth, all component

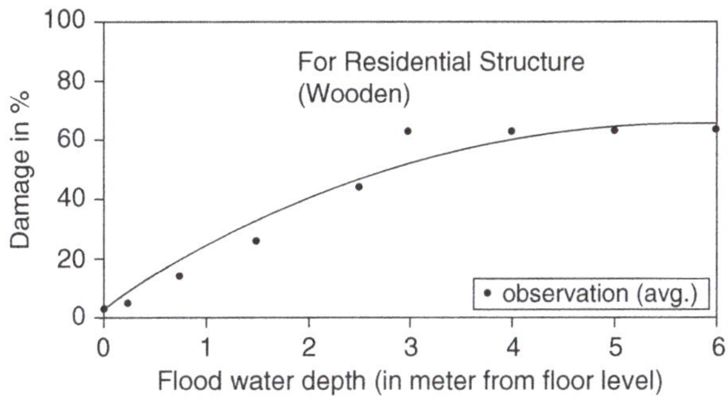

Fig. 2 Flood stage–damage function for residential structure (wooden) category [8]

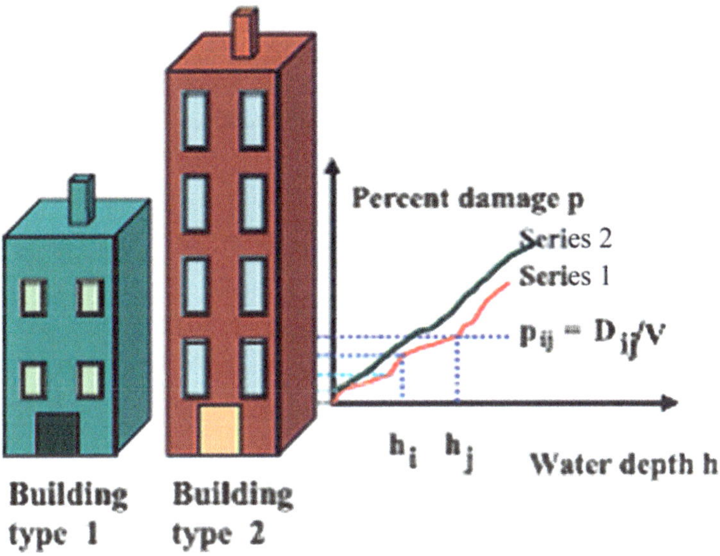

Fig. 3 Schematic of the local depth–percent damage relationships evaluation procedure [29]

categories damageable by water (the internal and external plaster, the textures, the paving tiles, the floors, and part of the electric appliances) were considered in order to compute the overall replacement costs. For each water depth, the percentages of damage were calculated by dividing the overall replacement cost with the estimated total replacement value of the building (structure plus contents) [29]. The procedure described is represented in Fig. 3.

2. *Commercial*: The Australian studies [31, 32] had produced stage–damage curves for commercial sectors. Commercial enterprises are classified by size and by value class. There are three sizes of classes: "Small" (<186 m^2) corresponds to the average high street shop, "medium" (186–650 m^2) to small supermarket, and for larger premises, the actual area (in m^2) was recorded. Each commercial

building was given a value class that indicates the susceptibility of the contents to flood damage. These are in the range of 1 (very low) to 5 (very high). The stage–damage curves form a matrix based on size and value class with average damages for each class given at five heights: 0.25, 0.75, 1.25, 1.75, and 2.00 m above floor level [26].

3. *Industrial*: According to Smith [26], loss assessment using stage–damage curves is inappropriate for industrial plants and they should be analysed using questionnaires. It is important to acknowledge that one single large industrial plant can be disposed to a direct flood damage that exceeds several hundred nearby dwellings subject to the same flood risk.

   A site questionnaire survey should be used to estimate damage to industrial and related properties [14]. The questionnaire method relies on the cooperation of local company management to provide information on the susceptibility of premises to physical damage and the likely magnitude of disruption to production. If information is incomplete or not forthcoming, estimates should be produced from either a similar type of premises within the area or from information previously collected. The example of depth–damage data curve of industry-related services for type 52 (vehicle services), 54 (contractors, merchants etc), and 55 (storage and wholesale establishment) is illustrated in Fig. 4.

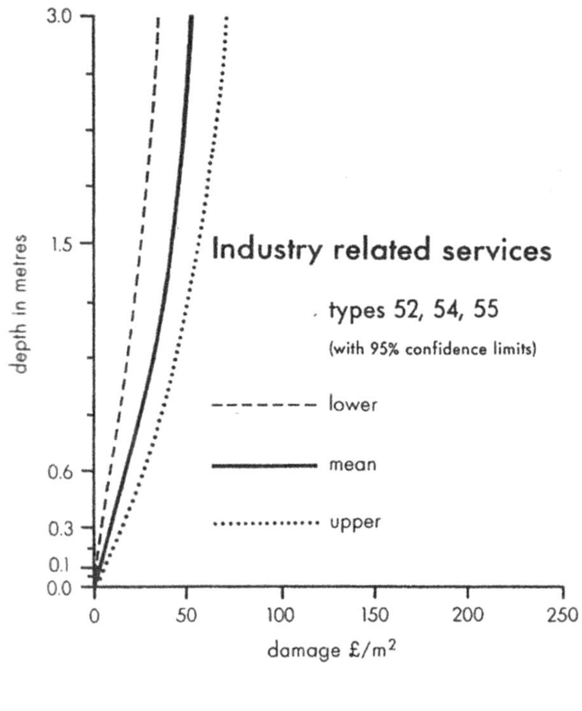

Fig. 4 Industry-related services depth/damage data [14]

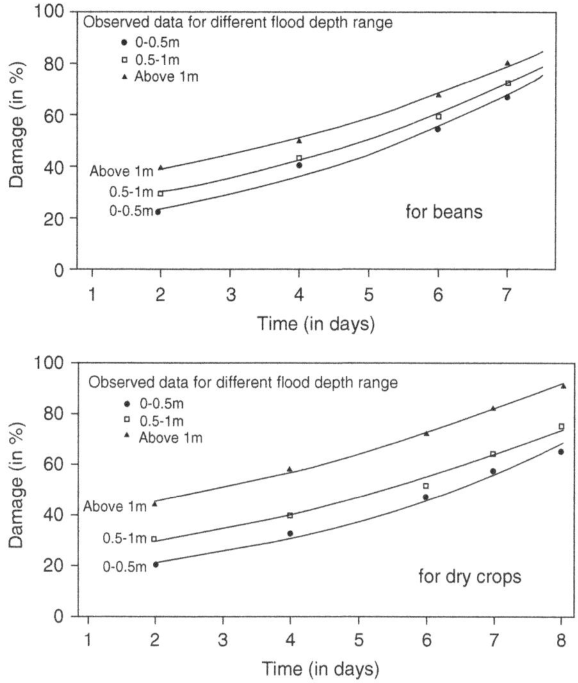

Fig. 5 Agricultural flood stage–damage function for beans and dry crops category [8]

4. *Agriculture*: Dutta et al. [8] formulated stage–damage functions for agriculture damage in Ichinomiya river basin, Japan. They considered two flood parameters: depth and floods duration. The agricultural stage–damage functions are categorized into eight categories, i.e. beans, Chinese cabbage, dry crops, melon, paddy, vegetable with root, sweet potato, and green leaf vegetables according to different flood depth range from 0 m to above 1 m. Example of stage–damage function curves formulated by [8] for agriculture product is shown in Fig. 5 for beans and dry crops category.

4 Development of Flood Stage–Damage Function Curve

In Malaysia, the pioneer study on flood damage estimation was conducted by Japan International Co-operation Agency (JICA), in the National Water Resources Study (NWRS) 1982 for Malaysia [33]. Later, in the year 2000, KTA Tenaga Sdn. Bhd., under the National Register of River Basin study commissioned by the Drainage and Irrigation Department (DID) Malaysia, carried out the updating of the flood damages that was completed by JICA in 1982. KTA Tenaga Sdn. Bhd. [29] proposed a two-tier approach, namely a rapid assessment method (RAM) and a detailed assessment method (DAM). The former is a first-level approximation which requires only basic information, whereas the latter requires extensive data collection and may be adopted for critical flood regions. However, the study by KTA Tenaga Sdn. Bhd. carried out exemplary detailed assessment following the methodology developed in Australia, which has limited applicability in the Malaysian context [15].

The fundamental steps in producing stage–damage curve were to collect damage data from actual flood events. However, the problem with this approach is that that kind of data is not available in Malaysia. Hence, synthetic stage–damage curves by the valuation surveys are the preferred option [26]. Synthetic approach develops standard flood damage information from a wide variety of sources, not necessarily related to specific flood events. This approach can be done by interview survey to gain flood damage information.

The development of questionnaire survey for different classes of building had been discussed by McBean et al. [17] and [26]. As stated by Smith [26], the development of questionnaire survey should include information on types of structure, description of room, household contents, location (either in basement, first floor, or second floor), the quality, and age of the structure and contents. The questionnaire survey also can include information on the height above the floor of each item and general information such as the household/commercial income, number of person in the houses/premises, and how often they experienced flooding.

Penning-Rowsell and Chatterton [14] conducted site survey for residential stage–damage data collection. The survey was divided into two parts: building fabric and inventory. The building fabric and inventory checklist for the site surveys is shown in Table 3. The details of type, total area or number, and quality

Table 3 Flood damage information required for stage–damage curve [14]

Category	Flood damage information required	
Residential	*Building fabric*: Plasterwork and wall finish Floors Joinery Decorations Path and paved areas Boundary/fence/garage/gates Main building	*Inventory*: Domestic appliances Heating appliances TV/hi-fi etc Furniture Personal effects Floor covering and curtains
Commercial	Damage to stock, damage to building fabric, equipment damage, clean-up coat, depth function	
Industrial	Name of company, type of business, total area of premises, total ground floor area of building only, height of floor level of buildings, plant and equipment, raw material and unfinished goods, finished goods (stocks if no processing is involved), total physical damage (estimate), average weekly turnover/output, ability to defer production/work, ability to transfer production/work	
Agricultural	Types of crops/land use, building fabric and clean-up cost, damage to stored crops, feedstuff, and fertilizers, damage to agricultural vehicles and movable equipment, damage to fixed equipment	

of building fabric and inventory had to be obtained during the surveys. The flood damage information required for the development of stage–damage function for commercial, industrial, and agricultural category as recommended by Penning-Rowsell and Chatterton [14] is illustrated in Table 3.

The questionnaires were sent to residents, companies, and farm and farm building owners according to their categories. A successfully completed questionnaire should yield answers which can be totalled to provide an overall damage figure.

Using the synthetic method, flood losses are calculated independently of a particular flood experience and independently of assumptions about damage averted by emergency actions. Although this method is preferable for a country with limited historical flood data, the disadvantages of the synthetic approach are the difficulty of incorporating every facet of flood damage into the standard data to allow for all flood event variables (velocity, effluent content, etc.). Also, the resulting data may not always be applicable to the area of concern, which might be particularly high quality or third rate. However, the advantages include not having to rely on the vagaries of historical data while still retaining the savings in resources arising from standard data rather than other approaches [14].

In this review, the assessment of flood damage is limited to considering only direct and indirect tangible damage, as the intangible damage assessment is very complicated and difficult to quantify due to its subjectivity [11]. The estimation of flood–damage function in previous study mostly concentrates on two variables, i.e. depth and duration of flood. Hence, it is suggested that other variables, such as velocity of flow, warning times and responses, and types, price, and different design of building, should also be taken into account in future research.

5 Conclusion

Assessment of flood damages is a fundamental step for the economic analysis of a flood control project. Moreover, the assessment of flood damage is gaining greater attention as nowadays flood risk management is becoming the dominant approach of flood control policies. One approach of flood damage assessment is to develop a flood–damage function, which relates flood damage to flood inundation parameters for different classes of assets. Synthetic method is the preferred method in constructing this for a developing country like Malaysia which has limited history/actual flood damage data. This presented review can be a starting point in producing Malaysia's very own flood loss estimation procedure that can be used as a tool for flood management practice.

Acknowledgement The authors acknowledge Water Research Alliance of Universiti Teknologi Malaysia for its financial support from research grant (Q.J130000.2409.00G08), Universiti Malaysia Pahang, and Drainage and Irrigation Department (DID) Malaysia for the research materials and support on the present works.

References

1. E.A. McBean, J. Gorrie, M. Fortin, J. Ding, R. Moulton, Flood depth-damage curves by interview survey. J. Water Resour. Plann. Manag. **114**(5), 613–634 (1989)
2. B. Merz, J. Hall, M. Disse, A. Schumann, Fluvial flood risk management in a changing world. Nat. Hazards Earth Syst. Sci. **10**, 509–527 (2010). doi:10.5194/nhess-10-509-2010
3. B. Merz, H. Kreibich, R. Schwarze, A. Thicken, Assessment of economic flood damage. Nat. Hazards Earth Syst. Sci. **10**, 1697–1724 (2010). doi:10.5194/nhess-10-1697-2010
4. A. Keizrul, Floods in Malaysia (20 January 2004), http://www.icharm.pwri.go.jp/publication/jan_20_22_2004_ws/pdf_output/abdullah.pdf
5. J.C. Ho, Coastal flood risk assessment and coastal zone management case study of Seberang Perai and Kuantan Pekan in Malaysia, Erasmus Mundus Master's Thesis in Insurance Against Natural Disasters, Joseph Henry Press, Washington, DC, 2009
6. U.G. Aliagha, T.E. Jin, W.W. Choong, M.N. Jaafar, Factors affecting flood insurance penetration in residential properties in Johor Malaysia. Nat. Hazards Earth Syst. Sci. Discuss. **2**, 3065–3096 (2014). doi:10.5194/nhessd-3065-2014
7. T. Andrews, G. Billen, H. Kim, A. Lui, C. Pollack, L. Simpson, D. Smith, S. Underhill, D, Whittle, The insurance of flood risks, in *16th General Insurance Seminar*, Institute of Actuaries of Australia, 9–12 Nov 2008
8. D. Dutta, S. Herath, K. Musiake, A mathematical model for flood loss estimation. J. Hydrol. **277**(2003), 24–49 (2003)
9. L.D. James, R.R. Lee, *Economics of Water Resources Planning*(McGraw-Hill, New York, 1971)
10. E. Kuiper, *Water Resources Project Economics* (Butterworths, London, 1971)
11. A. Lekuthai, S. Vongvisessomjai, Intangible flood damage quantification. Water Resour. Manag. **15**, 343–362 (2001)
12. D.J. Parker, C.H. Green, P.M. Thompson, *Urban Flood Protection Benefits: A Project Appraisal Guide* (Gower Technical, Aldershot, 1987)
13. K. Smith, R. Ward, *Floods: Physical Processes and Human Impacts* (Wiley, Chichester, 1998)
14. E.C. Penning-Rowsell,J.B. Chatterton, *The Benefits of Flood Alleviation: A Manual of Assessment Techniques* (Gower Technical, Aldershot, 1979)
15. W.K. Lee, I.N. Mohamad, Flood economy appraisal: An overview of the Malaysian Scenario, in *Proceeding of the International Civil and Infrastructure Engineering Conference 2013*, Part III (2014), pp. 263–274
16. D.L. James, B. Hall, Risk information for floodplain management. J. Water Resour. Plann. Manag. **112**(4), 485–499 (1986)
17. E.A. McBean, J. Gorrie, M. Fortin, J. Ding, R. Moulton, Flood depth damage curves by interview survey. J. Water Resour. Plann. Manag. **114**(6), 613–634 (1988)
18. D.J. Parker, The assessment of the economic and social impacts of natural hazards. Paper presented at the international conference on preparedness and mitigation for natural disaster 92, Reykjavic, Iceland, 28–29 May 1992
19. K.M.N. Islam, Impact of floods in Bangladesh, in *Floods, Routledge Hazards and Disaster Series*, ed. by D.J. Parker, vol. 1 (Routledge, London, 2000), pp. 156–171
20. D.I. Smith, Assessment of urban flood damage, in *Proceedings of Flood Plain Management Conference*, Australian Water Council, Canberra, Australia, 1981, pp. 145–180
21. UNSW, Evaluation Methodology of Flood Damage in Australia, Technical Project Report, 1981
22. MOC, Flood Damage Statistics in Japan, Technical Report, River Engineering Bureau, Ministry of Construction, Japan, in Japanese, 1996
23. USACE, National Economic Development Procedures Manual. USA, 1988
24. NTIS, Analysis of non-residential content value and depth-damage data for flood damage reduction studies. USA, 1996

25. D.J. Parker, *Floods*, vols. I and II. Routledge Hazards and Disasters Series (Routledge, London, 2000)
26. D.I. Smith, Flood damage estimation—A review of urban stage-damage curves and loss functions. Water SA **20**, 231–238 (1994)
27. R. Krzysztofowicz, D.R. Davis, Category-unit loss functions for flood forecast-response system evaluation. Water Resour. Res. **19**(3), 1476–1480 (1983)
28. G.F. White, Choice of adjustment to floods. University of Chicago, Department of Geography, Research Paper No. 93
29. E. Oliveri, M. Santoro, Estimation of urban structural flood damages: The case study of Palermo. J. Urban Water **2**(2000), 223–234 (2000)
30. S.J. Appelbaum, Determination of urban flood damage. J. Water Resour. Plann. Manag. **111** (3), 269–283 (1985)
31. D.I. Smith, P. Den Exter, M.A. Dowling, P.A. Jelliffe, R.G. Munro, W.C. Martin, *Flood Damage in the Richmond River Valley, New South Wales* (Centre for Resource and Environmental Studies, Australian National University, Canberra, 1979)
32. D.I. Smith, J.W. Handmer, M.A. Greenaway, T.L. Lustig, *Lesson and Losses from the Sydney Flood of August 1986* (Centre for Resource and Environmental Studies, Australian National University, Canberra, 1990)
33. KTA Tenaga Sdn. Bhd. Flood damage assessment of 26 April 2001 flooding affecting the Klang Valley and the preparation of generalised procedures and guidelines for assessment of flood damages (Final report). Volume 2: Guidelines and Procedures for the Assessment of Flood Damages in Malaysia. Drainage and Irrigation Malaysia, 2003

Is Farmer's Agricultural Income Dependent on Type of Irrigation Delivery System?

Noorul Hassan Zardari, Sharif Shirazi, Irena Naubi, and Siti Mariam Akilah Mohd Yusoff

Abstract The study was conducted to verify the principles of the century-old rotational irrigation water delivery system (locally known as the *warabandi* system) in Pakistan. Fair allocation of water and equitable distribution of water shortage (if any) are the two main principles of the warabandi system. Two watercourses (1-R and 3-R) in lower Indus River Basin were selected and 86 farmers were interviewed in a face-to-face survey. A questionnaire asking for their farm income, weekly water share, farming experience, water shortage, and other information was distributed to the farmers. We found that both principles of the *warabandi* system were not fully implemented in the surveyed watercourses. Unfair distribution of irrigation water to the farmers was observed within and across watercourses. Water shortage was also not fairly distributed between the farmers. The ineffectiveness of the *warabandi* principles was causing economical and social impacts on the local farmers. The farmer's income was dependent on the type of irrigation delivery system. More flexibility in irrigation water availability may increase the farm income.

Keywords Indus River basin • Irrigation allocation system • Water share • Water value

1 Introduction

The irrigation system in Pakistan (14.6 M ha) is the largest contiguous irrigation system in the world. It comprises the Indus and its major tributaries, three major storage reservoirs, 19 barrages/headworks, 44 canal commands, and some 100,000 watercourses. The total length of canals is about 56,073 km, with watercourses, farm channels, and ditches running another 1.6 million km in length [1, 2]. The

N.H. Zardari (✉) • S. Shirazi
Institute of Environmental and Water Resources Management, Faculty of Civil Engineering, Universiti Teknologi Malaysia, Skudai, Johor, Malaysia
e-mail: noorulhassan@utm.my

I. Naubi • S.M.A. Mohd Yusoff
Department of Hydraulics and Hydrology, Faculty of Civil Engineering, Universiti Teknologi Malaysia, Skudai, Johor, Malaysia

S.H. Abu Bakar et al. (eds.), *ISFRAM 2014*, DOI 10.1007/978-981-287-365-1_14

irrigation system uses barrages to divert water from rivers into canals. The main canals are designed for continuous operation at or near full capacity. Distributaries are the lowest level channels directly controlled by the Irrigation Department. The turnouts (*mogha*) are ungated modular outlets designed to deliver fixed amounts of water into watercourses. Each watercourse irrigates 60–250 ha (225 ha average) of land, usually cultivated by 10–150 farmers (average about 50 farmers) [3, 4]. Each farmer's water share is supposed to be proportional to the size of his land holding [5].

1.1 Problems with the Existing System

Since the introduction of the *warabandi* system, little effort has been made to see what has happened to the historical water share granted on the basis of land holdings, despite major political and socioeconomic changes [6]. Much has been said about the shortage of water resources in relation to crop water requirements, but little has been done to evaluate the effectiveness of water allocation or to find ways and means to improve distribution procedures. The thorough examination of basic rules of the *warabandi* system has shown many drawbacks in the existing delivery system, for example, the size of water share under the *warabandi* system is supposed to be exclusively decided on the basis of the size of the farmland. Water allocation based totally on a single criterion is unlikely to improve productivity from the scarce resource, as the water productivity depends on a variety of critical factors [7, 8]. Reference [6] indicates that the volume of water available to a farmer is independent of the stage of crop growth and the farmer is forced to either take his/her turn to receive water (whether needed or not), or to forego that water. This means that the system not only leads to wastage of water but also sometimes results in low crop yields due to over-irrigation. Likewise, the rigidity implicit in the *warabandi* system prevents farmers from maximizing economic and social net benefits from the scarce water [9, 10].

2 Methodology

2.1 Salient Features of the Study Area

The study was conducted in two watercourse commands, namely 1-R and 3-R. Both watercourses off-take from *Lundo* Distributary. 1-R watercourse is situated at the head of the distributary and 3-R watercourse at the middle of the distributary. *Lundo* Distributary off-takes from *Ali Bahar* Branch Canal of *Sukkur* Barrage Circle. The *Lundo* Distributary is 32 km long. The design discharge of the Distributary is 4.6 m^3/s with 13,360 ha of cultural command area (CCA), which is equivalent to

Table 1 Features of the surveyed watercourses

Number of watercourse	Design discharge (m^3/s)	Length (km)	CCA (ha)	No. of farmers interviewed	Average size of farm (ha)	Average size of weekly water share (m^3/ha)
1-R	0.067	3.1	330	46	4.8	134
3-R	0.025	2.4	166	40	4.1	86

224 $m^3/ha/week$ for the whole of the CCA. However, the average water allocation in the survey areas was between 86 and 134 $m^3/ha/week$. It shows about 50 % of water delivered to the Distributary was lost or unfairly allocated elsewhere. Three reasons have been initially found for such huge water losses: (1) Farmers on other watercourses were legally allocated more water; (2) Evaporation and seepage from the channels; or (and) (3) there is significant theft from the Distributary. Determining the magnitude of impact by each factor was beyond the scope of this study.

Farm and water management data were collected for both 2011–2012 crop seasons, i.e., *Rabi* (winter) and *Kharif* (summer). Crop survey data for both regions show that cotton was the main crop in *Kharif* and wheat in *Rabi*, with some area under sugarcane, banana, corn (maize), sorghum, millet, and fodder crops. The groundwater quality was fresh in 1-R region and marginal in 3-R region. A total of 86 farmers were interviewed and a questionnaire was completed for each. Information about the size of "weekly water share" ("water share" hereafter), proportion of allocated water actually received, the types of crops grown, and other farm and water management practices was collected. The farmers' fields were grouped into three categories, namely head, middle, and tail according to the distance from the watercourse outlet. Some details of watercourses are given in Table 1.

2.2 *Analysis of Weekly Water Share Data*

For this study, the allocated water share was calculated on the assumption that both watercourses were flowing with the design discharge. The average water allocation was very low (630 mm depth for 1-R and 470 mm depth for 3-R farms in each year) in both survey areas. The actual amount of water received by the farmers in the tail reach of the watercourse was probably smaller, as 20–25 % of allocated water share was likely to have been lost in both unlined watercourses. Studies on water losses in unlined watercourses by [11, 12] suggest that 20–25 % of water delivered to the watercourse is lost in the system before reaching the farm gate. This clearly shows that the structural system ensures that the main principle of *warabandi* system (i.e., equity in water distribution) fails in practice. (Note it was suggested above that at least 50 % of the water entering the Distributary never reaches the crops.)

Water share variation from watercourse to watercourse Weekly water share data were analyzed to determine equity in the water allocation. The average size of water share with maximum and minimum sizes of water share for both watercourses is shown in Table 2.

Table 2 Field survey results

Parameters		1-R watercourse	3-R watercourse
Proportion of farmers with less than high school education (%)		22	65
Proportion of farmers with farming experience of 15 years or less (%)		61	43
Farm location on watercourse (%)	Head	7	28
	Middle	28	37
	Tail	65	35
Average water share (m^3/ha)		134	86
Groundwater quality		Fresh	Marginal
Average size of farm (ha)		4.8	4.1
Proportion of farmers having resources of income other than agriculture (%)	None	56	65
	Own business	9	28
	Employed	35	7
Percent of farm area under cultivation (%)		86	65
Number of govt. tubewells		2	None
Number of private tubewells		36	None
Net crop income (RS/ha)		17,150[a]	11,365[a]
Value of a unit of water (RS/m^3)		3.9	6.9
Value of a unit of water for each location of farms (RS/m^3)	Head	NA	6.9
	Middle	4.31	8.3
	Tail	3.68	5.65
Water shortage faced (proportion of allocated water share) (%)		36	48
Water shortage faced by each location of farms (%)	Head	NA	34
	Middle	41	41
	Tail	33	57

[a]1 US dollar = 99 Pakistan Rupees (RS) (exchange rate in July 2012)

The average size of water share of all 1-R farms was 134 m^3/ha (13.4 mm depth of water per week). The average water allocation to 3-R farms was 86 m^3/ha (8.6 mm depth of water per week). The analysis shows that 50 % of 3-R farms have water share of less than 86 m^3/ha and all 1-R farms have water share above 86 m^3/ha. This uneven water allocation, in the watercourse commands, is shown in Fig. 1.

Water share variation within watercourse Unfair and inequitable allocation of scarce water does not only occur between but also within watercourses. Eighty-five percent of 1-R tail end farms have water share of less than 134 m^3/ha (i.e., average size of water share of all 1-R farms). The middle farms have greater share in canal water allocation compared to the tail end farms. Head farms were not considered, as the sample size was very small. In case of 3-R farms, none of the head farms had water share above the average size of water share for all 3-R farms. However, 75 % of farmers in the tail reach have 30 % higher water share than the

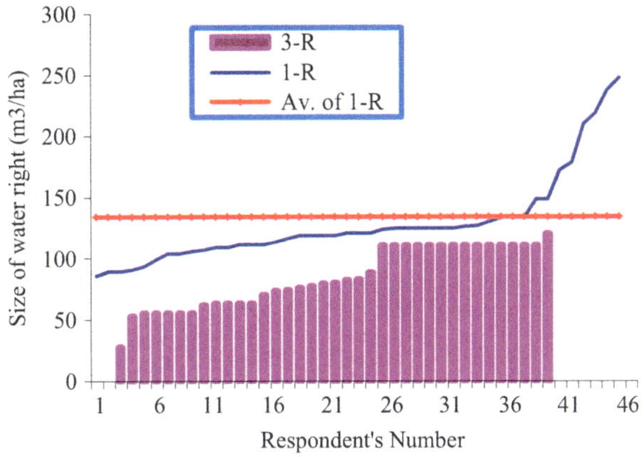

Fig. 1 Size of water share (plotted in increasing order) in 1-R and 3-R command areas

average. What was the philosophy behind such unequal allocations? The data in hand only give a single reason for such philosophy. The only resemblance between head and middle farms of 1-R and tail farms of 3-R was the equal size of farms. Medium to large size farms were dominant in those locations. So, one can say that the larger farms tend to attract a proportionally higher share of water.

2.3 Analysis of Water Shortage Data

Variation of water shortage between watercourses The second objective of the *warabandi* system, i.e., equitable distribution of any water shortage between all farms, was investigated in the survey areas. It was assumed that the parent channel (*Lundo* Distributary) remains closed for 6 weeks in each year for maintenance reasons. The farmers were entitled to receive only 46 weekly water turns instead of 52. In this chapter, the term "water shortage" is the proportion of "actual number of water turns received" to the "allocated number of water turns."

The collected data indicate that the 1-R farms missed 36 % of allocated water turns. This value for 3-R farms was 48 %. Hence 3-R farmers faced more water shortage than 1-R farmers (Table 2). This uneven distribution of water scarcity affects the farm area under cultivation and must have a huge effect on farm incomes. Table 2 shows that 65 % of 3-R command area was under cultivation, while for 1-R command it was 86 %. The farm income of 3-R farmers suffered more than for 1-R farmers. It was probably because the farm income was approximately proportional to the area under cultivation. The pattern of actual amount of water delivered to 1-R and 3-R is shown in Fig. 2.

Water shortage at different locations on same watercourse Analysis of the distribution of water shortage along each watercourse indicates that inequity in water shortage distribution does not stop at the outlet of the watercourse but also continues within watercourses (similarly to uneven water allocations). For example, the

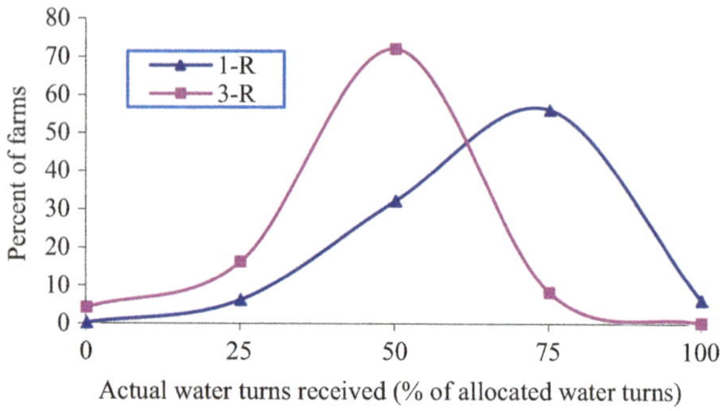

Fig. 2 Pattern of allocated water turns received by 1-R and 3-R farmers

farms located on 3-R watercourse shared average water shortage of 48 % in this manner: tail end farms did not receive 57 % of allocated water turns, middle farms missed 41 %, and head farms missed 34 %. This analysis demonstrates that the farmers in the tail reach of the watercourse missed a large portion of their water allocation compared to the head and middle farmers.

2.4 Economic Analysis from Survey Data

Farm benefits from water use In the earlier sections, two fundamental rules of the *warabandi* were analyzed and it was shown that those rules have been inadequately applied in the surveyed areas. In this section, the impacts of such irregularities on the farm income are estimated. The average annual crop income considering all 1-R farms was 17,150 RS/ha. This value for 3-R farms was 11,365 RS/ha (Table 2). It shows that the average farm income of 1-R farmers was 35 % more than the average farm income of 3-R farmers. The reasons for higher income by 1-R farmers would be numerous, but the more likely causes were their larger shares in water allocation and lesser share in water scarcity. Another reason for their high crop income was the availability of fresh groundwater, as 36 private and two government tubewells were augmenting canal water supplies where it was insufficient to meet crop water requirements. On the other hand, 3-R farmers had no access to groundwater supplies.

The impacts of the water shortage intensity on the crop income were not always linear. A linear relation between the crop income and water shortage only occurs where the proportion of supplied water to the allocated water was above 75 %. The average crop income of head and tail farms of 3-R watercourse was 12,798 RS/ha and 9,203 RS/ha, respectively. The value for the middle farms was 12,595 RS/ha. The data regarding the water allocation and the water shortage distribution show that the farmers in the tail reach of the watercourse were enjoying more water allocation, but were facing more frequent water shortages. The high water share allocated to the farmers in the tail reach of the watercourse does not appear to provide any significant advantage to them, as their income remains low compared

Fig. 3 Annual net crop income on 1-R and 3-R (plotted in increasing order)

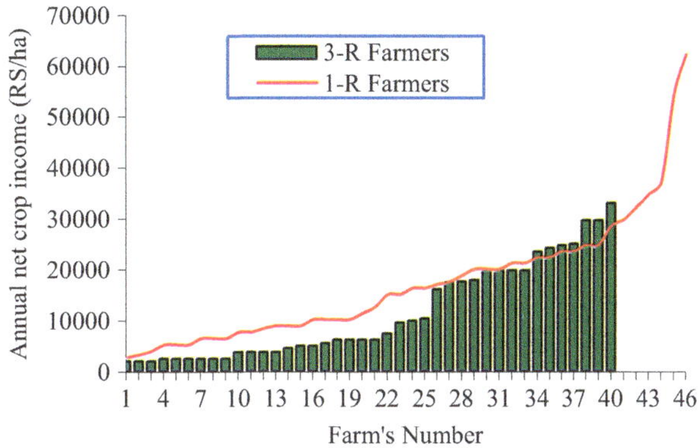

to the income of head and middle farms. Farm incomes fell progressively from head to tail of the watercourse. The only apparent reason for this trend was the distribution of water shortage between locations, as the intensity of water shortage was increasing from head to tail of the watercourse. Figure 3 shows annual net crop income of each watercourse.

Economic value of unit of water The economic value of a unit of water was calculated in order to find the efficient user(s) of irrigation water. Thirty-six private and two government-controlled tubewells were in the 1-R command area. On the other hand, not a single tubewell was found in 3-R command because of nonexistence of good quality groundwater. The amount of water delivered to the farms (only 1-R farms) by the private and the government tubewells was calculated on the basis of rough data available from the farmers and the tubewell operators about the tubewells' annual operational hours. It was calculated that each tubewell was annually operating for 170 h. The average operational cost of a single tubewell was determined as 61 rupees per hour. Each year, on average, each tubewell was adding 10,300 m^3 of groundwater into the canal water supplies. This means a total amount of 370,770 m^3 of groundwater was put into use in the data collection period, which was almost equal to the total amount of water delivered by 3-R watercourse or 30 % of total amount of water delivered by 1-R watercourse. The value of a unit of water calculated for each watercourse and location are shown in Table 2. The average net value of 1 m^3 of water consumed by 1-R and 3-R farmers was 3.90 rupees and 6.90 rupees, respectively. Similarly, the value of a unit of water for farms situated at different locations on the same watercourse varies significantly. The farms located at the middle of 1-R and 3-R were more efficient in water than other farms. Therefore, if the data about water share and value of a unit of water could be obtained from a large sample of watercourses, the optimal size of water share could possibly be determined.

3 Factors Controlling Farm Incomes

It was calculated that the income of 3-R farmers was 35 % lower than the income of 1-R farmers. Many factors are thought to be affecting 3-R farmers' income. The most likely factors appear to be: the majority of 3-R farmers fall in the "low education" category, they have small share in canal water allocation, most of the farmers have no other source of income, they have no other resource available for irrigation, and they missed more water turns. The inverse of this may be true for 1-R farmers.

Those variables were compared in order to determine the most likely variables that have affected 3-R farmers' income. The high level of farmer's education did not contribute to farm income, as the majority of farmers at the tail of 3-R watercourse were highly educated but were less efficient in water use. The majority of farmers at the tail of 1-R watercourse had other sources of income (own business, employed, etc.) but were not more efficient in generating farm income as compared to those on the head and the middle of the watercourse. So, it does not appear that this factor has any large impact on farm income. However, the analysis of water shortage and water share data produces significant conclusions. Twenty-five percent shortfall in allocated water share does not have significant impacts on the farm income. When the intensity of water shortage increases beyond 25 %, farm income falls rapidly compared to the proportional increase in the water shortage. It was also found that having a larger share in canal water allocation produces higher farm income. Therefore, two critical factors, which most likely have affected 3-R farmers' income, were their smaller share of canal water supplies and the higher degree of water shortage they suffered. The impact on farm income by other factors was minor.

4 Conclusions

The survey results showed that the basic principles of the *warabandi* were not fully operational. Varying rates of water allowances between watercourses and within each watercourse is one example of its ineffectiveness in the real world. Moreover, the variation in the distribution of water shortage between stakeholders particularly contributes to *warabandi's* inefficiency. The inequities in the *warabandi* affect farmers' economic and social outcomes. Farmers having lands on 3-R watercourse were double-handicapped, as they have a lower share in canal water supplies compared to 1-R farmers and have marginal quality groundwater in their locality, which is unfit for raising crops. The theoretical principles of *warabandi* do not provide equal opportunities to all farmers in generating farm benefits. The farms located on 1-R watercourse benefited from the available fresh groundwater resource. The amount of water extracted from the fresh aquifer was equal to 30 % of the watercourse supplies. This extra water contributed to the farms' income but produced lower net return from a unit of water. We conclude if the improving of land use efficiency is the aim of the delivery system, then the *warabandi* produces significant results with higher overall farm income. However, if improving water use efficiency is the target, the *warabandi* system will not produce satisfactory outcomes.

Acknowledgment This study was partly funded by the Ministry of Higher Education (MOHE) Malaysia and the Universiti Teknologi Malaysia (UTM) under GUP Grant Tier-1 with Vot No. 08H43 and FRGS Grant Vot No. 4 F539. This study was also supported by the Asian Core Program of the Japanese Society for the Promotion of Science (JSPS).

References

1. D.H. Murray-Rust, B. Lashari, Y. Memon, Water distribution equity in Sindh Province, Pakistan, Extended project on farmer managed irrigated agriculture under the National Drainage Program (NDP). Working Paper No. 9, International Water Management Institute, Colombo, Sri Lanka, 2000
2. S. Khan, R. Tariq, C. Yuanlai, J. Blackwell, Can irrigation be sustainable? Agr. Water. Manag. **80**, 87–99 (2006)
3. M.A. Chaudhary, A.Y. Robert, Irrigation water allocation institutions: Pakistan's warabandi system, in *Social, economic and institutional issues in third world irrigation management*, ed. by R.K. Sampath, A.Y. Robert. Papers in water policy and management, No. 15 (Westview Press, London, 1989)
4. M. Nakashima, Water users' organization for institutional reform in Pakistan's irrigation sector. J. Int. Dev. Stud. **9**, 79–93 (2000)
5. N.H. Zardari, I. Cordery, Estimating the effectiveness of rotational delivery system: A case study from Pakistan. Irrigat. Drain. **59**, 277–290 (2010)
6. D.J. Bandaragoda, S. Rehman, Warabandi in Pakistan's canal irrigation systems: Widening gap between theory and practice. IWMI Country Paper-Pakistan No. 7, International Water Management Institute, Colombo, Sri Lanka, 1995
7. N.H. Zardari, I. Cordery, Water productivity in a rigid irrigation delivery system. Water Resour. Manag. **23**, 1025–1040 (2009)
8. D.J. Bandaragoda, Design and practice of water allocation rules: Lessons from warabandi in Pakistan's Punjab. Research Report No. 17, International Water Management Institute, Colombo, Sri Lanka, 1998
9. M.A. Bhatti, W.K. Jacob, Irrigation allocation problems at tertiary level in Pakistan. ODI-IWMI, Irrigation management Network Paper 90/3C, Overseas Development Institute, London, 1990
10. X. Cai, C. Ringler, M.W. Rosegrant, Does efficient water management matter? Physical and economic efficiency of water use in the River Basin. Discussion Paper No. 72. Environment and Production Technology Division, International Food Policy Research Institute, Washington, DC, 2001
11. M. Latif, S. Sarwar, Proposal for equitable water allocation for rotational irrigation in Pakistan. Irrigat. Drain. Syst. **8**, 35–48 (1994)
12. P.Z. Kirpich, Z.H. Dorota, W.S. Stuart, Problems of irrigation in developing countries. J. Irrigat. Drain. E-ASCE **125**, 1–6 (1999)

Variable Parameter Muskingum Discharge Routing Method for Overland Flow Modeling

Ravindra V. Kale and Muthiah Perumal

Abstract A distributed physically based variable parameter Muskingum discharge routing (VPMD) method, which is derived directly from the Saint-Venant equations (SVE) considering uniform lateral flow, is presented for generating runoff from the overland flow plane. The operational performance of the method is carried out using the observed rainfall-runoff data (Izzard data) available in literature based on various evaluation measures. Besides, the simulation results are also evaluated by comparing with the corresponding solutions of the SVE obtained using the explicit numerical scheme. The study demonstrates that the proposed VPMD method is advantageous over the currently used numerical solution methods for overland flow simulation because of its unconditional numerical stability and high accuracy with larger degree of flexibility in the selection of computational spatial and temporal grid sizes leading to efficient and inexpensive computations. These capabilities imply its suitability for coupling with various land surface schemes available for meso- and microscale catchment modeling studies.

Keywords Runoff hydrograph • Overland flow • Variable parameter Muskingum • VPMD • Saint-Venant solutions

1 Introduction

Overland flow is the beginning of the hydrological runoff process by which the precipitation fallen on the land surface is transported down the slope as a sheet flow before draining into different forms of streams. Incorporation of overland flow component is an essential step towards developing a spatially distributed physically based modeling tool. Suitable physically based catchment models based on the

R.V. Kale (✉)
Research Management and Outreach Division, National Institute of Hydrology Roorkee, Roorkee 247667, Uttarakhand, India
e-mail: ravikale2610@gmail.com

M. Perumal
Department of Hydrology, Indian Institute of Technology Roorkee, Roorkee 247667, Uttarakhand, India
e-mail: p_erumal@yahoo.com

S.H. Abu Bakar et al. (eds.), *ISFRAM 2014*, DOI 10.1007/978-981-287-365-1_15

basic equations governing the flow process of rainfall-runoff transformation phenomenon are more useful to deal with the gauged, less-gauged, and ungauged field problems [1, 2].

As the analytical solution of the full Saint-Venant equations (SVE) is difficult, various researchers have used explicit and implicit numerical schemes to solve the full SVE for the simulation of runoff from overland flow planes and channels (e.g., [3–5]). However, the operational use of the SVE is severely hindered by the requirement of high computational skill, extensive topographical data, and knowledge of initial and boundary conditions under unsteady flow conditions. Further, the stability of the explicit numerical methods depends on the dynamic wave celerity, even for the flow cases that have negligible inertia [6].

In order to overcome the limitations of SVE, many researchers (e.g., [7–9]) suggested that a greater degree of sophistication can be achieved, without ignoring fundamental mechanism, by evolving simplified approximations to the full SVE for describing the overland flow process. On the basis of simplifications of the full SVE, two different models have been evolved: (1) the kinematic wave model (KW) and (2) the diffusion wave (DW) or the zero-inertia model. A comprehensive review of literature reveals that for the simulation of runoff from a basin as response to complex rainfall pattern occurring over complex catchment geometry, the modelers have to rely on various numerical schemes, i.e., the finite difference, finite volume, and finite element methods for the solutions of the SVE and their variants. Such numerical schemes employed in the solutions require the use of very short temporal and spatial grid sizes in order to ensure the stability of the results [10]. To overcome the shortcomings of the various numerical schemes used for arriving at the solutions of the SVE and its variants, another well-known simplified method available for channel and overland flow modeling is the variable parameter Muskingum–Cunge (VPMC) method. The VPMC method has been applied extensively for routing flow in channel and natural rivers [11–13]. However, only limited attempts have been made to verify the application of the Muskingum–Cunge method for overland flow modeling [14–17].

In the VPMC method, the weighting parameter θ and the travel time K of the Muskingum method have been linked to the channel and flow characteristics based on the concept of matching the numerical diffusion with the hydraulic diffusion described by the diffusion coefficient. However, the physical justification of the VPMC method based on the concept of matching the physical and numerical diffusion is questionable [18] on the pretext that since this method is based on the concept of matching the numerical diffusion with the Hayami's physical diffusion equation [19], the method should work for the entire range of the physical diffusion equation. However, it is a well-known fact that the VPMC method is only applicable for a small range [20] of the physical diffusion equation. Therefore, Perumal [21, 22] proposed a variable parameter Muskingum discharge routing (VPMD) method which is based on sound physical and mathematical considerations. Although the theories proposed by Cunge [23] and Perumal [21, 22] attempt to give physical interpretation to the classical Muskingum routing method, the interpretation given in the latter references is able to explain all the features of the classical Muskingum method, including the formation of the well-known "initial dip" in the beginning of the Muskingum solution. A study made by Perumal and Sahoo [20] on the applicability criteria clearly brought out that the VPMD method has a wider applicability range than that of the VPMC method for channel routing, which is attributed to the physical basis of the development of the VPMD method. The VPMD method preserves volume more accurately than the VPMC method [24] when it is applied within its applicability range. Thus, the use of the VPMD method is more useful for solving unsteady channel flow problems [25, 26]. But, the potential capabilities of these methods to solve the overland flow problem remain unexplored.

This study briefly presents the development of a simple and computationally inexpensive and numerically stable physically based method to simulate overland flow following the approach used in the development of the variable parameter Muskingum method proposed by Perumal [21]. The performance of this method is carried out using the observed rainfall-runoff data (Izzard data) available in literature. Besides, the simulation results are also compared with the solutions of the SVE obtained using the explicit numerical scheme.

2 Overland Flow Model Formulation

The overland flow can be described by the well-known SVE consisting of the continuity and momentum equations by assuming that the flow occurring over a plane surface is one dimensional. In many situations, it has been found that the continuity equation and the simplified form of the momentum equation obtained by eliminating and curtailing some of its terms are found to describe flow very well for many practical situations [27]. Perumal [21, 22] made one such simplification and used the continuity equation and the simplified momentum equation to develop the VPMD method.

2.1 Concept of the Proposed Overland Flow Model

During the steady-state flow condition in an overland flow strip, there exists a one-to-one relationship between the flow depth or the cross-sectional area of the flow and the discharge at any location defining the steady-state flow-depth relationship at that location. But, during unsteady flow situation, there exists no unique relationship between the flow depth and the discharge at any location, but the same unique relationship exists between the flow depth at a given section and the corresponding steady discharge, occurring somewhere downstream from that section. This condition is depicted in Fig. 1 which defines the Muskingum routing reach considered for the overland flow study. This figure clearly shows that within the subreach of length Δx, at any instant of time t, the flow depth, y_M, at the middle of the reach is uniquely

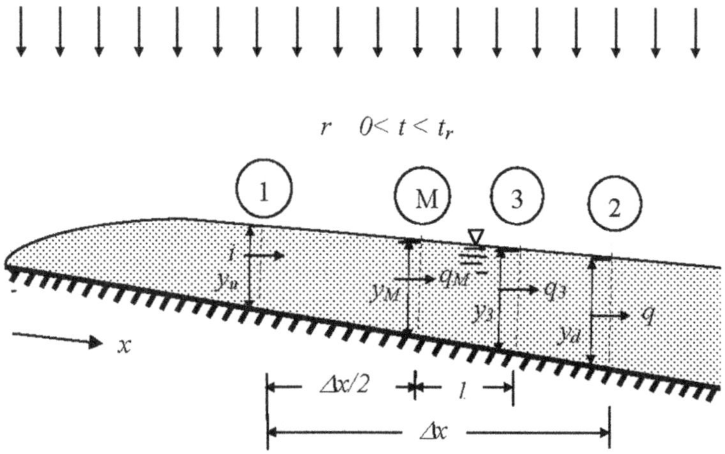

Fig. 1 Schematic sketch of the VPMD method governing the overland flow over an impervious surface

related with the discharge q_3 passing at a location l downstream from the midsection of the reach.

2.2 Governing Equations

The VPMD method for overland flow modeling is derived from the full SVE considering lateral flow. These equations applied to a unit width of the overland flow plane are expressed by the continuity and momentum equations, respectively, as [9]

$$\frac{\partial q}{\partial x} + \frac{\partial y}{\partial t} = q_{\mathrm{L}} \tag{1}$$

$$s_f = s_0 - \left(\frac{\partial y}{\partial x}\right) - \left(\frac{1}{g}\right)\left(\frac{\partial v}{\partial t}\right) - \left(\frac{v}{g}\right)\left(\frac{\partial v}{\partial x}\right) - \left(\frac{q_{\mathrm{L}} v}{g y}\right) \tag{2}$$

where q = flow rate per unit width of the flow plane ($\mathrm{L}^2\,\mathrm{T}^{-1}$), y = flow depth (L), s_f = energy slope ($\mathrm{L\,L}^{-1}$), g = acceleration due to gravity ($\mathrm{L\,T}^{-2}$), v = flow velocity ($\mathrm{L\,T}^{-1}$), s_0 is the bed slope ($\mathrm{L\,L}^{-1}$), q_{L} = net rate of lateral inflow per unit width ($\mathrm{L}^2\,\mathrm{T}^{-1}$) expressed as

$$q_{\mathrm{L}} = r - I_{\mathrm{a}} \tag{3}$$

where r = the rate of rainfall per unit area occurring for a duration t_r and I_{a} = the rate of loss due to abstraction. An order of magnitude analysis of (2) reveals that the momentum due to lateral inflow or outflow has a negligible effect on the flow dynamics [27, 28], and, hence, the term $(q_L v/gy)$ is neglected in the analysis.

2.3 Assumptions

The assumptions used in the derivation of this VPMD method are:

(a) A uniform thin sheet flow of runoff developing over an overland flow plane resembles to that of a shallow water unsteady flow in a wide prismatic rectangular cross-section channel reach.
(b) There is lateral inflow due to rainfall and/or outflow due to the interception, and infiltration loss is uniform over the computational reach.
(c) The momentum due to lateral inflow and/or outflow is assumed to be negligible.
(d) The slope of the water surface $(\partial y/\partial x)$, the slope due to local acceleration $(1/g)(\partial v/\partial t)$, and the slope due to convective acceleration $(v/g)(\partial v/\partial x)$ all remain constant at any instant of time in a given routing reach. This assumption implies that the friction slopes s_f is constant over the computation reach of length Δx and, accordingly, the flow depth variation is approximately linear over this reach Δx.
(e) The multiple of the derivatives of flow and section variables with respect to both time and space are negligible.
(f) At any instant of the time during unsteady flow, the steady flow relationship is applicable between the flow depth at the middle of the computational reach and the discharge passing somewhere downstream of it.

Assumptions (d–f) are employed in the VPMD routing method for channel flow routing without considering lateral flow [21].

2.4 Derivation

The discharge at any section of the unit width of the overland flow plane reach may be expressed as:

$$q = vy \tag{4}$$

Differentiating (4) after incorporating velocity v expressed by the Manning's friction law with reference to x and invoking assumption (d) of constant s_f over the length Δx, the resulting expression may be written as

$$\frac{\partial q}{\partial x} = (5/3)v\frac{\partial y}{\partial x} \tag{5}$$

Using (5), the celerity of flow wave may be expressed as

$$c = \frac{\partial q}{\partial y} = \frac{5}{3}v \tag{6}$$

Due to the assumption of linearly varying flow depth over a computational reach, the celerity of the overland flow governed by (6) is not unique, as in the case of the KW model [14], for the same flow depth occurring in the rising and falling limbs of the runoff hydrograph.

Differentiation of (5) with reference to x and invoking assumption (e) that the multiples of derivatives of flow and section variables with reference to x and t are negligible, the resulting expression can be written as

$$\frac{\partial^2 q}{\partial x^2} = 0 \tag{7}$$

Equation (7) implies that the discharge is also varying linearly over the computational reach considered.

The use of (1), (4), (5), and (6) and also the assumption of negligible momentum due to lateral inflow contribution enables one to arrive at the equation for the local acceleration term, which can be expressed as

$$\frac{1}{g}\frac{\partial v}{\partial t} = -\frac{10}{9}\frac{\partial y}{\partial x}F_{rp}^2 \tag{8}$$

where F_{rp} = Froude number of the overland flow expressed as $F_{rp} = v/\sqrt{gy}$.

Further, the use of (4) and (5) leads to the expression for the convective acceleration term which can be written as

$$\frac{v}{g}\frac{\partial v}{\partial x} = \frac{2}{3}\frac{\partial y}{\partial x}F_{rp}^2 \tag{9}$$

Using (8) and (9) in (2) leads to the expression for friction slope (s_{f}) as

$$s_{\mathrm{f}} = s_0\left[1 - \frac{1}{s_0}\frac{\partial y}{\partial x}\left\{1 - \frac{4}{9}F_{rp}^2\right\}\right] \tag{10}$$

The discharge q_{M} at the middle of the considered computational flow reach of Δx and after eliminating higher order terms of the $\frac{1}{s_0}\frac{\partial y}{\partial x}\left\{1 - \frac{4}{9}F_{rp}^2\right\}$ in the binomial series expansion of (10) leads to the expression for q_{M} as

$$q_{\mathrm{M}} = q_3 - \frac{q_3\left(1 - (4/9)F_{rp}^2\right)_{\mathrm{M}}}{2s_0(5/3)v_3} \frac{\partial q}{\partial x}\bigg|_3 = q_3 - l\frac{\partial q}{\partial x}\bigg|_3 \tag{11}$$

in which

$$q_3 = \frac{1}{n} y_{\mathrm{M}}^{5/3} s_0^{1/2} \tag{12}$$

where the subscripts M and 3, respectively, denote the midsection and section "3" of the computational reach of length Δx. In (11), the discharge q_{M} is expressed in terms of normal discharge, q_3, which is uniquely related to y_{M} by the steady-state relationship. Therefore, the adjunct term of $\partial q/\partial x|_3$ denotes the length l of the location of section (3) downstream from the midsection of the computational reach, at which the normal discharge q_3 corresponding to the flow depth y_{M} passes during unsteady flow.

Finally, the approximate convection-diffusion (ACD) equation governing the overland flow can be given as

$$\frac{\partial q}{\partial t} + c\frac{\partial q}{\partial x} = cq_{\mathrm{L}} \tag{13}$$

This simplified governing equation is used in the development of the VPMD method for overland flow modeling and is referred to as the approximate convection-diffusion (ACD) equation. Note that the form of (13) is similar to the form of the kinematic wave equation expressed by Ponce [14]. Although, the ACD equation and the kinematic wave equation resemble each other because of the same form, the ACD equation is capable of modeling the overland flow in an approximate manner in the transition range of the unsteady overland flow governed by the diffusion and kinematic waves (including the latter) in a way similar to the VPMD method developed for routing floods in channels [29]. The basis behind this inference is that the right hand side of (13) does not contain the diffusion term $\left(D\partial^2 q/\partial x^2\right)$ due to (7), but the diffusion coefficient $D \neq 0$. This enables the application of (13) for overland flow modeling in the range of lower order shallow diffusive flood waves, including the kinematic waves.

Application of the governing equation as given in (13) of the ACD wave to section "3", as depicted in Fig. 1, leads to the expression

$$\frac{\partial q}{\partial x}\bigg|_3 = -\left(\frac{1}{c}\frac{\partial q}{\partial t}\right)\bigg|_3 + q_{\mathrm{L}} \tag{14}$$

The application of the ACD equation as in (13) leads to the routing equation for overland flow modeling in the similar form as given by Ponce [14] using the

Muskingum–Cunge method and can be expressed on the space-time computational grid as

$$q_2 = C_1 i_2 + C_2 i_1 + C_3 q_1 + C_4 q_L \Delta x \tag{15}$$

where q_2 = the outflow discharge from the computational reach at current time level; i_2 = the inflow discharge to the computational reach at current time level; i_1 = the inflow to the computational reach at the previous time level; q_1 = the outflow runoff discharge from the computational reach at the previous time level; and q_L = average lateral inflow rate within the computational reach over Δt. The routing coefficients C_1,C_2,C_3, and C_4 are given as

$$C_1 = \frac{-K\theta + 0.5\Delta t}{K(1-\theta) + 0.5\Delta t} \quad C_2 = \frac{K\theta + 0.5\Delta t}{K(1-\theta) + 0.5\Delta t}$$
$$C_3 = \frac{K(1-\theta) - 0.5\Delta t}{K(1-\theta) + 0.5\Delta t} \quad C_4 = \frac{\Delta t}{K(1-\theta) + 0.5\Delta t} \tag{16}$$

The parameters K and θ can be expressed as

$$K = \frac{\Delta x}{c_3} \tag{17}$$

$$\theta = \frac{1}{2} - \frac{q_3\left(1 - (4/9)F_{rpM}^{\,2}\right)}{2 s_0 c_3 \Delta x} \tag{18}$$

The detailed explanations about derivations (13–18) and step-by-step procedure for overland flow simulation are given in the study by Kale [30].

3 Field Verification with Izzard's Data

The classical overland flow laboratory experiments conducted by Izzard in USA during 1942–1943 were among the first systematic study carried out on the overland flow process under controlled conditions [31]. Even after more than 70 years have passed, these research experiments continue to remain as the basic study for model verification of the newly developed overland flow models. Therefore, the performance of the VPMD method under complex rainfall pattern is investigated by comparing its solution with the eight selected overland flow events of the Izaard's experimental study (as reported by Maksimovic and Radojkovic [32]). The data utilized in the present study were collected from two different experimental plots each having a width of 1.83 m and a length of 21.95 m and each characterized by different surfaces to define two uniform roughness values, namely (1) smooth asphalt surface and (2) dense blue grass turf sod placed directly upon roofing paper with transverse slats. Considering typical variability of the rainfall pattern, the runoff events of 34, 35, 36, and 50 corresponding to smooth asphalt surface and the runoff events of 301, 302, 318, and 319 corresponding to the plane characterized by dense blue grass turf sod as reported by Maksimovic and Radojkovic [32] were considered for simulation using the proposed method.

The various plane characteristics and rainfall conditions corresponding to these selected events used for the verification of the developed method are summarized in Table 1. In order to evaluate the Manning's roughness coefficient required to simulate the runoff hydrograph, the practice of matching the observed hydrograph closely using the simulated hydrograph is adopted [33, 34]. Following the approach of Wong and Lim [35], the Manning's n is computed for each runoff event based on the sensitivity analysis tests carried out by varying n from 0.010 to 0.018 for the asphalt experimental surface, while from 0.21 to 0.38 for the dense blue grass sod experimental surface by assessing the reproduction performance using the Nash–Sutcliffe efficiency (η_q). The obtained optimum Manning's roughness coefficients with the event number written within the adjacent brackets are 0.016 (34), 0.017 (35), 0.015 (36), 0.015 (50), 0.36 (301), 0.27 (302), 0.25 (318), and 0.38 (319). These values are close to those recommended values (0.010–0.013 for the asphalt plane and 0.38–0.63 for the dense blue grass turf) by Woolhiser [33] and Engman

Table 1 Details of Izzard's experimental events considered for the VPMD method application

Surface cover, length of planes, and constant slopes (s_0)					
Asphalt, 21.95 m, 0.005			Dense blue grass turf, 21.95 m, 0.01		
Event	Manning's n	Rainfall intensity r (cm/h) versus time t (s)	Event	Manning's n	Rainfall intensity r (cm/h) versus time t (s)
34	0.015	9.756, $0 \leq t \leq 420$	301	0.28	9.678, $0 \leq t \leq 1{,}680$
		0, $t > 420$		0.36[a]	0, $t > 1{,}680$
35	0.015	4.752, $0 \leq t \leq 600$	302	0.28	4.65, $0 \leq t \leq 1{,}800$
		9.474, $600 \leq t \leq 1{,}080$			0, $t > 1{,}800$
		0, $t > 1{,}080$			
36	0.015	9.702, $0 \leq t \leq 480$	318	0.28	4.524, $0 \leq t \leq 1{,}320$
		0, $480 \leq t \leq 540$			9.096, $1{,}320 \leq t \leq 1{,}980$
		9.702, $540 \leq t \leq 780$			0, $t > 1{,}980$
		0, $t > 780$			
50	0.015	4.8, $0 \leq t \leq 120$	319	0.28	9.348, $0 \leq t \leq 1{,}140$
		9.6, $120 \leq t \leq 420$		0.38[a]	0, $1{,}140 \leq t \leq 1{,}440$
		0, $420 \leq t \leq 540$			9.348, $1{,}440 \leq t \leq 1{,}890$
		9.6, $540 \leq t \leq 840$			0, $t > 1{,}890$
		0, $840 \leq t \leq 960$			
		4.8, $960 \leq t \leq 1{,}320$			
		0, $t > 1{,}320$			

[a]Additional Manning's n used to verify the effect of increased roughness. Note that Manning's n for these events are chosen based on the sensitivity analysis test

[34]. Therefore, a uniform value of Manning's roughness coefficient of $n = 0.015$ for asphalt plane and a value of $n = 0.28$ for the dense blue grass turf were used in the simulation of all the events considered in this study. However, the differences between the employed and the calibrated roughness values are high for the events 301 and 319, and therefore, the corresponding calibrated roughness values for these events are also used in the performance verification of the VPMD method. Runoff simulations using the VPMD method were obtained for two different sets of spatial and temporal grid sizes, viz., $\Delta x = 0.51$ m and $\Delta t = 1$ s; $\Delta x = 2.195$ m and $\Delta t = 10$ s.

The typical runoff hydrograph simulation results by the proposed method are shown along with the observed hydrographs and the corresponding numerical solutions of the SVE as shown in Fig. 2. It can be inferred from these results that the VPMD method has the ability to reproduce the observed data accurately; besides, it matches perfectly with the solution of the SVE for all the events under considerations. It is interesting to note that, particularly, for the cases of rainfall-runoff events on asphalt plane, *Izzard* observed a sudden increase of discharge immediately after the cessation of rain as shown in Fig. 2 (Events 36 and 50). This spike observed in the discharge hydrograph is due to sudden reduction of the surface roughness caused by the termination of the impact of the rain drops on the surface of the overland flow ([31], as quoted by Maksimovic and Radojkovic [32]). However, for dense blue grass turf, such an instant rise of runoff discharge was not observed when rainfall ceased.

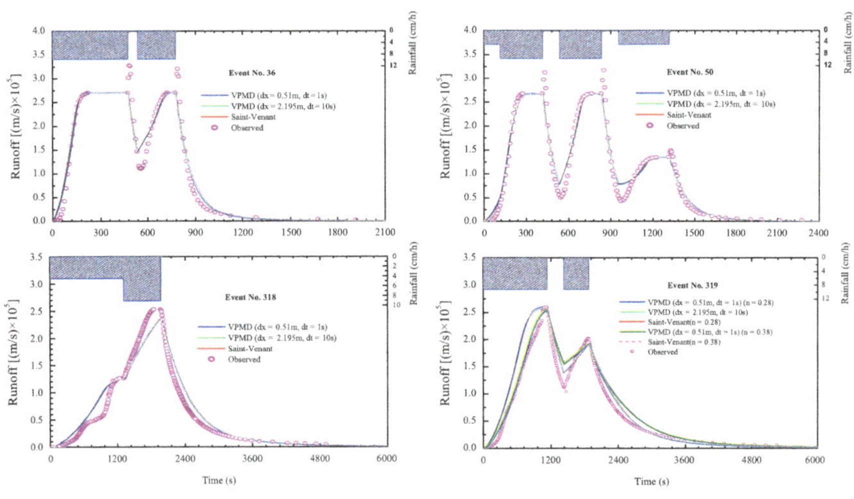

Fig. 2 Izzard's experimental hydrographs and the corresponding simulated hydrographs of the VPMD and SVE methods

4 Conclusions

This study presents the development of the VPMD routing method for runoff generation from the overland flow planes. The performance of the VPMD routing method under complex rainfall pattern is investigated by comparing its solutions with the simulated runoff hydrograph for Izzard's data. The study demonstrates that the simulated runoff hydrographs by the VPMD method are in good agreement with the observed data and SVE solution. The close reproduction of the runoff hydrographs recorded at the downstream end of the overland flow plane and that simulated by the SVE solutions at the end of the overland flow plane demonstrates that the VPMD method can be confidently used for catchment modeling studies. The ability of the VPMD method to provide wide flexibility in the selection of temporal and spatial grid sizes for computational purposes remains to be tested. VPMD method provides wide flexibility in the selection of temporal and spatial grid sizes.

Acknowledgment R. V. Kale is very thankful to Sh. R. D. Singh, Director, National Institute of Hydrology, Roorkee, India, and Dr. V. C. Goyal, Scientist F and Head, RMOD, NIH, Roorkee, for providing an opportunity for working on this manuscript.

References

1. V.P. Singh, D.A. Woolhiser, Mathematical modelling of watershed hydrology. J. Hydrol. Eng. ASCE **7**(4), 270–292 (2002)
2. T.S.W. Wong, Physically based approach in hydrology—What is the benefit? J. Hydrol. Eng. ASCE **11**(4), 293–295 (2006)
3. D.A. Woolhiser, J.A. Liggett, Unsteady one-dimensional flow over a plane: The rising hydrograph. Water Resour. Res. **3**(3), 753–771 (1967)
4. M.B. Abbott, J.C. Bathurst, J.A. Cunge, P.E. O'Connell, J. Rasmussen, An introduction to the European Hydrological System—Systeme Hydrologique Europeen, "SHE", 1. History and philosophy of a physically-based, distributed modeling system. J. Hydrol. **87**, 45–59 (1986)
5. J.S. Bathurst, Physically-based distributed modeling of an upland catchment using the Système Hydrologique Europèen. J. Hydrol. **87**, 79–102 (1986)
6. M.G. Ferrick, Analysis of river wave types. Water Resour. Res. **21**(2), 209–220 (1985)
7. E.M. Morris, D.A. Woolhiser, Unsteady one-dimensional flow over a plane: Partial equilibrium and recession hydrographs. Water Resour. Res. **16**, 355–360 (1980)
8. J.R. Richardson, P.Y. Julien, Suitability of simplified overland flow equations. Water Resour. Res. **30**(3), 665–671 (1993)
9. V.P. Singh, *Kinematic Wave Modelling in Water Resources: Surface-Water Hydrology* (Wiley, New York, 1996). 1424 pp
10. D.L. Johnson, A.C. Miller, A spatially distributed hydrologic model utilizing raster data structures. Comput. Geosci. **23**(3), 267–272 (1997)
11. Natural Environment Research Council (NERC), Flood routing studies, in Flood Studies Report, vol. III (Institute of Hydrology, Wallingford, 1975)
12. V.M. Ponce, V. Yevjevich, Muskingum-Cunge method with variable parameters. J. Hydraul. Div. ASCE **104**(HY12), 1663–1667 (1978)

13. V.M. Ponce, P.V. Chaganti, Variable parameter Muskingum-Cunge method revisited. J. Hydrol. **162**(3–4), 433–439 (1994)
14. V.M. Ponce, Diffusion wave modeling of catchment dynamics. J. Hydraul. Eng. ASCE **112**(8), 716–727 (1986)
15. V.M. Ponce, A.C. Klabunde, Parking lot storage modelling using diffusion waves. J. Hydrol. Eng. ASCE **4**(4), 371–376 (1999)
16. S. Orlandini, R. Rosso, Diffusion wave modeling of distributed catchment dynamics. J. Hydrol. Eng. ASCE **1**(3), 103–113 (1996)
17. S. Orlandini, R. Rosso, Parameterization of stream channel geometry in the distributed modeling of catchment dynamics. Water Resour. Res. **34**(8), 1971–1985 (1998)
18. M. Perumal, Discussions on "Assessment and review of the hydraulics of storage flood routing 70 years after the presentation of the Muskingum method", by Kousis, A. D., in Hydrol. Sci. J., 54(1), 43–60. Hydrol. Sci. J. **55**(8), 1431–1441 (2010)
19. S. Hayami, On the propagation of flood waves. Bull. Disaster Prev. Res. Inst. Kyoto Univ. Jpn. **1**(1), 1–16 (1951)
20. M. Perumal, B. Sahoo, Applicability criteria of the variable parameter Muskingum stage and discharge routing methods. Water Resour. Res. **43**(5), 1–20 (2007). doi:10.1029/2006WR004909. W05409
21. M. Perumal, Hydrodynamic derivation of a variable parameter Muskingum method—Part 1: Theory and solution procedure. Hydrol. Sci. J. IASH **39**(5), 431–442 (1994)
22. M. Perumal, Hydrodynamic derivation of a variable parameter Muskingum method—Part 2: Verification. Hydrol. Sci. J. IASH **39**(5), 442–458 (1994)
23. J.A. Cunge, On the subject of a flood propagation method (Muskingum method). J. Hydraul. Res. **7**(2), 205–230 (1969)
24. M. Perumal, B. Sahoo, Volume conservation controversy of variable parameter Muskingum-Cunge method. J. Hydraul. Eng. ASCE **134**(4), 475–485 (2008)
25. M. Perumal, P.E. O'Connell, K.G. Ranga Raju, Field applications of a variable-parameter Muskingum method. J. Hydrol. Eng. ASCE **6**(3), 196–207 (2001)
26. M. Perumal, K.B. Shrestha, U.C. Chaube, Reproduction of hysteresis in rating curves. J. Hydraul. Eng. ASCE **130**(9), 870–878 (2004)
27. F.M. Henderson, *Open Channel Flow* (Macmillan, New York, 1966). 522 pp
28. P.S. Eagleson, *Dynamic Hydrology* (McGraw Hill, New York, 1970). 462 pp
29. M. Perumal, K.G. Ranga Raju, Approximate convection-diffusion equations. J. Hydrol. Eng. ASCE **4**(2), 161–164 (1999)
30. R.V. Kale, Overland flow modeling using approximate convection-diffusion equations, Ph.D. thesis submitted to Indian Institute of Technology, Roorkee, August, 2010
31. C.F. Izzard, The surface-profile of overland flow. Eos Trans. AGU **25**(VI), 959–968 (1944)
32. Ć. Maksimović, M. Radojković, *Urban Drainage Catchments—Selected Worldwide Rainfall-Runoff Data from Experimental Catchments* (Pregamon, Oxford, 1986), pp. 313–330
33. D.A. Woolhiser, Simulation of unsteady overland flow, in *Unsteady Flow in Open Channels*, ed. by K. Mahmood, V. Yevjevich, vol. II (Water Resources Publications, Fort Collins, CO, 1975), pp. 485–508
34. E.T. Engman, Roughness coefficients for routing surface runoff. J. Irrigat. Drain. Eng. ASCE **112**(1), 39–53 (1986)
35. T.S.W. Wong, C.K. Lim, Effect of loss model on evaluation of Manning roughness coefficient of experimental concrete catchment. J. Hydrol. **331**, 205–218 (2006)

Part IV
Water Quality

Detection of Pathogenic Bacteria in Flood Water

Zummy Dahria Mohamed Basri, Zulhafizal Othman, and Marfiah Ab. Wahid

Abstract End of year, flood normally hits the eastern coast of the Peninsular Malaysia. The natural hazard usually causes property damage and loss of lives. However, pathogenic bacteria in flood water should be taken into account to ensure that it does not affect human health during flood. This study shows that three types of bacteria were found in flood water in Pahang. *Shigella flexneri*, *Salmonella typhimurium*, and *Escherichia coli*, which are pathogenic, can affect the human health if the hygiene factors are not taken care. *E. coli* was found in flood water with an average 671 cfu/ml. While Salmonella were found in Paloh Hinai, Kg. Batu Dua Belas, and Chini with 25, 50, and 38 cfu/ml, respectively. An average of 233 cfu/ml *Shigella* was found in flood water. Overall the flood water quality fulfilled the National Water Quality Standards for recreational use with body contact. However, more data of the actual flood situation are needed to be able to confirm the actual microbes present in flood water.

Keywords Pathogenic bacteria • Flood • Water

1 Introduction

Floods in Asia represent the most significant natural hazard [1]. Usually the northeast monsoon (November to March) is largely responsible for the wide flood on the east coast of peninsular Malaysia. Flood human death cause analysis stated that the most of flood related deaths was mainly caused by drowning, which accounts 44 % of flood deaths [2]. Flood losses are high [3] and it impacted thousands of people in Malaysia [4]. Either the impact is direct like homeless and

Z.D.M. Basri (✉)
Faculty of Civil Engineering, UiTM, Shah Alam, Malaysia
e-mail: zummy.d@gmail.com

Z. Othman
Faculty of Civil Engineering, UiTM, Pahang, Malaysia
e-mail: zulhafizal445@pahang.uitm.edu.my

M. Ab. Wahid
Universiti Teknologi MARA, Selangor, Malaysia
e-mail: marfi851@salam.edu.uitm.my

S.H. Abu Bakar et al. (eds.), *ISFRAM 2014*, DOI 10.1007/978-981-287-365-1_16

Table 1 Survival time of pathogenic bacteria

Bacteria	Survival time	Reference
E. coli	Over 28 days	[8]
	Up to 260 days	[9]
Salmonella typhimurium	132 h	[10]
Shigella flexneri	Up to 18 h	[11]

destruction of property or loss of life due to disease. The environmental impacts of extreme flooding are complex, interesting, and largely unused in policy making [5].

Flood water contained many pollutants including pathogenic bacteria and consequently increased the human health risks [6]. Sources of pathogenic bacteria come from sewer, washout from surface, and open channel. Previous study classified the water-borne diseases into four groups such as Gastrointestinal diseases, Hepatitis A and E, Respiratory and skin infections, and Leptospirosis [7].

Shigella flexneri, *Salmonella typhimurium*, and *Escherichia coli* are the most types of bacteria found in flood water. Table 1 shows several studies conducted to determine the survival rate of *E. coli*, *Salmonella typhimurium*, and *Shigella flexneri* in water environment.

1.1 E. coli

E. coli are gram-negative, non-sporeforming *bacilli*, which are approximately 0.5 μm in diameter and 1.0–3.0 μm in length. *E. coli* is an inhabitant of the colon of human and warm-blooded animals, and always used as an indicator for fecal contamination [12]. The main symptoms of *E. coli* are bloody diarrhea, stomach cramps, and nausea with vomiting.

1.2 Salmonella typhimurium

Salmonella are zoonotic pathogens and can be transferred between humans and other animals. Salmonellosis is an infection of a bowel caused by the *Salmonella* bacteria. It is a major cause of bacterial enteric illness in both humans and animals [13, 14]. Spread can occur by eating food and drinking water that has been contaminated by very small amounts of feces from infected people. *Salmonellosis* continues to be a major public health problem worldwide and contributes to negative economic impacts due to the cost of surveillance investigation, treatment, and prevention of illness [15].

1.3 Shigella flexneri

There has been increased recognition in recent years of the importance of *Shigella* as an enteric pathogen with global impact [16]. *Shigella* species are a common cause of bacterial diarrhea worldwide, especially in developing countries. *Shigella* can survive transit through the stomach since they are susceptible to acid than other bacteria. *Shigellosis* is the third most common enteric bacterial infection in the United States. Although infection is typically self-limiting, empiric treatment is often prescribed [17].

Lack of information about health may cause infectious diseases easily transmitted to humans during and after flood. This study was conducted to investigate the occurrence of pathogenic bacteria in flood water during monsoon in Pahang.

2 Methodology

2.1 Selected Area

Three flood areas were selected in Pahang state. The areas are located at Chini, Paloh Hinai, and Kg. Batu Dua Belas. Three sampling sites were selected to see the concentration of pathogenic bacteria in the flood water in different places. Figure 1 shows Kampung Batu Dua Belas which is located near Penur River and 5 km from the sea. The geographical coordinates are 3°38′15.3″N 103°18′13.5″E.

Fig. 1 Kampung Batu Dua Belas (*white mark area*)

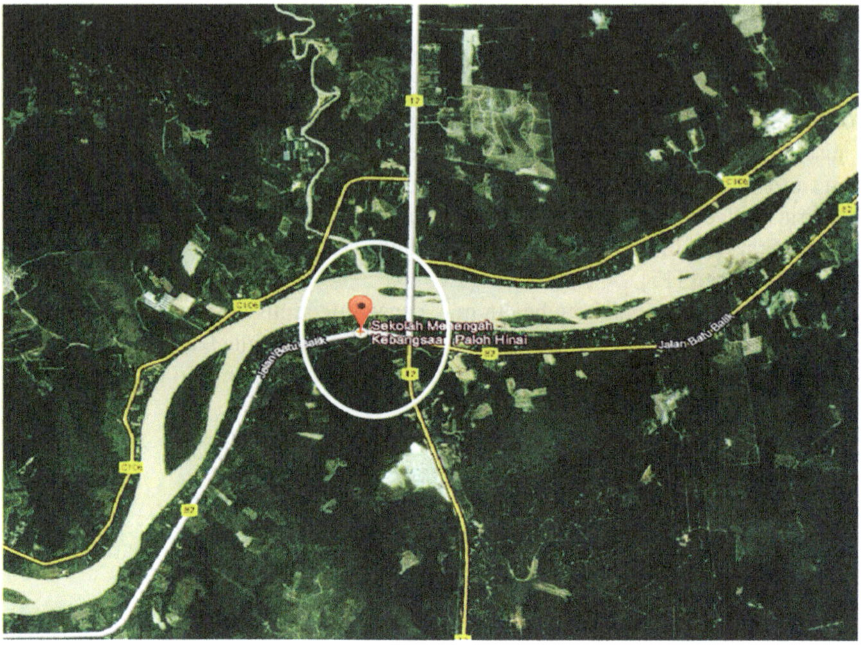

Fig. 2 Paloh Hinai (*white mark area*)

Figure 2 shows the second area of sampling. Kampung Paloh Hinai is situated in Pahang and its geographical coordinates are 3°29′0″ North, 103°6′0″ East. The location is extremely near to the Pahang River.

Another area affected by flooding is Chini, as shown in Fig. 3. The area is situated near Pahang River and Chini Lake in central Pahang. The coordinates are 3.4333° N, 102.9167° E. Chini River, which drains from the lake, flows directly into Pahang River.

2.2 *Meteorological and Hydrological Data*

The Pahang River basin has an annual rainfall of about 2,136 mm, a large proportion of which is brought by the North-East Monsoon from mid October to mid of January [18], with the strong winds and rough sea during the monsoon. The flood occurred due to rising of sea level, heavy rainfall, and overflow of river water. Annual rainfall varies from approximately 1,700–2,800 mm within the basin (mean annual rainfall obtained from 10-year rainfall data of 1999–2008 is 2,136 mm.

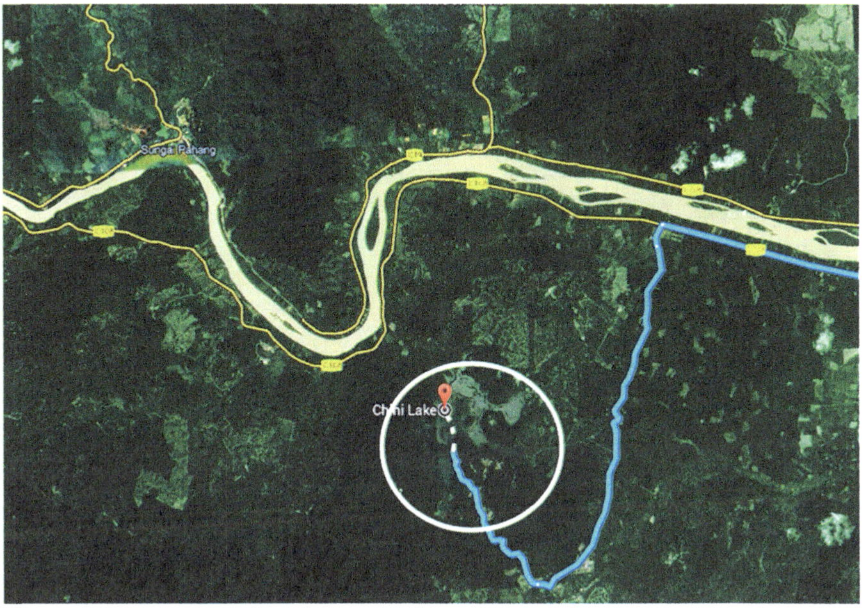

Fig. 3 Chini (*white mark area*)

2.3 Water Quality Measurement

The selected parameter of water quality such as temperature, dissolved oxygen, pH, total dissolved solid, and salinity were selected for on site measurement. The measurement was conducted by using water quality probe (HORIBA).

2.4 Microorganisms Analysis

Samples were collected from three areas affected by flooding in Pahang, such as Kg Batu Dua Belas, Chini, and Paloh Hinai. For each place, there were two points of sample collecting. All the water samples were collected and preserved following the Standard Method Procedure. Then each sample was cultured onto xylose lysine deoxycholate (XLD), thiosulfate citrate bile salt (TCBS), and cetrimide agar plates (Thermo Fisher Scientific). 10 μl loop was used to streak the sample on the agar. The plates were incubated at 37 °C for 24 h. After 24 h, the plates were removed from incubator and bacteria were counted.

There are a few pathogenic bacteria predicted in flood water, such as *Salmonella* spp., *Shigella*, *Lestospira*, *Campylobacter jejuni*, *Enterococci* spp., *E. coli*, and *Legionella*, but in this study only *Shigella flexneri*, *Salmonella typhimurium*, and *Escherichia coli* presented in the water sample. By using XLD agar plates,

Salmonella typhimurium showed 1–2 mm pink colonies with black centers on the agar. 1–2 mm pink colonies referred to *Shigella flexneri*. Yellow colonies on agar indicate the presence of *E. coli*. While TCBS agar showed the occurrence of *Vibrio alginolyticus* with 2–4 mm yellow colonies and *Vibrio parahaemolyticus* with 2–3 mm green colonies. Colorless colonies on the Cetrimide agar showed appearance of *Pseudomonas aeruginosa*.

3 Result and Discussion

3.1 Flood Water Quality

Table 2 shows the result from on-site measurement at selected area.

Water samples showed the temperature ranged between 26.41 and 27.9 and pH showed ranging between 6.82 and 7.04. According to [19], pH 6.5–8.5 classified as preferable for water. pH may affect habitat species by altering the species distribution, while high temperature may reduce the oxygen level in the water. Flash flood showed the temperature ranged between 13 and 15 °C and pH between 7.7 and 8.1 [20]. One of the ways to assess health risk of flood water is to compare flood water quality data to bathing water quality standards [21].

Because such biological activity takes place in waters, it is essential that DO levels are measured on-site. DO showed ranging between 3.8 and 11.60 mg/l. When there are too many bacteria or aquatic animals in the area, they may overpopulate, using DO in great amounts. But to protect early life stage, the DO should be more than 5 mg/l [22]. It shows that flood in Chini and Paloh Hinai is not suitable for habitat due to low DO in water.

During the sampling, minimum and maximum values of TDS were recorded at Kg. Batu Dua Belas (18 mg/l) and Chini (51 mg/l), respectively. This may happen due to natural sources and runoff from the flood flow. Water with a TDS $<$ 1,200 mg/l normally had an acceptable value [23]. Higher TDS could adversely influence the taste of water.

Salinity showed reading between 0 and 0.03 from the water sample. This specific parameter is of interest only in tidal waters or in other surface waters where there may be infiltration of seawater. The presence of a high salt content may render

Table 2 Water quality parameters in flood water

Parameter	1 (Chini)	2 (Paloh Hinai)	3 (Kg Batu Dua Belas)
Temperature (°C)	27.9	26.41	26.88
DO (mg/l)	3.8	4.4	11.60
pH	7.03	7.04	6.82
TDS (mg/l)	51	48	18
Salinity (ppt)	0.03	0.03	0.00

water unsuitable for domestic, agricultural, or industrial use, or may affect its suitability for shellfish [19].

3.2 *Pathogenic Bacteria in Flood Water*

Figure 4 shows the measured data from flood water samples taken from the selected areas. *E. coli* are the most abundant bacteria in flood water. Water usually is tested for fecal contamination by isolating *Escherichia coli* from a water sample. *E. coli* is called an indicator organism because as *E. coli* is a natural inhabitant of the human digestive tract, its presence in water indicates that the water is contaminated with fecal material [24]. Urban flood water was found to be fecally contaminated, which was demonstrated by elevated concentrations of fecal indicator bacteria 10^3–10^7 cfu/l [25]. According to EPA [19] criteria of recreational water (full body contact), the value of *E. coli* presented in this flood water (an average of 671 cfu/ml) exceeded the limit (1 cfu/ml).

This type of pathogenic bacteria exists very much in the flood water, as flood water has gone through all the sewage tank, animal waste, and surface water. Feces were known have very high concentrations of *E. coli*. These bacteria are of definite fecal origin (human and animal), and they are excreted in vast numbers. Their presence in flood water is proof that fecal contamination has occurred, and it is therefore a definite indication of the risk that pathogens may be present. The reading may be affected by the number of residents and activities that exist in the vicinity of the flood area. Therefore, the presence of *E. coli* was higher than other bacteria.

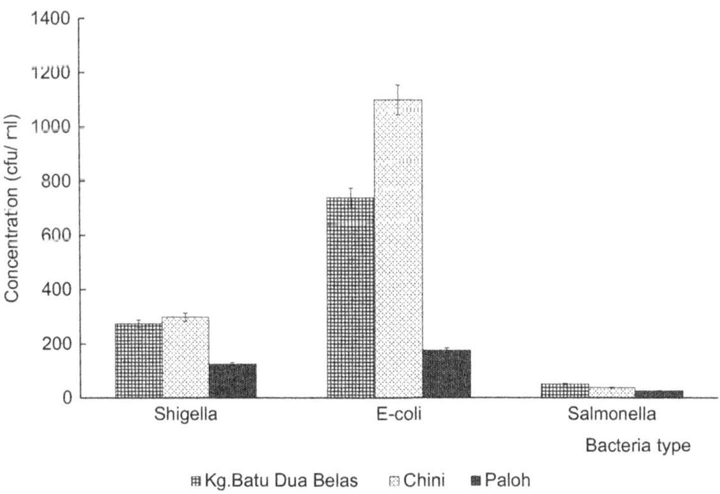

Fig. 4 Concentration of pathogenic bacteria in flood water

A number of *Salmonella* detected were not as many as *E. coli*, but the presence of these bacteria should be concerned by humans because of their effects to human health. *Salmonella* represents a major health risk [26], and these pathogenic microorganisms cause diseases such as typhoid fever and paratyphoid fever. From the sampling area, less than 50 cfu/ml of *Salmonella* was detected in each sampling area.

The amount of *Shigella* spotted in flood water should not be underestimated. Shigellosis is one of the most common diarrheal diseases in human worldwide [27]. An amount of 275, 300 and 125 cfu/ml of *Shigella flexneri* was detected in Kg. Batu Dua Belas, Chini, and Paloh Hinai, respectively. *Shigella flexneri* cells are not only able to survive but can grow extensively at warmer temperatures and reach a population size of about 108–109/g or ml within 6–18 h, and the data shows that more than enough to cause infection to humans [11].

4 Conclusion

Overall the flood water quality fulfilled the National Water Quality Standards for recreational use with body contact. From the result, it was indicated that pathogenic bacteria such as *Shigella flexneri*, *Salmonella typhimurium*, and *Escherichia coli* were found in flood waters. An average 671 cfu/ml of *E. coli* that is detected in flood waters can cause illness to human because the limit is 1 cfu/ml. However, more data of the actual flood situation are needed to be able to confirm the actual microbes present in flood water.

Acknowledgment Authors thank *Dana Kecemerlangan* (RIF) 04/2012: 600-RMI/DANA 5/3/ RIF (9/2012), Research Management Institute, and Universiti Teknologi MARA for financial support in this study.

References

1. P. Gourbesville, J. Batica, J.Y. Tigli, S. Lavirotte, G. Rey, D.K. Raju, Flood warning systems and ubiquitous computing. La Houille Blanche **6**, 11–16 (2012)
2. R. Osti, T. Nakasu, Lesson learned from southern and eastern Asian urban floods: From a local perspective. J. Flood Risk Manag. doi: 10.1111/jfr3.12107 (2014)
3. W.C. Ngai, Increasing flood risk in Malaysia: Causes and solutions. Disaster Prev. Manag. **6** (2), 72–86 (1997)
4. P.Y. Julien, A.A. Ghani, N.A. Zakaria, C.K. Chang, Case study: Flood mitigation of the Muda River, Malaysia. J. Hydraul. Eng. **136**, 251–261 (2010)
5. T.H. John, D.S. Jose, Environmental effects of extreme floods, in *U.S.-Italy Research Workshop on the Hydrometeorology, Impacts and Management of Extreme Floods*, 1995
6. J.A.E.T. Veldhuis, F.H.L.R. Clemens, G. Sterk, B.R. Berends, Microbial risks associated with exposure to pathogens in contaminated urban flood water. Water Res. **44**, 2910–2918 (2010)

7. K. Aldermen, L.R. Turner, S. Tong, Flood and human health: A systematic review. Environ. Int. **47**, 37–47 (2012)
8. S.A. Wilks, H. Mitchels, C.W. Keevil, The survival of Escherichia coli 0157 on a range of metal surfaces. Int. J. Food Microbiol. **105**(3), 445–454 (2005)
9. K.P. Flint, The long-term survival of *Escherichia coli* in river water. J. Appl. Bacteriol. **63**, 261–270 (1987)
10. N. Mezrioui, B. Baleux, M. Trousselier, A microorganism study of the survival of *E. coli* and *Salmonella typhimurium* in brackish water. Water Res. **29**(2), 459–465 (1995)
11. M.S. Islam, M.K. Hasan, S.I. Khan, Growth and survival of Shigella flexneri in common Bangladesh foods under various conditions of time and temperature. Appl. Environ. Microbiol. **59**(2), 652–654 (1993)
12. J.M. Willey, L.M. Sherwood, C.J. Woolverton, *Prescott, Harley and Klein's Microbiology*, 7th edn. (McGraw Hill, New York, 2008)
13. F.W. Brenner, R.G. Villar, F.J. Angulo, R. Tauxe, B. Swaminathan, *Salmonella* nomenclature. J. Clin. Microbiol. **38**(7), 2465–2467 (2000)
14. S. Mirzaie, M. Hassanzadeh, I. Ashrafi, Identification and characterization of Salmonella isolates from captured house sparrows. J. Vet. Anim. Sci. **34**(2), 181–186 (2010)
15. C.F. Pui, W.C. Wong, L.C. Chai, R. Tunung, P. Jeyaletchumi, M.S. Noor Hidayah, A. Ubong, M.G. Farinazleen, Y.K. Cheah, R. Son, Salmonella: A foodborne pathogen. Int. Food Res. J. **18**, 465–473 (2011)
16. K.L. Kotloff, J.P. Winickoff, B. Ivanoff, J.D. Clemens, D.L. Swerdlow, P.J. Sansonetti, G.K. Adak, M.M. Levine, Global burden of Shigella infections: Implication for vaccine development and implementation of control strategies. Bull. World Health Organ. **77**, 651–666 (1999)
17. M.R. Wong, V. Reddy, H. Hanson, Antimicrobial resistance trends of Shigella serotypes in New York City 2006–2009. Microb. Drug Resist. **16**, 155 (2010)
18. A.A. Ghani, C.K. Chang, C.S. Leow, N.A. Zakaria, Sungai Pahang digital flood mapping: 2007 flood. Int. J. River Basin Manag. 10, 139–148. doi: 10.1080/15715124.2012.680022 (2012)
19. EPA, Bacterial Water Quality Standards for Recreational Waters (Freshwaters and Marine Waters) Status Report, 2003
20. E. Fouilland, A. Trottet, C. Bancon-Montigny, Impact of a river flash flood on microbial carbon and nitrogen production in a Mediterranean Lagoon (Thau Lagoon, France). Estuar. Coast. Shelf Sci. **113**, 192–204 (2012)
21. G. Sterk, J.A.E. Ten Veldhuis, F.L.H.R. Clemens, B.R. Berends, Microbial risk assessment for urban pluvial flooding, in *11th International Conference on Urban Drainage*, Edinburgh, Scotland, UK, 2008. Unpublished
22. EPA, Ambient Water Quality Criteria for Dissolved Oxygen, Water Clarity and Chlorophyll a for the Chesapeake Bay and Its Tidal Tributaries, 2003
23. B. Mansouri, S.P. Moussavi, K. Salehi, J. Salehi, Assessment of Birjand flood plain water quality by physico-chemical parameters analysis in Iran. J. Adv. Environ. Health Res. **1**(2), 1–11 (2013)
24. G.B. Jacquelyn, *Microbiology. International Student Version*, 8th edn. (Wiley, Hoboken, NJ, 2013)
25. H.D. Man, I. Leenen, F.V. Knapen, A.M.D.R. Husman, Potential health risk associated with exposure to microorganisms in urban floods, in *Novatech*, 2013, pp. 1–4
26. S.Y. Ong, F.L. Ng, S.S. Badai, A. Yurvec, M. Alam, Analysis and construction of pathogenic island regulatory pathways in Salmonella enterica serovar typhi. J. Integr. Bioinformatics **7**(1), 145 (2010)
27. H. Zhang, P. Liu, Y. Feng, F. Yan, Fate of antibiotics during wastewater treatment and antibiotic distribution in the effluent-receiving waters of the Yellow Sea, northern China. Mar. Pollut. Bull. **73**, 282–290 (2013)

Stormwater Treatment Using Porous Rock Matrices

Marfiah Ab. Wahid, Ismacahyadi Bagus Mohd Jais, Ana Miraa Mohd Yusof, and Nurul Amalina Ishak

Abstract Stormwater carries a broad mix of toxic chemicals, bacteria, sediments, fertilisers, oil and grease to nearby water bodies, which makes it polluted for other use. Under Malaysia climate, abundance stormwater runoff can be collected, treated and then reused for other purposes depending on the treated water quality. For the purpose of the study, a site in UPM Serdang was identified and selected to house a treatment unit with stormwater feed system using porous rock matrices as treatment media. The rock matrix is originally used as landscape tool due to its aesthetic value. However, the porous nature of the rock matrix has potential to remove stormwater impurities via biological treatment. Thus, the objective of this study is to evaluate the treatment efficiency of the porous rock matrices as treatment media. The research methodologies involve design, build and commission of the treatment unit followed by water quality test for selected parameters. The treatment unit achieved the following removal efficiency for total suspended solid (TSS) (57–72 %), turbidity (15–87 %), colour (55–87 %) and dissolved oxygen (DO) (2–17 %). pH value shows an increasing trend of 7–40 %. Nevertheless, the result for total dissolved solid (TDS) and electrical conductivity (EC) fluctuates throughout the treatment process.

Keywords Stormwater treatment • Porous rock media • Treatment efficiency

1 Introduction

Due to the increase in impervious surfaces throughout urbanised catchment, there has been an increase in the volumes of stormwater runoff into the waterways, which is carrying larger amounts of pollutants including sediments, heavy metals, gross pollutants, excess nutrients, suspended solids, polycyclic aromatic hydrocarbons,

M.A. Wahid (✉)
Universiti Teknologi MARA, Selangor, Malaysia
e-mail: marfi851@salam.uitm.edu.my

I.B. Mohd Jais • A.M. Mohd Yusof • N.A. Ishak
Faculty of Civil Engineering, UiTM, Shah Alam, Selangor, Malaysia
e-mail: ismac821@salam.uitm.edu.my; ana_miraa@yahoo.com; nurulamalinaishak@gmail.com

S.H. Abu Bakar et al. (eds.), *ISFRAM 2014*, DOI 10.1007/978-981-287-365-1_17

pesticides, herbicides, faecal indicator bacteria, pathogens and others [1, 2]. According to Chiew and Vase [3], in urban stormwater runoff, the pollutant loadings, concentrations of nutrients and suspended solids are generally much higher than the runoff from unimpaired and rural areas.

Polluted surface runoff, when collected and appropriately treated can be distributed back to the environment as a resource of non-potable water for urban centres, irrigation, industrial, domestic and stock uses across the world [4]. Stormwater reuse system is typically using a surface detention pond or other water storage system to capture the water and recycle that water for non-potable needs. Many detention ponds are designed on site-by-site basis, but their watershed-wide performance and effectiveness cannot be assured [5]. For reuse purpose, treatment system is needed if the quality of stormwater is critical.

Treatment of stormwater means the reduction and removal of pollutants from the water [6]. The collected water is treated to reduce pathogen and pollution levels before meeting additional requirements of end users. Thus, improved and cost-effective treatment technologies are often needed to reduce the adverse impacts of surface water quality [7]. Design and implementation of stormwater treatment strategies often involved a substantial investment in the construction of stormwater treatment measures [8]. Depending on the nature of the stormwater pollutants being targeted, different types of treatment measures are being used, i.e. detention ponds, wet lands, swales, infiltration system, etc.

Porous media is widely used in filter designs during wastewater pre-treatment stage. Gravel filters were found to be very effective for removal of sediments and heavy metals under all water level regimes, even as the system clogged over time [9]. It is also established that percolation through a natural soil profile or separate filters containing natural or engineered porous media is the most frequently used treatment system [10]. For instance, porous media is widely used in trickling filters to provide biological treatment for municipal and industrial wastewaters [11].

2 Methodology

2.1 Site Background

The site is located within UPM Serdang with coordinates of latitude 3°0′18″ and longitude 101°42′9″. Stormwater from a retention pond is currently channelled to irrigate the greenhouses planted with rock melons. Raw water is pumped from the pond and filtered by sand filters before distributing to the greenhouses. Figure 1 shows satellite image of the site.

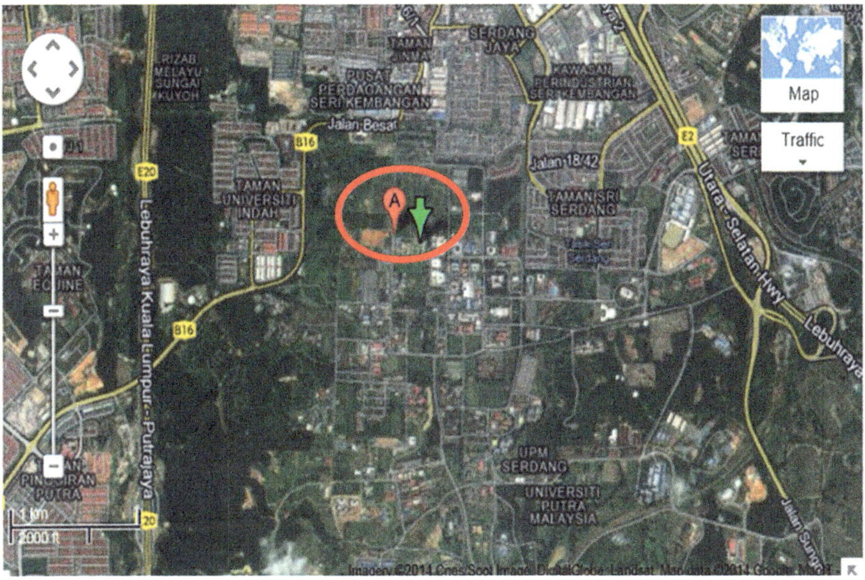

Fig. 1 Satellite image of site UPM, Serdang

2.2 Conceptual Design

The existing irrigation system consists of pumping stormwater from the detention pond and passing it through sand filters. The filtered water is then distributed to the greenhouses to irrigate the rock melons. For the purpose of this study, an in situ continuous flow treatment unit is proposed and executed. Both raw water and sand-filtered water will be fed to the treatment unit having a porous rock slab as treatment media under two treatment modes. The schematic diagram is shown in Fig. 2. Stormwater flows into the treatment tank from top and intercepts with the porous rock slab. The treated stormwater is discharged out via a constant length pvc pipe and collected in a collection tank. Treatment is done under a continuous flow with constant head condition.

2.3 Tank Design

The storage tank is designed to store 100 l of water to last up to 7 days to achieve a flow rate of 0.6 l/h. The water level will be maintained at a depth of 240 mm from the bottom of the tank before discharging out via the pvc pipe to achieve the constant head condition. This gives a retention time of approximately 1 day (24 h). The treatment tank capacity is 300 mm × 300 mm × 200 mm. The treatment

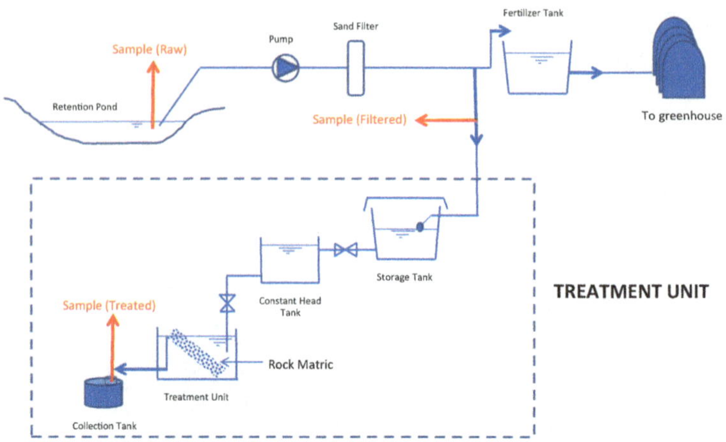

Fig. 2 Schematic diagram of treatment unit

Fig. 3 Treatment tank

tank is constructed using Perspex material reinforced with wood frames as shown in Fig. 3.

2.4 Treatment Mode

The treatment process is carried out in two modes, i.e. Mode 1 using filtered water and Mode 2 using raw water direct from the pond. Initially, the raw water from the pond was not able to be supplied directly to the treatment unit. As such the sand-filtered water which is tapped to the treatment tank is taken as the influent fed to the treatment unit. However, after several samplings collected under Mode 1 treatment, the site managed to organise manual collection of the raw water from the lake and feed it directly to the treatment unit. Therefore, the results consist of Modes 1 and 2 samples. Each mode is completed in approximately 1 month duration.

2.5 Porous Rock Media

Granular grains of size Grade 10 (~5 mm diameter) is selected for the purpose of this study. Two types of binding chemicals with specific ratio combination are churned together with the grains in a cement mixer before being compacted in a rectangular mould. The dimension of the porous rock slab to suit the treatment tank is 300 mm × 288 mm × 50 mm.

A full slab of the porous rock is positioned to cover the whole span of the rectangular length of the perspex tank as shown in Fig. 4. The top end of the rock matrix will lean against the top of the tank wall. The bottom end of the rock matrix will sit on the bottom of the tank at ~60° angle. The full height of rock slab will be submerged in the water at all time during testing. The rock matrix is positioned in such a way that the water will flow down through the slab and discharged out via the pvc pipe outlet.

2.6 Binding Chemicals

The rock matrix is packed by churning together the granular grains with two different chemical solutions. The chemicals used are as follows: (1) BASF Elastopave T 6551/201 CB and (2) BASF Elastopave* T 6551/102 CA.

2.7 Commissioning of Treatment Unit

The treatment unit was commissioned by adopting the following procedures: (1) Fill up the storage tank with influent. (2) Filling in the constant head tank. (3) Position rock matrix in the test tank. (4) Fill up the treatment tank with raw water/influent (5) Regulate desired flow rate by adjusting valve at discharge pipe. (6) Allow for desired retention time. (7) Collection of effluent via collection tank. (8) Sampling starts at T1 and repeated at specified interval. (9) In situ physical characteristic check (using mobile water quality metre). (10) After completion of the first mode, drain all water, back wash porous rock slab and repeat procedure for Mode 2 treatment.

2.8 Sampling

Sampling was done weekly for Mode 1 treatment and twice weekly for Mode 2 treatment. Three (3) points of samples were taken, i.e. raw water, filtered water

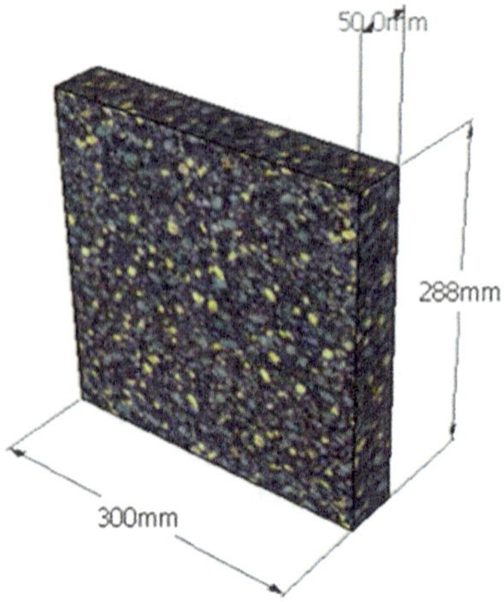

Fig. 4 Porous rock media

and treated water as per schematic diagram in Fig. 3. Three litres of samples at each point were taken for laboratory analysis.

2.9 Laboratory Test and Data Collection

Wastewater is characterised in terms of its physical, chemical and biological composition [11]. The data obtained from the laboratory test will illustrate the concentrations of each parameter within the system and how much removal can be achieved after treatment with the porous rock media. In this study, the parameters in Table 1 were measured and analysed using standard laboratory method [12] and in situ HORIBA mobile water quality metre.

2.10 Limitations

There are several limitations that influenced the overall result of this study. The following limitations shall be addressed and taken into consideration in concluding the research:

1. There was a water shortage during the period of sampling due to the prolonged draught, thus leading to water interruption during the treatment period. Thus, treatment flow is not fully continuous.

Table 1 Water quality parameters (standard method test)

Water quality parameter	Abbreviation	Remark
Total suspended solids	TSS	Laboratory test
Total dissolved solids	TDS	In situ test probe
Turbidity	NTU	In situ test probe
Colour		Laboratory test
Temperature	°C	In situ test probe
Conductivity	EC	In situ test probe
pH	pH	In situ test probe
Dissolved oxygen	DO	In situ test probe

2. It is not possible to carry out a real-time/online water quality test on the samples due to site set-up. The collected and stored samples do not represent the 'real-time' composition of the samples.
3. The treatment unit at site is subjected to fluctuation of surrounding environment like temperature, sunlight, small insects and so forth, which influence data reliability.

3 Result and Discussion

The porous rock matrices are currently manufactured and supplied for landscape purposes due to its aesthetic values. The applications among others are for pedestrian walkways, residential driveways, outdoor benches/seatings, drainage gratings, assembly pavement and car parks as can be seen in Fig. 5. The porous nature of the rock matrices permits water to pass through it and allows the water to infiltrate directly into the soil.

To evaluate a more true performance of the porous rock media, only Mode 2 result is discussed in this chapter. Table 2 summarises the water quality test result.

3.1 Total Suspended Solid

Fig. 6 shows high removal in TSS before and after treatment with the rock matrix. TSS concentration in raw water ranging from 4 to 34 mg/l and reduced to a range of 0–9 mg/l. The removal efficiency achieved is as high as 57–72 %. The trend of the result is expected as the treatment unit provides 24 h retention time which is able to reduce the TSS by gravity settling process.

Fig. 5 Various applications of porous rock matrices

3.2 Total Dissolved Solids

As shown in Fig. 7, there is a mixture of trending for total dissolved solids (TDS) in treated water. In two occasions, the reduction of the TDS after treatment is about 10–20 %, but three occasions detected an increase in TDS ranging from 0.5 to 60 %. The increase could be attributed to the rock media dissolving material or dissolving of material due to absence of oxygen in the treatment tank. To verify the trending of TDS, a test was carried out by soaking the rock media in distilled water for about 2 weeks. The average TDS value detected is 6.5 mg/l coming from the porous rock media.

3.3 Turbidity

As shown in Fig. 8, turbidity removal after treatment is as expected as TSS since the nature of these two parameters is similar. The result shows high removal efficiency ranging from 58 to 100 %. The turbidity is mostly reduced to <10 mg/l after

Table 2 Results under Mode 2 treatment

	Raw	Out	% Removal
TSS (mg/l)			
19/4/14	34.3	9.3	72.82 %
23/4/14	N/A	N/A	N/A
27/4/14	22	9	57.73 %
30/4/14	4.0	0.0	100.00 %
7/5/14	10.0	9.0	10.00 %
Turbidity (NTU)			
19/4/14	26.1	10.6	59.46 %
23/4/14	7.3	2.8	61.70 %
27/4/14	88.4	37.1	58.03 %
30/4/14	78.0	0.0	100.00 %
7/5/14	6.4	2.0	69.22 %
EC (μS/cm)			
19/4/14	83.8	86.9	−3.70 %
23/4/14	75.8	79.5	−4.88 %
27/4/14	89.0	72.0	19.10 %
30/4/14	91.7	142.0	−54.85 %
10/5/14	79.6	75.9	4.65 %
DO (mg/l)			
19/4/14	13.3	9.9	25.79 %
23/4/14	8.7	8.8	−1.73 %
27/4/14	14.3	15.7	−9.66 %
30/4/14	13.1	13.8	−5.66 %
7/5/14	8.8	9.9	−12.97 %
TDS (mg/l)			
19/4/14	39.5	39.7	−0.51 %
23/4/14	36.4	37.5	−3.02 %
27/4/14	54.3	43.3	20.26 %
30/4/14	57.0	92.0	−61.40 %
10/5/14	38.8	34.6	10.82 %
Colour (PtCo)			
19/4/14	107.7	91.7	14.86 %
23/4/14	55.0	32.3	41.27 %
27/4/14	190.0	89.0	53.16 %
30/4/14	47.0	0.0	100.00 %
7/5/14	46.0	45.0	2.17 %
pH			
19/4/14	4.9	6.8	−39.59 %
23/4/14	7.1	7.4	−4.67 %
27/4/14	5.7	5.0	12.17 %
30/4/14	5.1	5.5	−7.48 %
7/5/14	6.7	9.2	−37.46 %

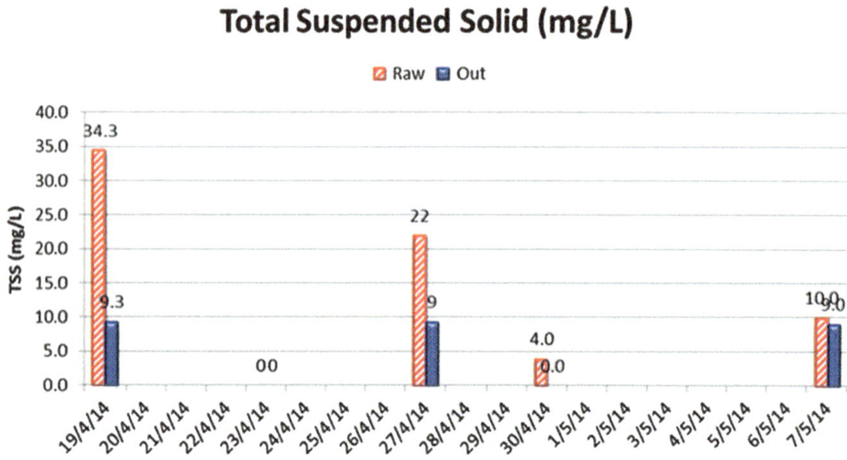

	19/4/14	23/4/14	27/4/14	30/4/14	7/5/14
Raw	34.3	0	22	4.0	10.0
Out	9.3	0	9	0.0	9.0

Fig. 6 TSS concentration

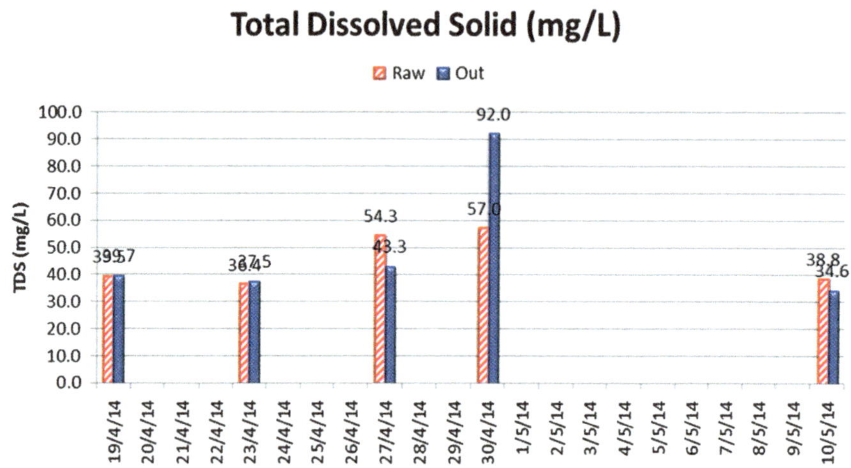

	19/4/14	23/4/14	27/4/14	30/4/14	10/5/14
Raw	39.5	36.4	54.3	57.0	38.8
Out	39.7	37.5	43.3	92.0	34.6

Fig. 7 Total dissolved solid (TDS) concentration

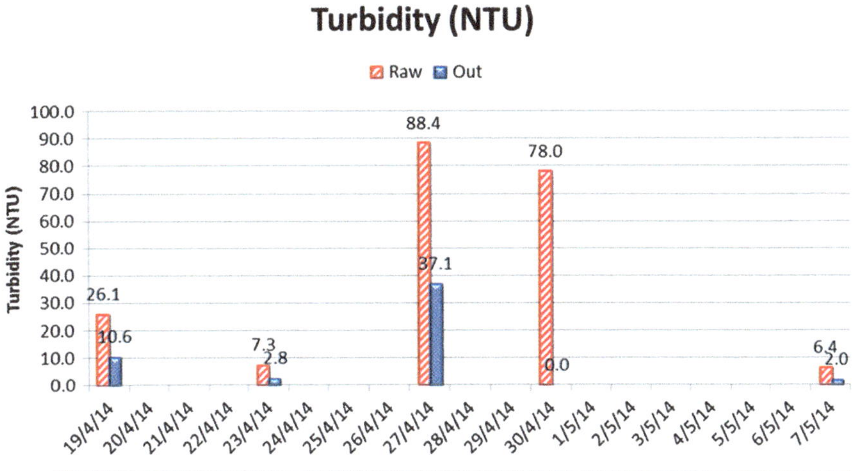

	19/4/14	23/4/14	27/4/14	30/4/14	7/5/14
Raw	26.1	7.3	88.4	78.0	6.4
Out	10.6	2.8	37.1	0.0	2.0

Fig. 8 Turbidity concentration

treatment. There is a wide range of TSS in raw water from 7 to 88 mg/l due to different pollutants loading on each occasion. The removal trend is expected as the treatment unit provides 24 h retention time, which is able to reduce the turbidity by gravity settling process.

3.4 *Colour*

The colour of the water is mainly contributed by the suspended solids which are measured under TSS and turbidity. As shown in Fig. 9, as expected, the treatment unit achieved high removal efficiency on most occasions ranging from 40 to 100 %. This is due to the settling process in the treatment tank that will remove the colour.

3.5 *Electrical Conductivity*

Since the electrical conductivity (EC) is a measure of capacity of water to conduct electrical current, it is directly related to the concentration of salts dissolved in water and, therefore, to TDS concentration. Thus, similar result to TDS is expected, i.e. the conductivity value fluctuates with time. As shown in Fig. 10, the result shows a decrease in EC in two occasions (5–20 %) and an increase of EC in three occasions (4–55 %). The increase in conductivity after treatment might be due to

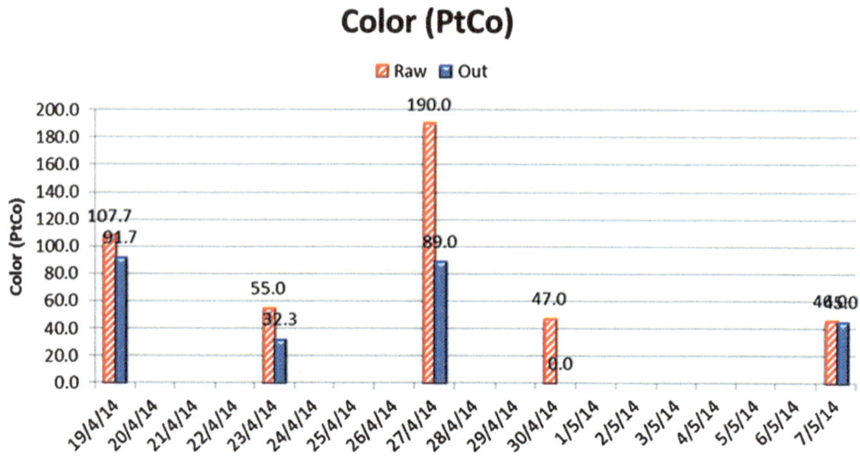

	19/4/14	23/4/14	27/4/14	30/4/14	7/5/14
Raw	107.7	55.0	190.0	47.0	46.0
Out	91.7	32.3	89.0	0.0	45.0

Fig. 9 Colour concentration

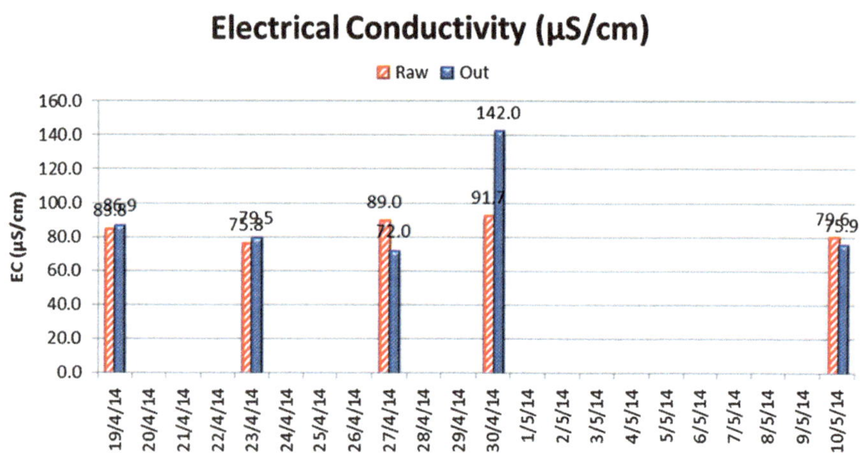

	19/4/14	23/4/14	27/4/14	30/4/14	10/5/14
Raw	83.8	75.8	89.0	91.7	79.6
Out	86.9	79.5	72.0	142.0	75.9

Fig. 10 Electrical conductivity

the release of dissolved solids into the water when it flows through and over the porous rock media throughout the treatment period. To verify the increase of conductivity, a test was carried out by soaking the rock media in distilled water for about 2 weeks. The averaged conductivity reading recorded is 15.84 μS/cm, which is coming from the rock media.

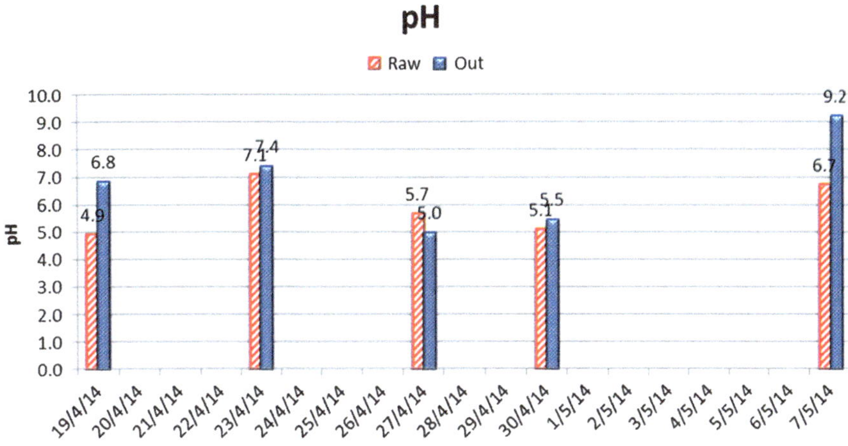

	19/4/14	23/4/14	27/4/14	30/4/14	7/5/14
Raw	4.9	7.1	5.7	5.1	6.7
Out	6.8	7.4	5.0	5.5	9.2

Fig. 11 pH concentration

3.6 pH

Generally, pH value has increasing trend throughout the treatment as shown in Fig. 11. However, the increase of pH is still within the permitted raw water limit (pH 5–9) in accordance with National Water Quality Standard for Malaysia [13]. There is a tendency for the pH reading to change from acidic towards neutral value. This could be due to some removal of suspended solids that carry acidic particle or oxidation of acidic matter. The increase of pH may also be contributed by the chemical bonding of rock media, which is used to bind together the loose granular pebbles. To further verify the result, a test was carried out by soaking the rock media in distilled water for about 2 weeks. The average pH reading recorded is 7.42, a slight increase coming from the porous rock media.

3.7 Dissolved Oxygen

As shown in Fig. 12, overall result shows an increase of dissolved oxygen (DO) ranging from 2 to 26 %. The increase of DO after treatment could be due to several factors. The treatment tank is considerably small (~14 l capacity), which

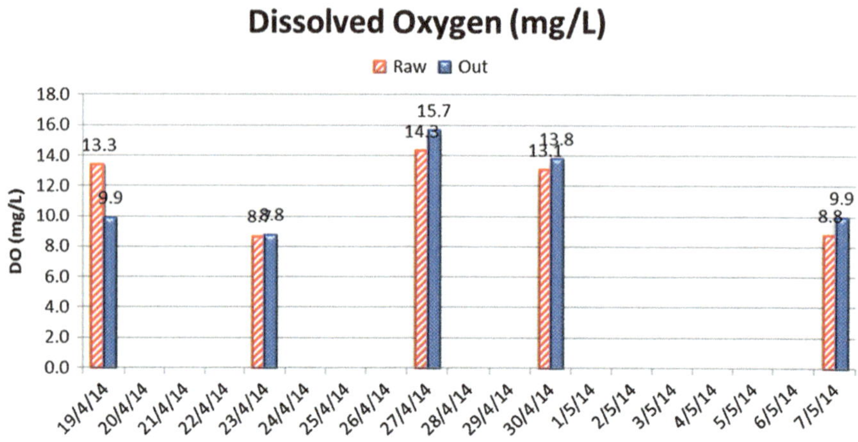

	19/4/14	23/4/14	27/4/14	30/4/14	7/5/14
Raw	13.3	8.7	14.3	13.1	8.8
Out	9.9	8.8	15.7	13.8	9.9

Fig. 12 Dissolved oxygen (DO)

allows more oxygen to penetrate through certain depth of the tank. Moreover, the samples taken might interact with oxygen during transportation and storage at the laboratory. Lower storage temperature can also increase the DO concentration.

4 Conclusion

From test result, the porous rock matrices have potential to improve the water quality of the stormwater from the retention pond with significant removal efficiency of physical properties such as total suspended solid (TSS), turbidity and colour. However, further test is recommended to verify the reading for total dissolved solid (TDS) and electrical conductivity (EC). The fluctuation for both TDS and EC readings might be contributed by the reaction of the porous rock matrices with the stormwater throughout the treatment period. Due to some limitations during the study, further analysis can be carried out to verify if higher treatment efficiency can be achieved under aerobic condition and under controlled environment.

As far as water quality standard is concerned, the treated water quality at least complies with National Water Quality Standards for Malaysia (NWQS) under Class IV (defined water quality required for major agricultural irrigation activities which may not cover minor applications to sensitive crops [13]).

Acknowledgment Authors express utmost gratitude to Dana Kecemerlangan (RIF) 04/2012 : 600-RMI/DANA 5/3/RIF (9/2012), Research Management Institute, Universiti Teknologi MARA, for financial support for this study.

References

1. M. Schwecke, B. Simmons, B. Maheswari, Sustainable use of stormwater for irrigation case study. Environmentalist **27**, 51–61 (2007)
2. M.N. Chong, S. Jatinder, A. Rupak, T. Janet, G. Wolfgang, E. Beate, T. Simon, Urban stormwater harvesting and reuse: A probe into the chemical, toxicology and microbiological contaminants in water quality. Environ. Monit. Assess. **185**, 6645–6652 (2012)
3. H.S.F. Chiew, J. Vase, Nutrient loads associated with different sediment sizes in urban stormwater and surface pollutants. Environ. Eng. **4**(130), 391–396 (2004)
4. S. Begum, M.G. Rasul, R.J. Brown, *Stormwater Treatment and Reuse Techniques: A Review* (WSEAS, Greece, 2008)
5. D. Su, X. Fang, Z. Fang, *Effectiveness and Downstream Impacts of Stormwater Detention Ponds Required for Land Development* (American Society of Civil Engineers, Rhode Island, 2010)
6. UNEP, Wastewater and stormwater treatment, Online (2013), available at: http://www.unep.or.jp/ietc/publications/freshwater/sb_summary/4.asp
7. D.B. Hackett, J. Schenk, M. Stinson, *Evaluating Innovative Stormwater Treatment Technologies Under the Environmental Technology Verification (ETV) Program* (EPA, Washington, DC, 2009)
8. T.H.F. Wong, T.D. Fletcher, H.P. Duncan, G.A. Jenkins, Modelling urban stormwater treatment—A unified approach. Ecol. Eng. **2**(27), 58–70 (2006)
9. B.E. Hatt, T.D. Fletcher, A. Deletic, Treatment performance of gravel filter media: Implications for design and application of stormwater infiltration systems. Water Res. **41**, 2513–2524 (2007)
10. T.K. Stevik, K. Aa, G. Ausland, J.F. Hanssen, Retention and removal of pathogenic bacteria in wastewater percolating through porous media: A review. Water Res. **38**, 1355–1367 (2004)
11. Metcalf & Eddy, Wastewater constituents, in *Wastewater Engineering, Treatment and Reuse*, ed. by G. Tchobanoglous (McGraw-Hill, New York, 2004), pp. 29–76
12. HACH Company, *Water Analysis Handbook*, 7th edn. (Hach Company, Loveland, CL, 2008)
13. DOE, *Malaysia Environmental Quality Report* (DOE, Kuala Lumpur, 2009)

Removal of Oil from Floodwater Using Banana Pith as Adsorbent

Ramlah Mohd Tajuddin and Noor Sa'adah Abdul Hamid

Abstract Oily water has become one of the environmental problems which can cause clogging and pollution. One way to remove the oil is by using adsorption methods where activated carbon is one of the effective adsorbents. The activated carbon is quite expensive, thus natural adsorbent studies were conducted. This study's objective was to determine the ability of banana pith as natural adsorbent in removing oil from oily wastewater. The performance of banana pith was studied by using dosage and contact time, while the banana pith itself was characterized using FT-IR and SEM methods. By using mathematical models, Langmuir and Freundlich isotherms were used to determine the adsorption pattern. Results indicate that both raw and modified banana piths have an ability to act as adsorbents. Treated banana pith showed the highest removal when compared with raw banana pith. The conclusion for these studies shows that banana piths have an ability to remove oil from water and treated banana pith gives a higher removal when compared with raw banana pith.

Keywords Adsorption • Banana pith • Natural adsorbent

1 Introduction

Oily water causes problems to the environment. Various sources can contribute to oily water, such as flood, food processing industries, farmhouses, and slaughterhouses. Oil can contaminate water in two different forms: emulsified oil or free oil [1]. High concentration of oil in water can cause clogging due to the deposited oil and grease inside the sewer system. According to the research by Bridges [2], the overflow will increase the potential of water pollution and increase the potential exposure of pathogens toward the environment.

Various studies have been conducted in searching the best ways to remove oil in the water. Adsorption is one of the common methods used in removing pollutants. According to Wahi et al. [3], the advantages of using adsorption methods are good

R. Mohd Tajuddin (✉) • N.S.A. Hamid
Department of Civil Engineering, Universiti Teknologi Mara, Shah Alam, Selangor, Malaysia
e-mail: ramlah2007@gmail.com; saadahamid@gmail.com

S.H. Abu Bakar et al. (eds.), *ISFRAM 2014*, DOI 10.1007/978-981-287-365-1_18

oil removal efficiency, low cost, and low processing cost. But there are also some disadvantages in using adsorption methods, such as it is labor intensive and poor removal of fine emulsions [3]. Adsorption is a process when the molecules of sorbate concentrate at the surface of adsorbent without penetrating into the adsorbent [3]. In recent years, there were various studies on the use of natural sorbent as the adsorption materials.

According to Wahi et al. [3], the use of natural organic and inorganic sorbents can give a higher adsorption, which can adsorb between 3–15 and 4–20 times of adsorbent weight. However, natural sorbents tend to have a higher water uptake.

According to Wahi et al. [3], this is due to the low hydrophobicity or water repelling ability of the sorbent, which causes the uptake of oil at the surface. This study will focus more on the characterization of raw banana pith (RBP) and NaOH-treated banana pith (TBP) adsorbent and also to see the performance between raw and treated banana plants as adsorbent.

2 Materials and Methods

2.1 Materials Preparation and Characterization

Treated and untreated banana pith was used as an oil adsorbent material to remove oil from wastewater discharged from Lecker Food Sdn. Bhd. The oil was tested using method SPE10300. For characterization of the banana pith, Scanning Electron Microscopy (SEM) was used for physical characterization and Fourier Transform Infrared (FT-IR) for chemical characterization.

2.2 Adsorption Studies

For adsorption studies, two different parameters were used to determine the performance of banana pith as adsorbent. The parameters are contact time and dosage. For contact time, fixed amount of sorbent (5.0 g) and 500 mL of wastewater were mixed at various contact times (15–75 min). For dosage adsorption experiment, the dosages of sorbent were varied (1–9 g) using fixed volume of wastewater (500 mL) and time (60 min). Experiments were conducted in duplicate and the mean values were used in the calculation. The amount of removal was calculated using percentage removal of oil from the wastewater using Eq. (1):

$$\text{Percentage removal } (\%) = \frac{C_0 - C_e}{C_0} \quad (1)$$

From the equation, C_0 and C_e are the concentrations of oil and grease in the initial and equilibrium in g/L.

3 Result and Discussion

3.1 Wastewater Characterization

Table 1 shows the result for the wastewater characteristic. From the experimental data, the concentration of oil in wastewater is 148.79 g/L.

3.2 Study on Physical and Chemical Characterization of Adsorbent

To give a higher removal, the morphology structure of adsorbent should contain hollow structures. These hollow structures will give a large surface area for adsorption surface; thus the adsorption process will increase [3]. Figures 1 and 2 show the structure of both RBP and TBP under SEM. From the figure, the morphology structure of RBP and TBP shows changes after the treatment process.

As for chemical characteristic study, Fig. 3 shows the FT-IR spectra for both RBP and TBP before the adsorption process.

Plant wax is commonly present in the plant, according to the study by Abdullah et al. [4]; the plant wax usually consist of *n*-alkanes, alcohols, and carbonyl group. This statement was also supported by Lim and Huang [5], where the plant wall generally consists of alkanes, ketones, alcohols, and esters. Table 2 shows the summary of functional group that is present in RBP and TBP. From the table, the carbonyl group in TBP is missing. It can be justified that the absence of carbonyl group will result in the destruction of plant wax since carbonyl group is one of the components inside plant wall.

3.3 Performance Study for Banana Pith as an Adsorbent

The contact time study is to find the desirable time for the highest oil removal for the performance study experiment. For contact time, the adsorption processes were divided into two phases as stated in the study by Ibrahim et al. [1]: primary rapid phase and slow phase of adsorption. The primary rapid phase can be seen when the adsorption capacity increased with an increase in contact time of adsorption up to 30 min for RBP and 15 min for TBP. For the slow phases, it starts after the primary

Table 1 Wastewater characteristic

Parameter	Value
Temperature	25.9 °C
pH	6.39
Oil and grease	148.79 g/L

Fig. 1 SEM for RBP at ×1.0 k

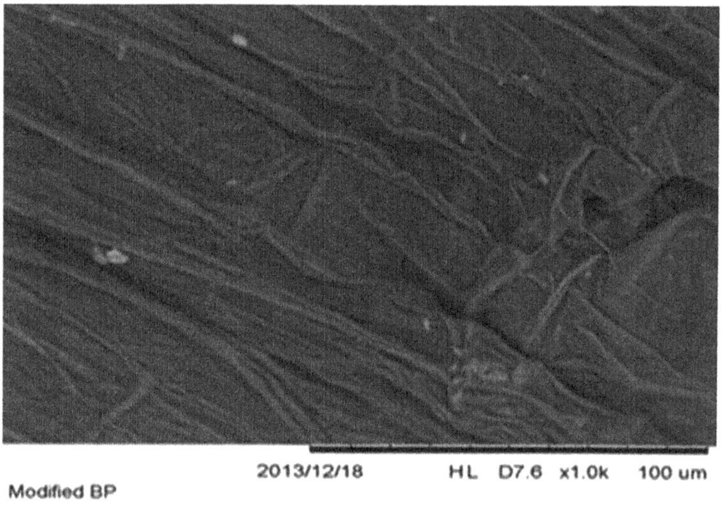

Fig. 2 SEM for TBP at ×1.0 k

rapid phase. The rapid phase of adsorption is due to the abundance of adsorption sites at the adsorbent surface, thus increasing the adsorption [1]. With the increase in time, the adsorption sites are slowly filled with oil molecules, thus reducing the adsorption sites which can be used to explain the slow phase of adsorption.

For the removal of oil using RBP, the trends for the removal of oil are increased rapidly for minutes 15 until minutes 30. After 30 min of contact time, the percentage removal of oil using RBP starts to give a percentage removal of 96 % for both

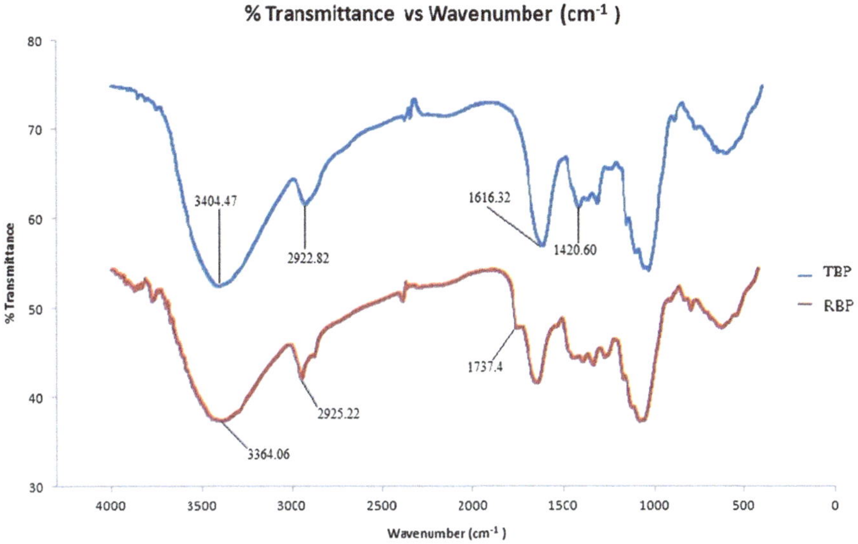

Fig. 3 FT-IR transmittance vs. wavenumber chart

Table 2 Functional group for raw and NaOH treated banana pith

Sample	Functional group			
	Alcohols (cm^{-1})	Alkanes (cm^{-1})	Carbonyl (cm^{-1})	Aromatics (cm^{-1})
Raw banana pith (RBP)	3,364.06	2,925.22 2,857	1,737.4	1,518.5
Treated banana pith (TBP)	3,404.47	2,922.82	–	1,420.60

30 and 45 min before the removal slightly increased to 97 %. But at the 75 min contact times, the percentage removal drops at 86 % removal for RBP. Based on Fig. 4 for TBP, it shows a rapid removal at minutes 15 with the percentage removal of 86 %. After that the percentage removal became the same for minutes 30 and 45 with the removal of 86 %. At minutes 60, the removal using TBP also showed an increase in percentage removal of 97 % but showed a lower removal at minutes 75 with the percentage removal of 79 %. From the figure, the highest percentage removal was achieved when using RBP with the highest removal at 97 %. TBP showed a lower percentage removal when compared with RBP, with the highest removal using TBP at 97 %. In terms of contact time, TBP showed a rapid removal at short time, when the removal is at 86 % at minutes 15, while for RBP at minutes 15, the percentage removal is only at 20 %.

For dosage performance study, the percentage removal of oil for banana pith is increased for both RBP and TBP as the dosage of adsorbent is increased. From

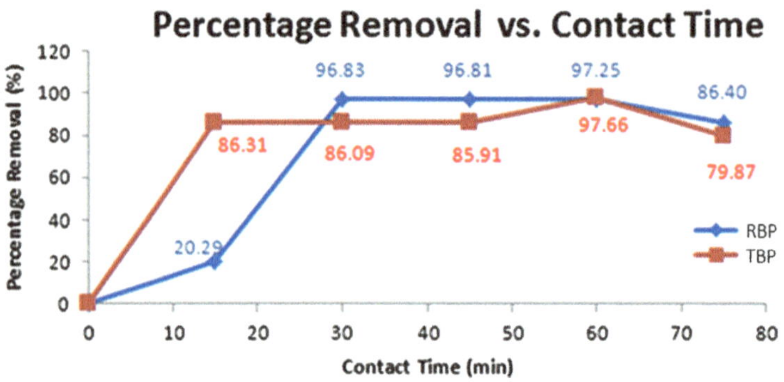

Fig. 4 Effect of contact time on the percentage of oil removal ($W = 5$ g; $V = 0.50$ L)

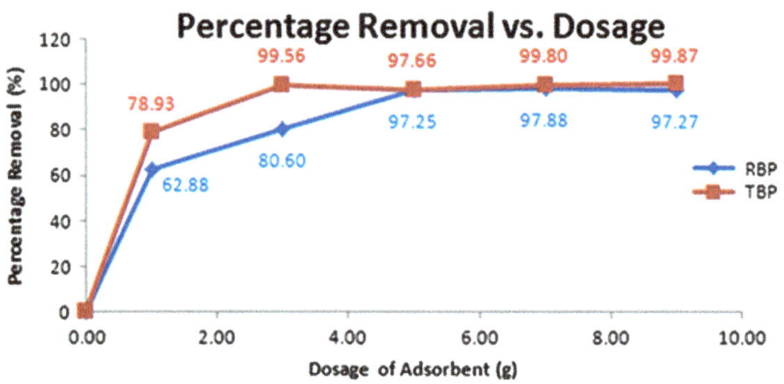

Fig. 5 Effect of dosage on adsorption of oil ($V = 0.50$ L)

Fig. 5, RBP shows the highest removal at dosage 7 g with the highest percentage removal at 97.88 %, while for TBP the highest percentage removal is 99.87 % at dosage 9 g. RBP showed an increase from 1 g of adsorbent until 5 g of adsorbent before the percentage removal became the same for 5, 7, and 9 g with the percentage of removal at 97 %.

For TBP, the percentage removal showed a rapid increase until 3 g removal where the percentage removal is at 99 %. This percentage removal is the same for both adsorbent dosage 7 and 9 g. But, for TBP at adsorbent dosage of 5 g, the percentage removal is slightly lower when compared with adsorbent dosage for TBP. The increase in adsorption efficiency is believed due to the availability of adsorption binding sites that are increased as the adsorbent dosage is increased [6]. This statement is also reported by Wahi et al. [3], where with an increase in active surface area, the sorption toward adsorbent site also increases.

Two different isotherms, Langmuir and Freundlich adsorption isotherms, were used to fit the adsorption data from the experiment. Figures 6, 7, 8, and 9 show the Langmuir and Freundlich adsorption isotherm graphs for both raw and treated

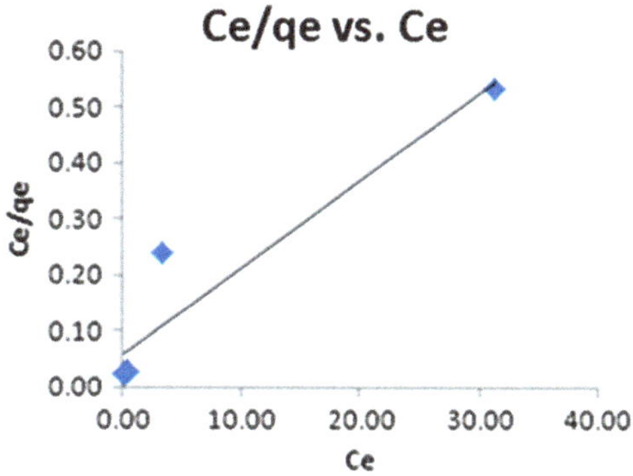

Fig. 6 Langmuir isotherm plots for RBP

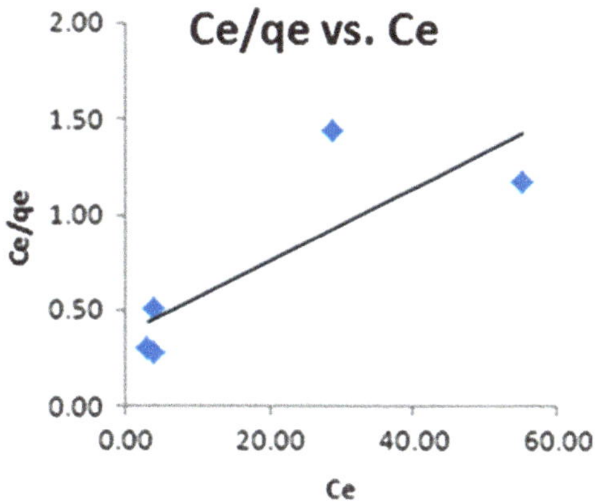

Fig. 7 Langmuir isotherm plots for TBP

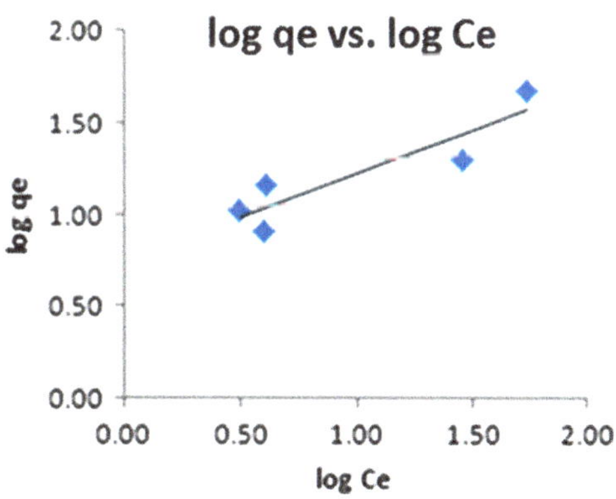

Fig. 8 Freundlich isotherm plots for RBP

Fig. 9 Freundlich isotherm plots for TBP

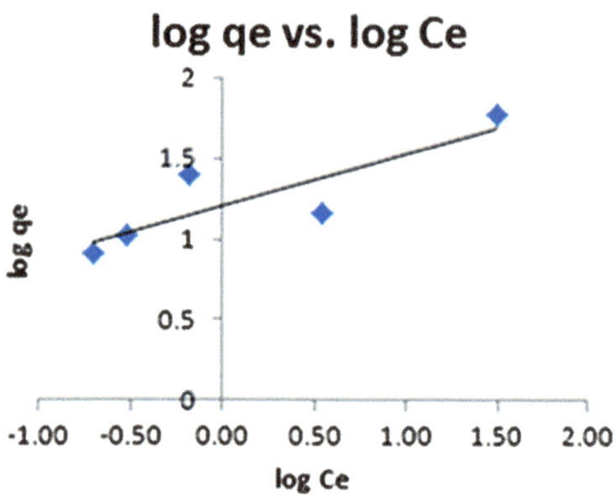

Table 3 Langmuir and Freundlich adsorption isotherm

Adsorbent	Langmuir constant		
	Q_0	b	R^2
Raw banana pith (RBP)	52.3560	0.0503	0.6686
NaOH-treated banana pith (TBP)	64.1026	0.2667	0.8964
	Freundlich constant		
	K_f	$1/n$	R^2
Raw banana pith (RBP)	5.6027	0.4697	0.8244
NaOH-treated banana pith (TBP)	16.2854	0.324	0.7416

banana piths. From the plotted graph, the linear equations from the graph are derived and the values of R^2 were obtained to determine which adsorption isotherm is more favorable for banana pith adsorption. The linear forms of Langmuir isotherm are given in Eq. (2), while the linear forms of Freundlich are given in Eq. (3).

$$\frac{C_e}{q_e} = \frac{C_e}{Q_o} + \frac{1}{Q_{ob}} \tag{2}$$

$$\log q_e = \log K_f + \frac{1}{n} \log C_e \tag{3}$$

C_e is the equilibrium concentration and q_e is the amount adsorbed during equilibrium. While Q_o and b are constants for Langmuir adsorption isotherm, K_f and n are constants for Freundlich adsorption isotherm. Table 3 shows the value of R^2 and constants for both Langmuir and Freundlich. Based on the R^2 value, the best fit adsorption isotherm is Freundlich isotherm, which is based on multilayer adsorption on heterogeneous surface of the sorbent [3]. It was also reported by Ahmad et al. [7], for the study of oil removal using rubber powder, where the Freundlich adsorption isotherm is more favorable when compared with Langmuir isotherm.

4 Conclusion

Based on the result of both physical and chemical characterizations, TBP structures for both physical and chemical structure were changed due to the treatment process. Treatment using NaOH can cause the disruption of plant wall and can be seen using the result of chemical characterization. The physical structure of TBP also showed the changes when all the narrow piths in RBP were missing. Based on the performance study, for contact time studies the highest removal is by RBP when compared with TBP, but TBP showed a rapid removal in short time when compared with RBP with the removal of 86 % rather than 20 % for RBP at minutes 15. For dosage, TBP showed the highest removal of 99 %, while RBP showed the highest removal at 97 %. Based on the adsorption isotherm data, the adsorption isotherm that is suitable to explain the adsorption behavior is Freundlich isotherm.

Acknowledgment The author would like to acknowledge and thank the valuable technical support given by the laboratory staff of Environmental Laboratory, Faculty of Civil Engineering, and Faculty of Dentistry and postgraduate students from Faculty of Pharmacy for their assistance during the experiment.

References

1. S. Ibrahim, H.-M. Ang, S. Wang, Removal of emulsified food and mineral oils from wastewater using surfactant modified barley straw. Bioresour. Technol. **100**(23), 5744–5749 (2009)
2. O. Bridges, Double trouble: health risks of accidental sewage release. Chemosphere **52**(9), 1373–1379 (2003)
3. R. Wahi, L.A. Chuah, T.S.Y. Choong, Z. Ngaini, M.M. Nourouzi, Oil removal from aqueous state by natural fibrous sorbent: an overview. Sep. Purif. Technol. **113**, 51–63 (2013)
4. M.A. Abdullah, A.U. Rahmah, Z. Man, Physicochemical and sorption characteristics of Malaysian Ceiba pentandra (L.) Gaertn. as a natural oil sorbent. J. Hazard. Mater. **177**(1–3), 683–691 (2010)
5. T.-T. Lim, X. Huang, Evaluation of hydrophobicity/oleophilicity of kapok and its performance in oily water filtration: comparison of raw and solvent-treated fibers. Ind. Crop. Prod. **26**(2), 125–134 (2007)
6. S. Ibrahim, S. Wang, H.M. Ang, Removal of emulsified oil from oily wastewater using agricultural waste barley straw. Biochem. Eng. J. **49**(1), 78–83 (2010)
7. A.L. Ahmad, S. Bhatia, N. Ibrahim, S. Sumathi, Adsorption of residual oil from palm oil mill effluent using rubber powder. Braz. J. Chem. Eng. **22**(3), 371–379 (2005)

Salinity Velocity Pattern in Estuary Using PIV

Nuryazmeen Farhan Haron, Wardah Tahir, Irma Noorazurah Mohamad, Lee Wei Koon, Jazuri Abdullah, and Natasya Anom Sheikh Aladin

Abstract Estuaries are bodies of water along the coasts that are formed when freshwater from rivers flows and mixes with saltwater from the ocean. The mixing process in estuary is due to the forces that forced the freshwater from river to interact with saltwater at the ocean. Thus, it will give effect to the mixing properties of the estuary. The objective of this research is to analyze the salinity mixing pattern using Particle Imaging Velocimetry (PIV). A laboratory investigation was conducted in a Perspex channel to observe the mixing process in estuary. The images of the mixing process had been captured using the PIV system. Five parameters had been analyzed: Reynolds Stress, Turbulence Intensity, U for speed, V for velocity, and Vorticity using Matlab. The results show that two layers of water had been established inside the Perspex channel where saltwater had been observed to be at the bottom of the channel due to difference in density between freshwater and saltwater. This showed typical salt-wedge estuary characteristics.

Keywords Mixing • Salinity • Estuary • Laboratory investigation • PIV

1 Introduction

Estuaries are semi-enclosed coastal bodies of water where freshwater and saltwater meet and mix [1, 2]. Based on mixing characteristics, estuaries can be classified as vertically mixed, slightly stratified, highly stratified, or saline-wedge. The freshwater from river is lighter than saltwater; thus it has a tendency to remain on top of the saltwater. Among previous studies on salinity, mixing pattern can be referred from [3–5].

In order to get a clear picture on how mixing processes occur in an estuary, computer modeling and physical modeling are great tools in providing the

This work was supported in part by the Ministry of Higher Education (MOHE) and Universiti Teknologi MARA (UiTM), Malaysia, under the Research Acculturation Grant Scheme (RAGS)

N.F. Haron (✉) • W. Tahir • I.N. Mohamad • L.W. Koon • J. Abdullah • N.A.S. Aladin
Flood Control Research Center, Faculty of Civil Engineering, Universiti Teknologi MARA, Shah Alam, Selangor, Malaysia
e-mail: neemzay@yahoo.com

S.H. Abu Bakar et al. (eds.), *ISFRAM 2014*, DOI 10.1007/978-981-287-365-1_19

mechanism such as particle imaging velocimetry (PIV) system. The PIV technique had been used for most of laboratory works such as the measurement of velocity in fluid mechanics [3] and the investigation of turbulent flows [4]. The system marked particles in the flow of a fluid and these marked particles are recorded on a video which are in the form of images after they are illuminated by light [5].

Therefore, this chapter presents the laboratory investigation to visualize estuary salinity mixing pattern using PIV. In order to obtain a clear picture on how mixing pattern in an estuary occurs, this research involves the designing and simulation of a simple estuary system by using a Perspex channel.

2 Methodology

2.1 Experimental Set Up

A Perspex channel with a length of 45 cm, width of 29.8 cm, and height of 30.5 cm had been used to represent a simple estuary system with PIV system. Freshwater and saltwater flow rates used were 0.0429 l/s and 0.132 l/s, respectively. Saltwater was mixed with red dye tracer. The entrance of the saltwater represents the condition in the event of heavy raining, in which an increase in the amount of rainfall leads to the rise in sea level, thus leading to the entry of saltwater from the coastal area into the downstream area of a river which contains freshwater. Figure 1 shows the experimental setup for freshwater and saltwater.

2.2 Setting Up of PIV System

PIV system component consists of seeding particles, light sources, image capturing, recording hardware, and image processing. Freshwater and saltwater, which were

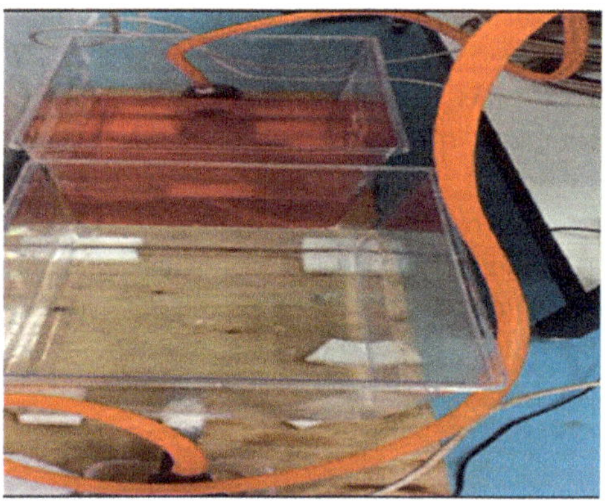

Fig. 1 Experimental setup

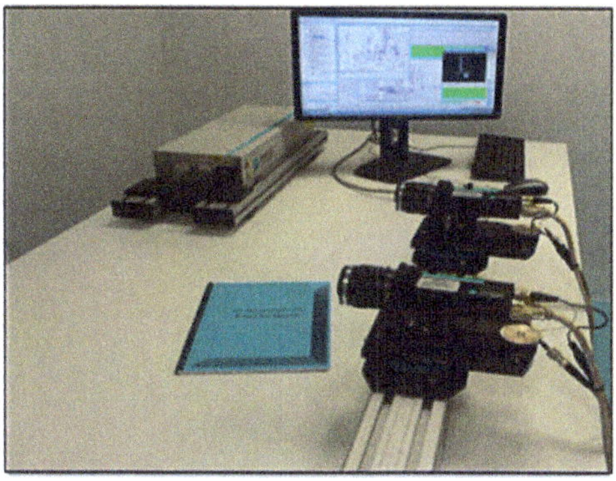

Fig. 2 The PIV system

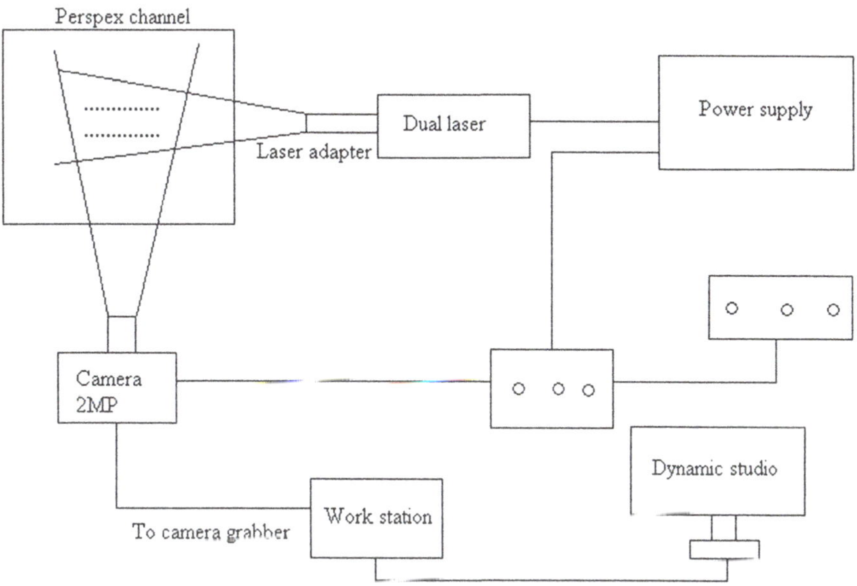

Fig. 3 The experimental layout

mixed with the seeding particles to ensure capturing of the mixing image, were introduced into the Perspex channel. Laser was used in this experiment as the light sources. Double frame-single exposures were used for the analysis during image capturing by charge-coupled devices (CCD). Figure 2 shows the PIV system.

Meanwhile, Fig. 3 shows the experimental layout. Alignment of the laser and the CCD camera together with the position of the Perspex channel is an important element to ensure capturing of the images is accurate.

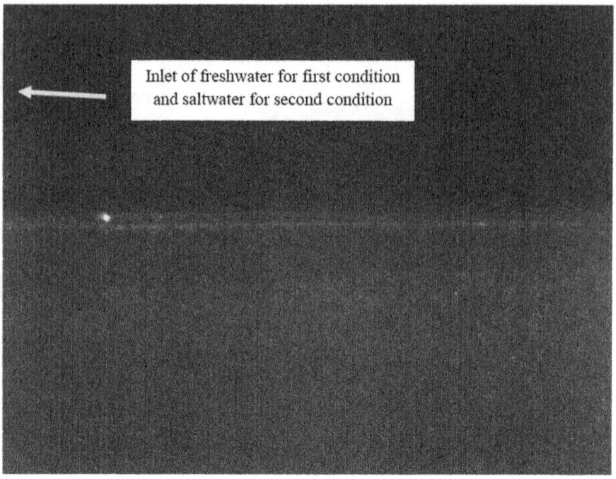

Fig. 4 Images of mixing freshwater and saltwater inside the Perspex channel

3 Results and Analysis

The CCD Cameras started capturing images when freshwater and saltwater were introduced into the Perspex channel. The images obtained from PIV showed the mixing of freshwater and saltwater (Fig. 4). Due to the difference in density between freshwater and saltwater, these two different densities of water do not usually mix together.

3.1 Reynolds Stress

The origin of Reynolds Stress is from Reynolds Number, which refers to the indication whether the fluid is flowing in steady or turbulent condition. The investigation of Reynolds Stress of a flowing fluid is important because it will identify and give a prediction on whether the flowing fluid is in a laminar or turbulent condition.

Figures 5 and 6 show the result of Reynolds Stress for freshwater into saltwater and saltwater into freshwater. Both of the figures show a high value of Reynolds Stress in the area of inlet for freshwater for the first condition and saltwater for the second condition. This is due to the forces of the flow rate that influences the value of Reynolds Stress where the higher the value of flow rate, the higher the value of Reynolds Stress will be. The case is different at the end area of the Perspex channel where the value of Reynolds Stress is lower since low momentums or force is produced around the area.

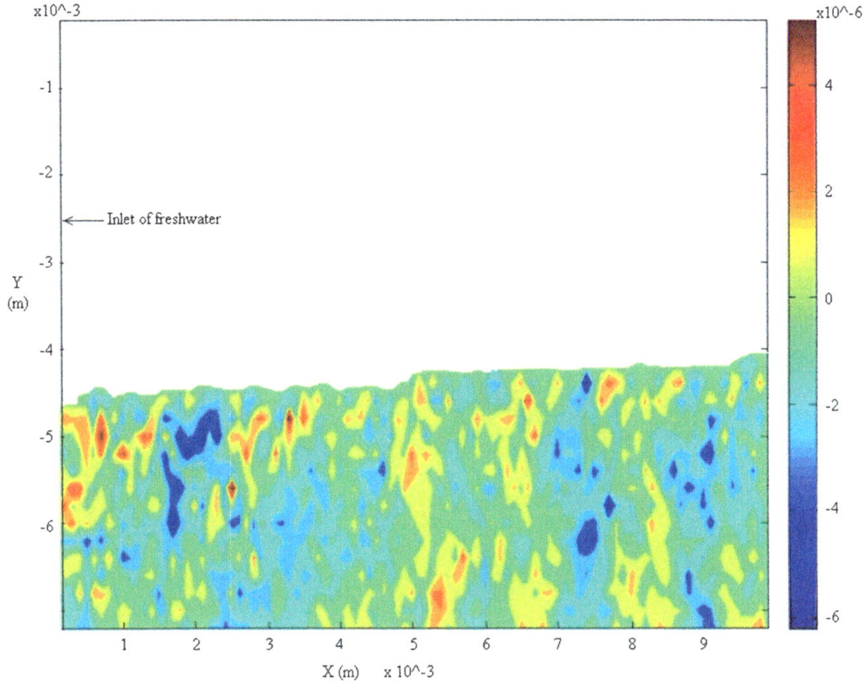

Fig. 5 Reynolds stress distribution of freshwater into saltwater

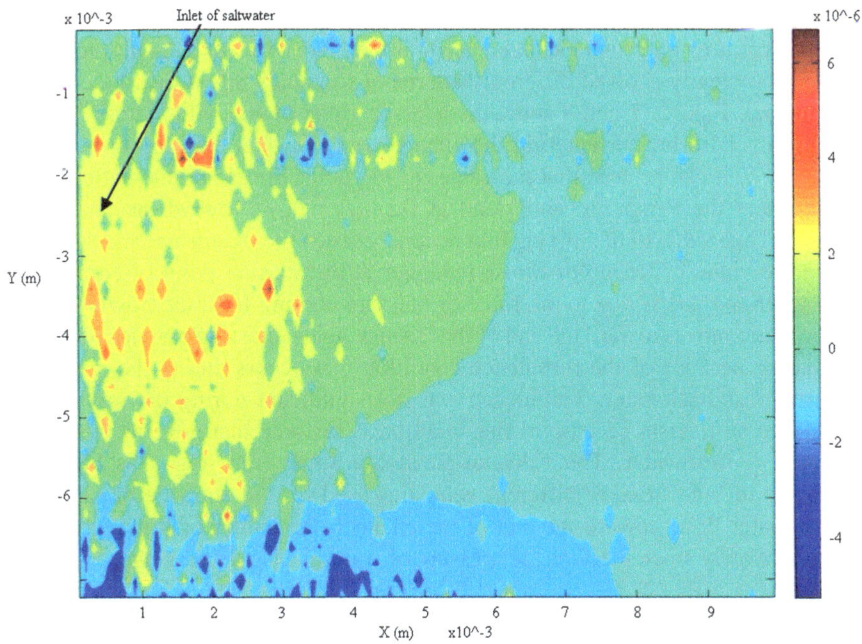

Fig. 6 Reynolds stress distribution of saltwater into freshwater

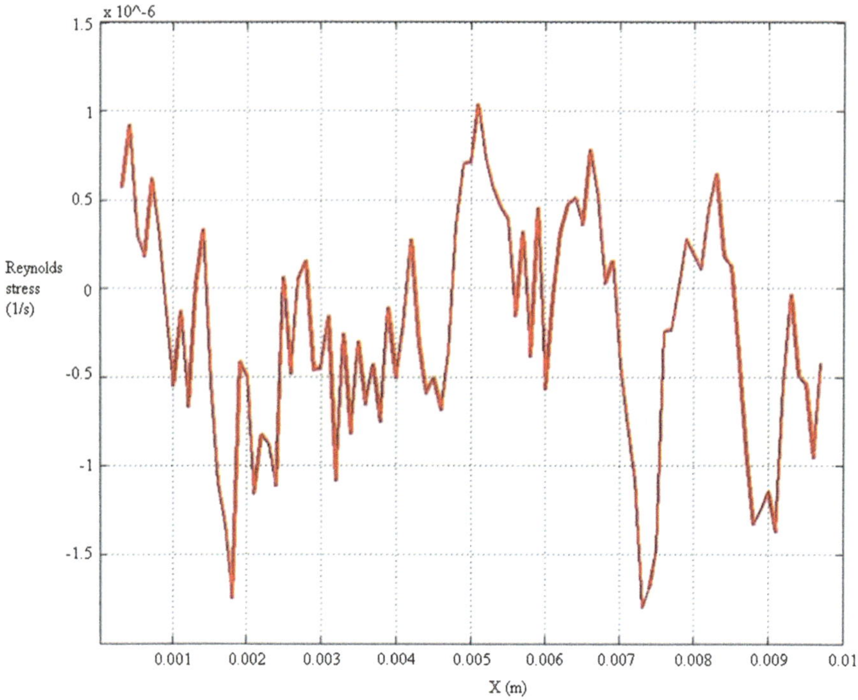

Fig. 7 Average value of Reynolds stress of freshwater into saltwater

Different flow rates that had been used for the first and second condition for this experiment clearly showed different patterns of Reynolds Stress distributed along the channel. Figures 7 and 8 show the average value of Reynolds Stress for both conditions of freshwater entering the channel and saltwater entering the channel. From Fig. 7, it can be seen that the values of Reynolds Stress are fluctuating as the water enters the Perspex channel and as the water flows toward the end of the channel. This is due to the mixing that occurs between the freshwater and saltwater where the value of Reynolds Stress is higher at the starting point of entrance of freshwater and this is due to the force or pressure coming from the flowing water. As the water moves toward the end of the channel, the values of Reynolds Stress are fluctuating because of the turbulence condition that occurs during the mixing of freshwater and saltwater. Turbulence was not significant during the mixing, but turbulence still exists because of the small press or force that governs during the entrance of freshwater. The value of Reynolds Stress from Fig. 8 is higher as compared to Fig. 7 due to different velocities used for the entrance of saltwater. As the water flows toward the end of the channel, the values of Reynolds Stress decrease slowly since the force or pressure of water is also decreasing.

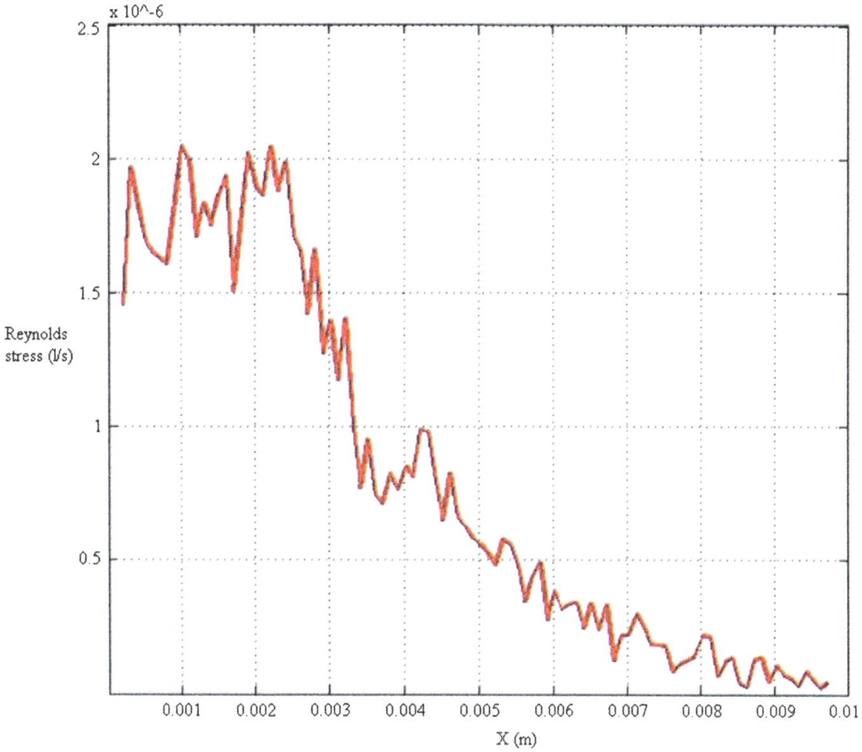

Fig. 8 Average value of Reynolds stress of saltwater into freshwater

3.2 Turbulence Intensity

Turbulence can be defined as rapid movement of fluid, air, or even any other type of fluid and it can be expressed as the root mean square (RMS) values of turbulent velocity fluctuations at a selected location with a specified time period, over the average of velocity at the same selected location with the same specified time [6].

Figures 9 and 10 show the turbulence intensity for both conditions of freshwater into saltwater and saltwater into freshwater. The turbulence intensity seems to occur in the area where freshwater for the first condition and saltwater for the second condition were being introduced into the Perspex channel. The occurrence of turbulence intensity was due to pressure or high force that was governed by the value of the flow rate at the valve. As the flow moves toward the end of the Perspex channel, the value of turbulence intensity gradually decreases as the amount of pressure or force decreased as well.

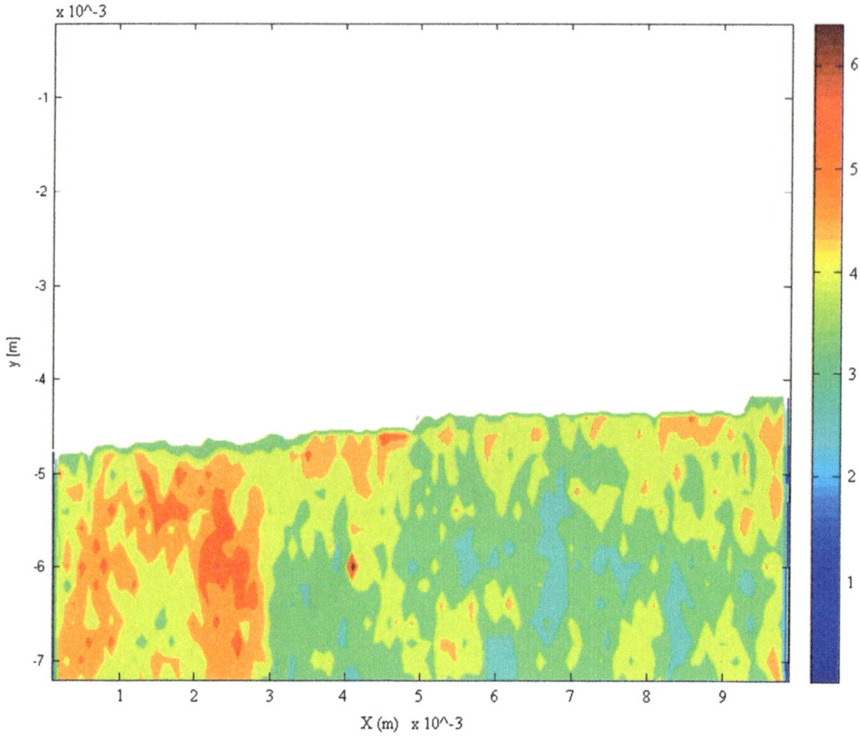

Fig. 9 Turbulence intensity distribution of freshwater into saltwater

Due to the difference in the value of the flow rate between freshwater and saltwater that were being introduced into the channel, the results can clearly be seen and differentiated, Figs. 11 and 12 show the average value of turbulence intensity for both conditions. From Fig. 11, the value of turbulence intensity increases and starts to decrease at 0.0025 m due to the force and pressure coming from the inflow of freshwater and starts to decrease as the water flows toward the end of the channel. Figure 12 shows a gradual decrement of turbulence intensity as saltwater enters the channel and moves toward the end of the channel.

3.3 U for Speed

U for speed refers to the measurement of speed in a flowing fluid. PIV measures the speed of fluid by measuring the distance over time taken to move particles into one another.

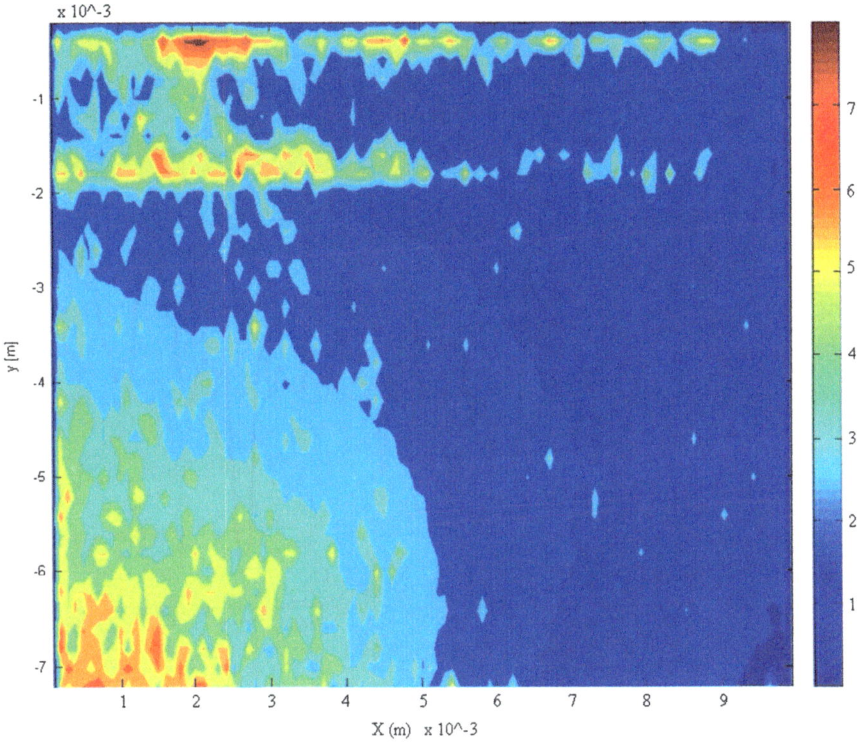

Fig. 10 Turbulence intensity distribution of saltwater into freshwater

Figure 13 shows a high value of speed during the entrance of freshwater, but the speed decreases as the water flows and leaves the entrance area, while Fig. 14 shows different behaviors, whereby as the water flows toward the end of the channel, the value of speed increases and higher value is shown at the bottom of the channel. This is due to the difference in the value of flow rate used for the entrance of freshwater and saltwater; thus it explains the difference between Figs. 13 and 14, where the value of speed gradually decreases as the mixing water moves toward the end of the channel which can be seen in Fig. 13. The condition is different in Fig. 14 where the value of speed gradually increases as it moves toward the end of the channel due to the force or pressure occurring at the entrance of saltwater that leads to the high value of speed, even though the mixing water is flowing toward the end of the channel. Also, due to the high value of flow rate, the impact of saltwater entering the Perspex channel has led to the high value of speed at the bottom of the channel in Fig. 14.

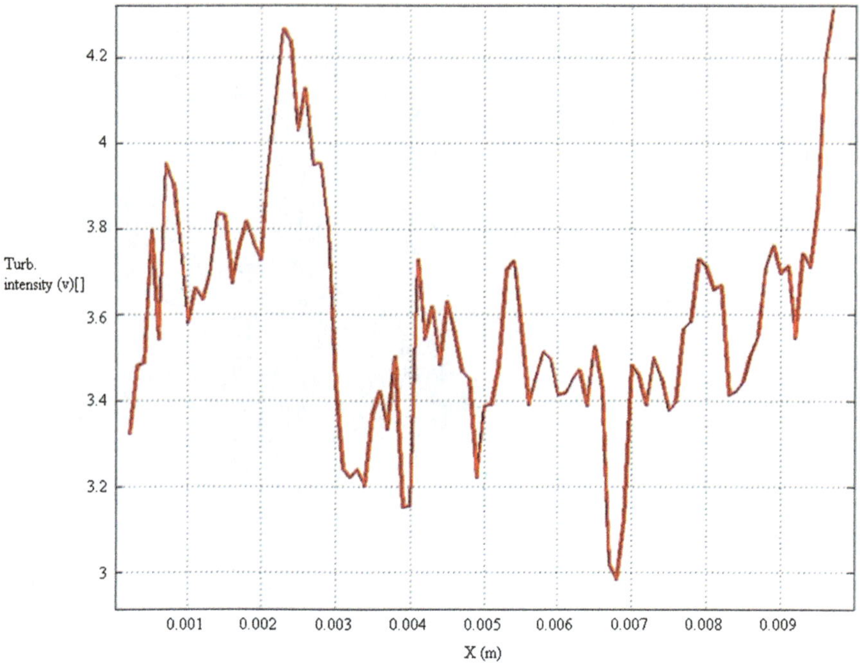

Fig. 11 Average value of turbulence intensity of freshwater into saltwater

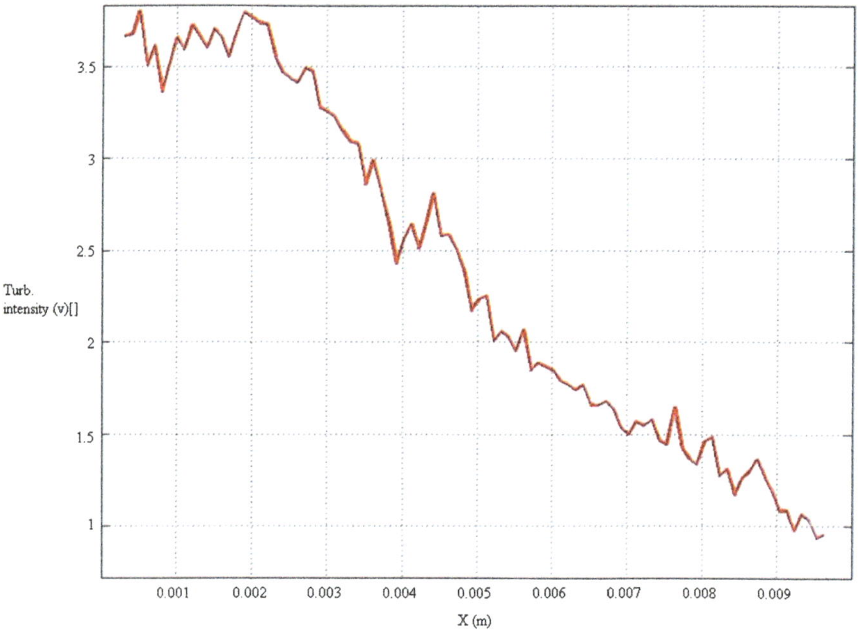

Fig. 12 Average value of turbulence intensity of salt water into fresh water

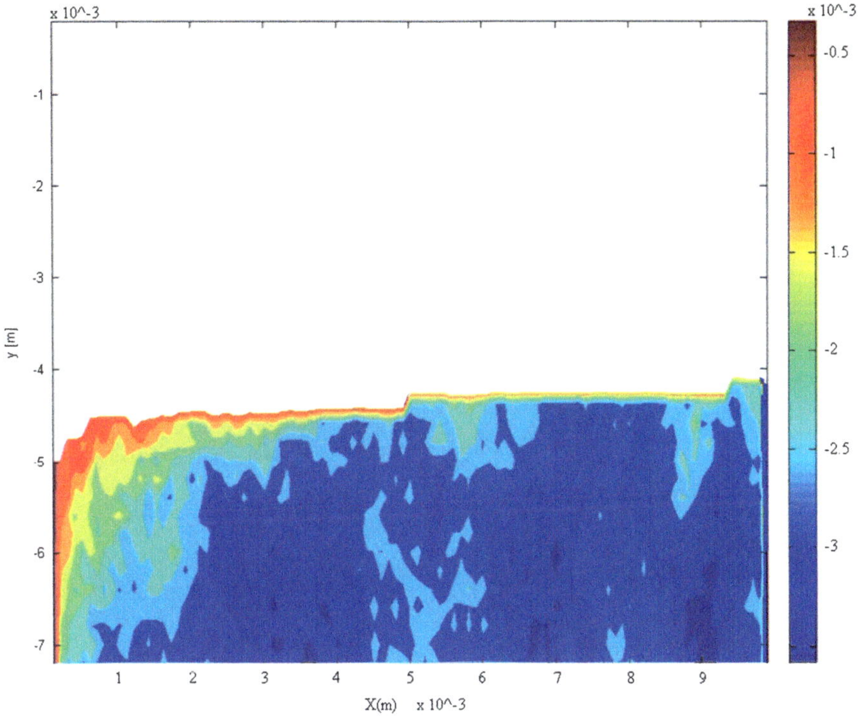

Fig. 13 Speed distribution of freshwater into saltwater

Figures 15 and 16 represent the average value of speed for both conditions and the figure reflects the condition of speed presented in a contour form as shown in Figs. 13 and 14.

3.4 V for Velocity

V for velocity is the measurement of change in position at a point in a flowing fluid. As freshwater enters the Perspex channel, the high value of velocity is created in the area of the entrance and decreases as the water flows toward the end of the channel. Freshwater with a small value of flow rate shows a gradual decrement of velocity as the water becomes slower which can be seen in Fig. 17.

The value of velocity in Fig. 17 does not change abruptly as compared to Fig. 18, which shows abrupt changes in velocity as saltwater is introduced into the channel.

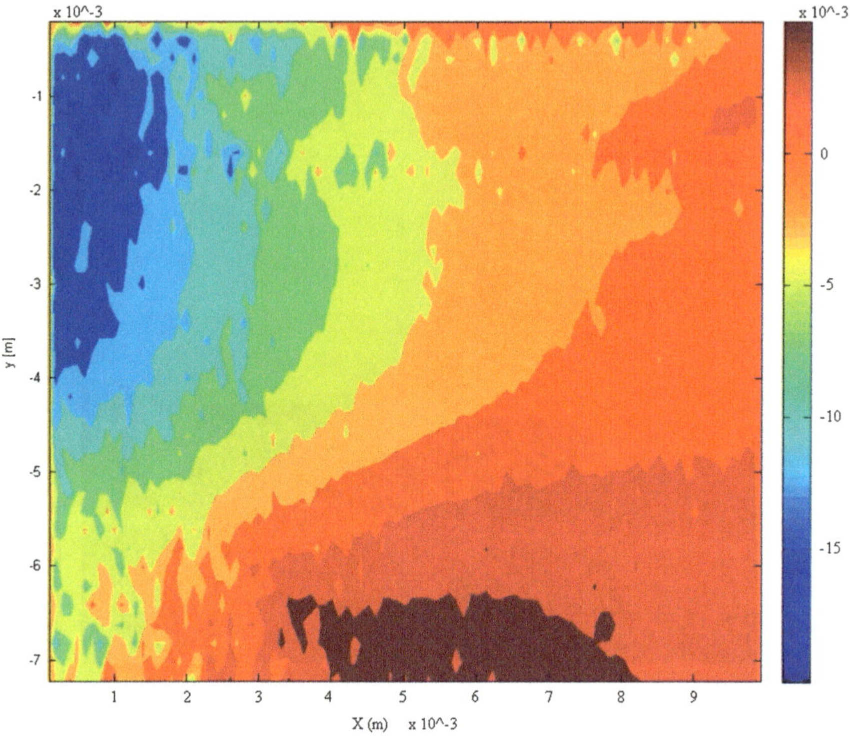

Fig. 14 Speed distribution of saltwater into freshwater

The abrupt changes are due to value of velocity for saltwater which is slightly higher as compared with the first condition which is the introduction of freshwater into saltwater.

Figures 19 and 20 show the average value of velocity for both conditions of freshwater entering the channel and saltwater entering the channel. The velocity of freshwater entering the channel fluctuates as the water flows along the channel due to the occurrence of force being created as the water flows along the channel. As shown in Fig. 20, the value of the velocity is high at the entrance area of saltwater and starts to decrease as the mixing water flows along the channel. This is due to the decreasing force which usually occurs at the entrance of the channel due to the moving water.

3.5 Vorticity

Vortices are defined as a spinning movement of a mass of fluid or air or as a vector field that determines the rotation at any point in a fluid. Vorticity that occurs in

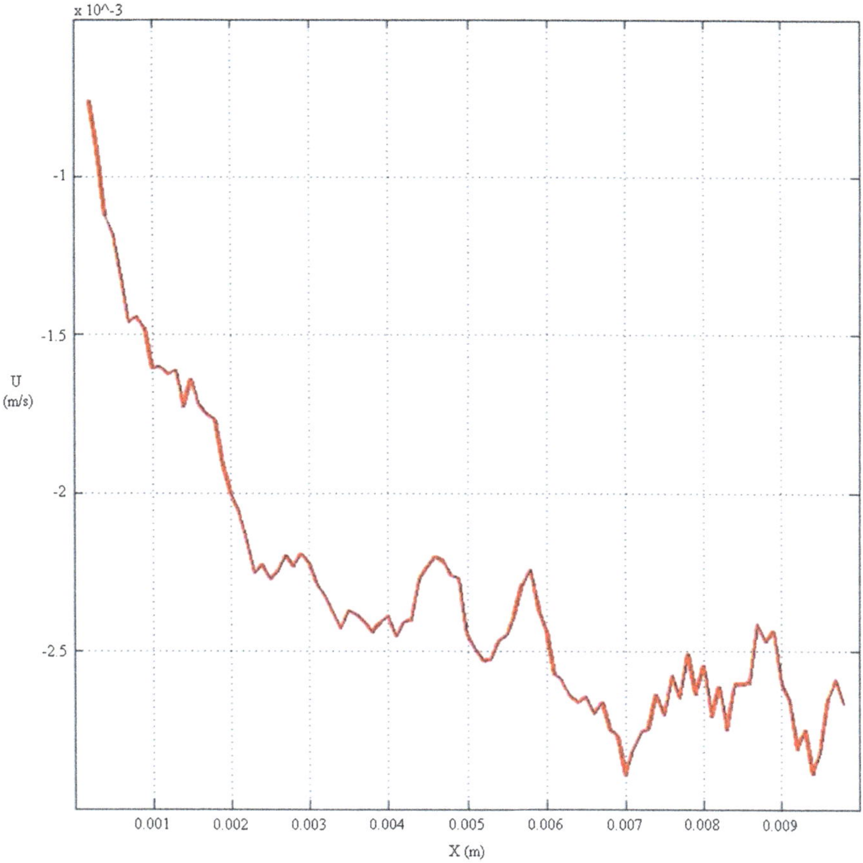

Fig. 15 Average value of speed for freshwater into saltwater

moving fluid during mixing of freshwater and saltwater can be seen in Figs. 21 and 22. All the figures show that vorticity is created at the edge of the channel where the inlets of freshwater and saltwater were located and this is the area where circulation or rotation of fluid occurs as two different densities of fluid meet and collide among themselves.

Vorticity increases gradually as it moves and leaves the core area. These explain the formation of vorticity which can be seen inside the figures where the value of vorticity is high as it moves or flows leaving the area of inlet where circulation occurs. As the fluid flows toward the end of channel leaving the inlet area, the values of vorticity approach to zero as there is no rotation of fluid occurring along

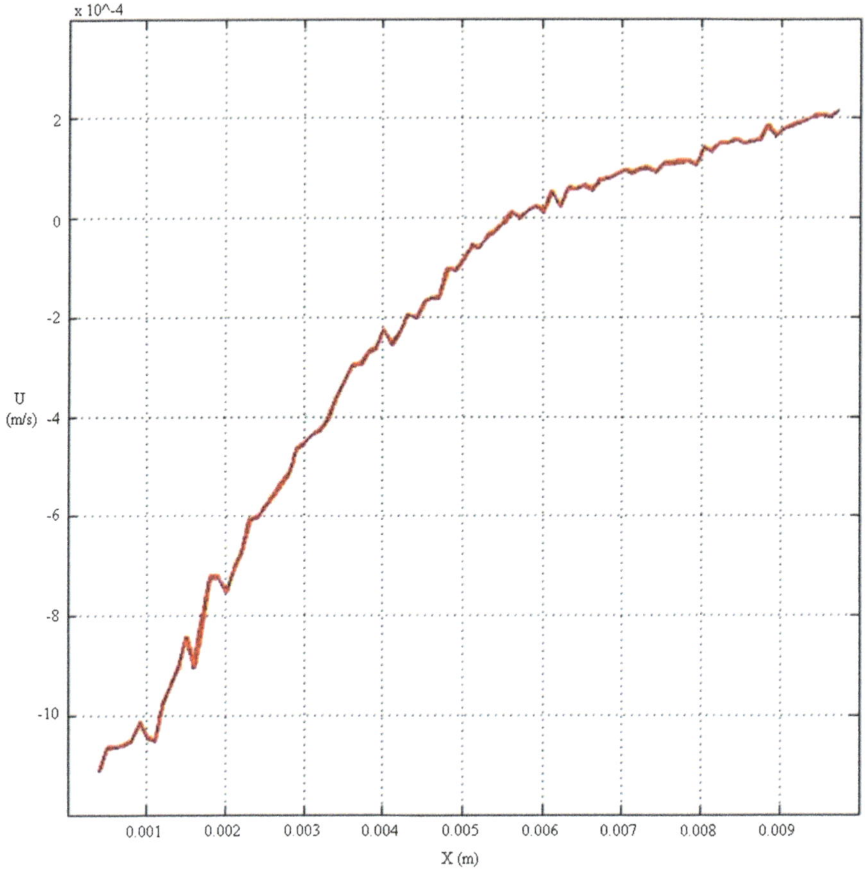

Fig. 16 Average value of speed of saltwater into freshwater

the movement. The quantitative value of the vorticity can be seen in Figs. 23 and 24, which show the average value of vorticity for freshwater entering the channel and saltwater entering the channel.

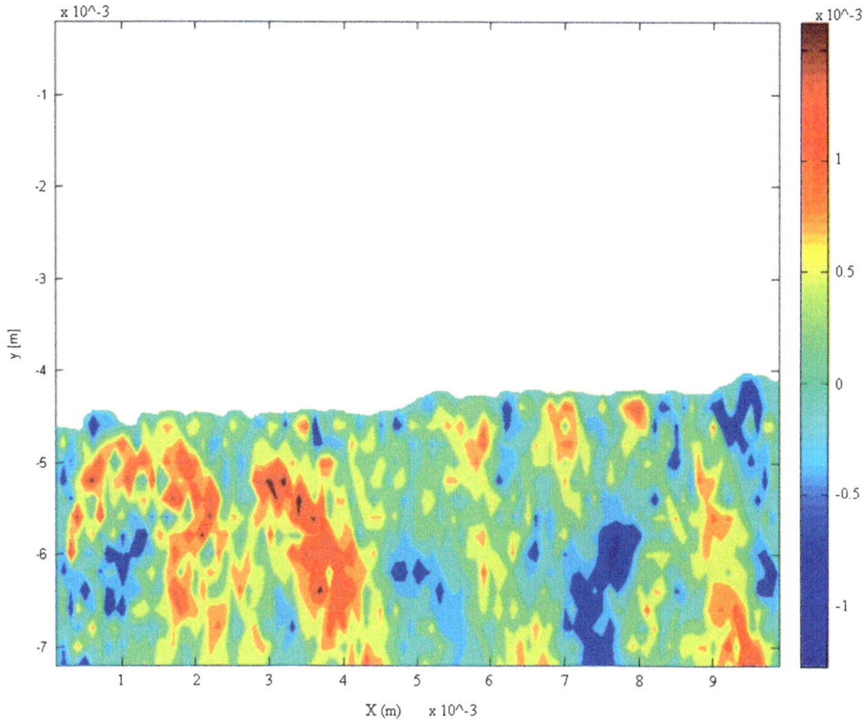

Fig. 17 Velocity distribution of fresh water into salt water

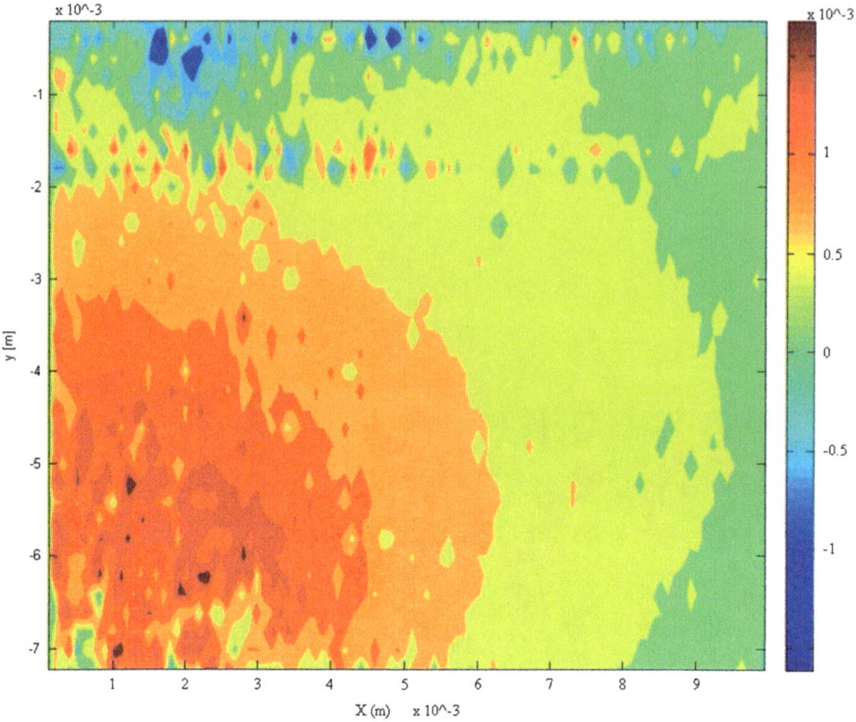

Fig. 18 Velocity distribution of saltwater into freshwater

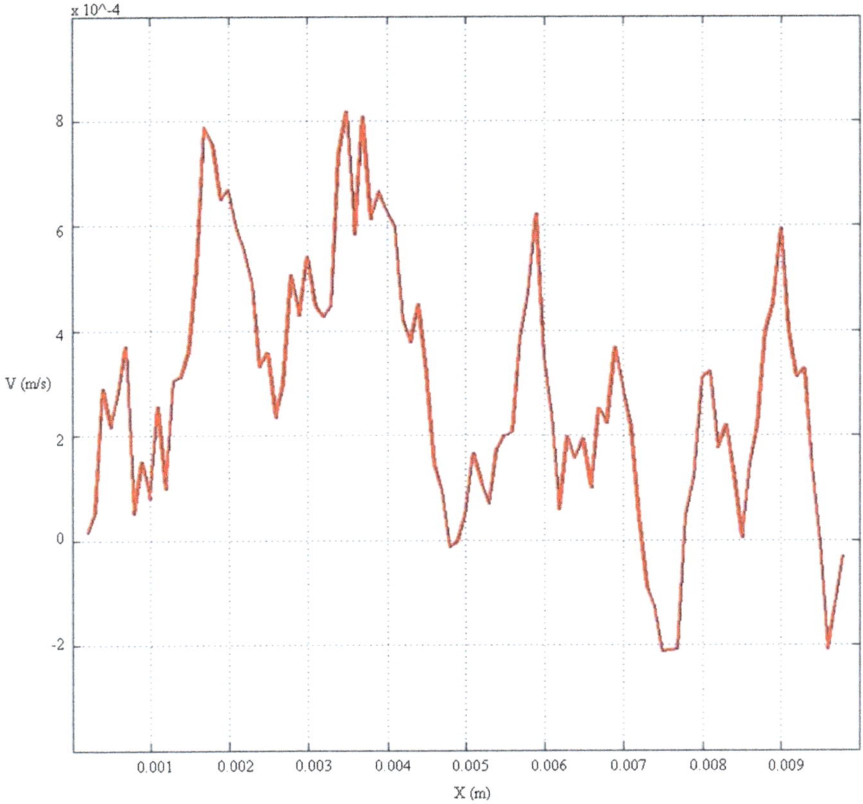

Fig. 19 Average value of velocity of freshwater into saltwater

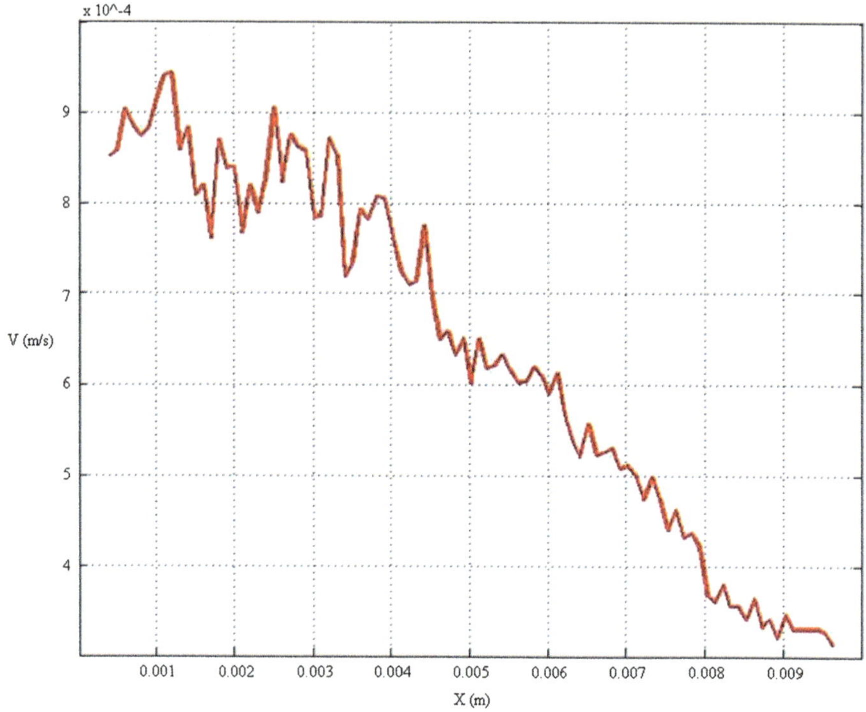

Fig. 20 Average of velocity of saltwater into freshwater

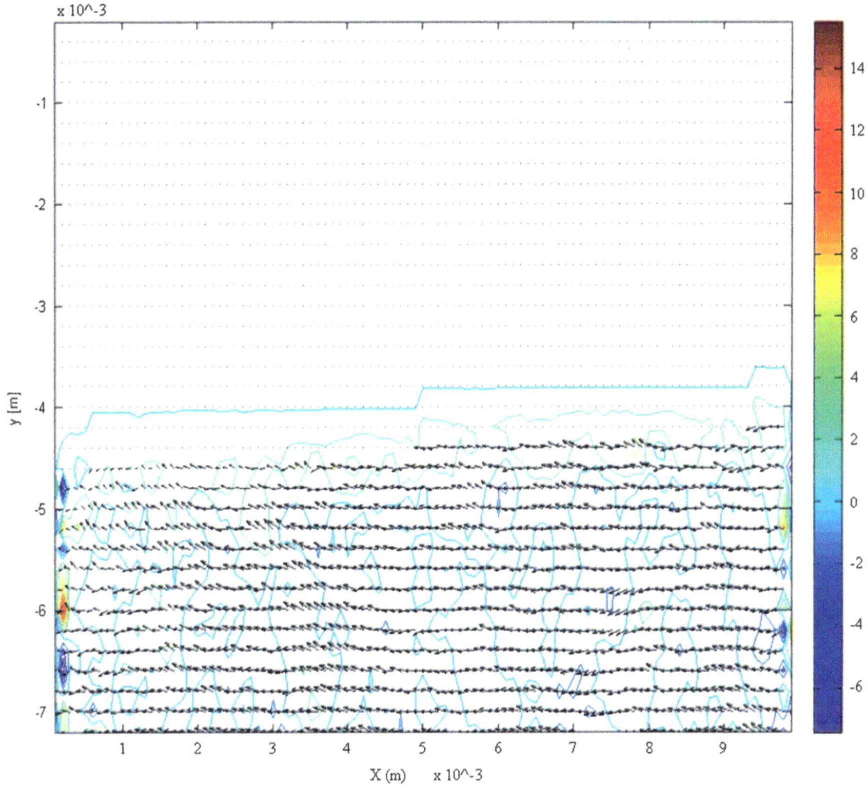

Fig. 21 Vorticity distribution of freshwater into saltwater

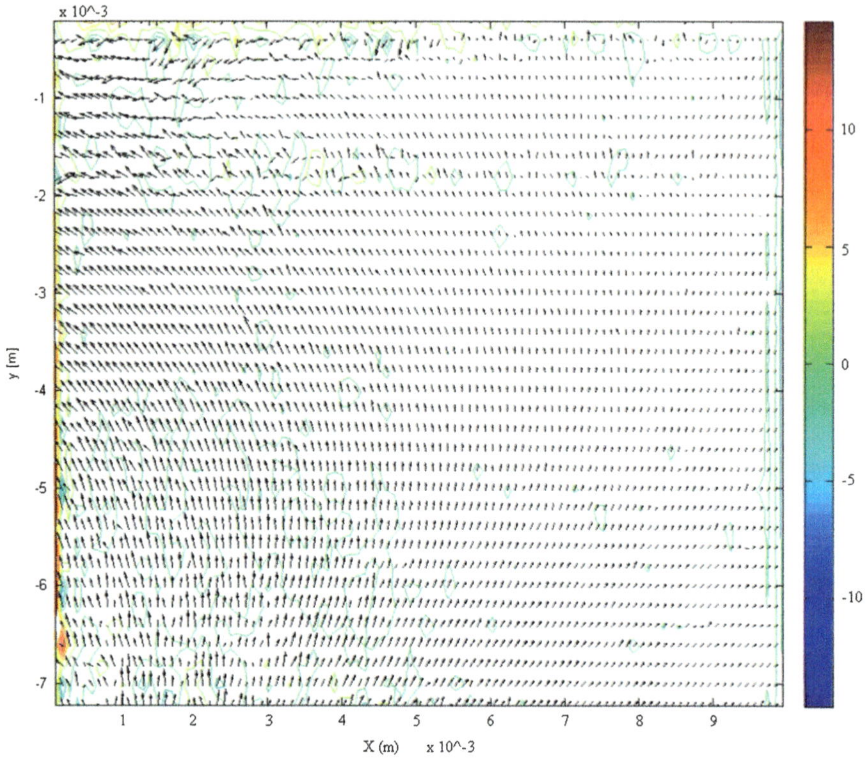

Fig. 22 Vorticity distribution of saltwater into freshwater

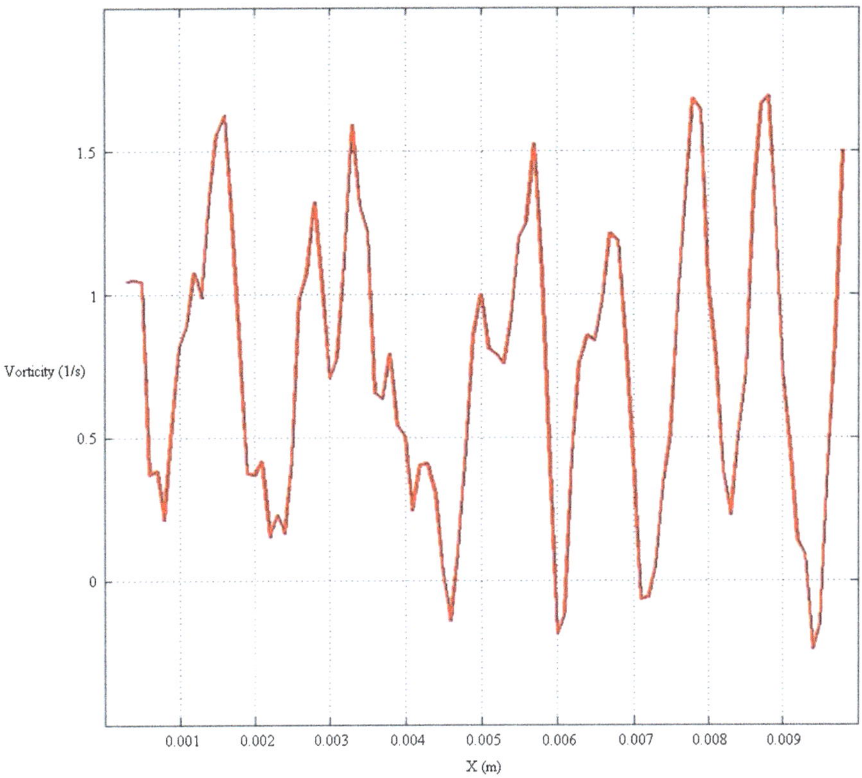

Fig. 23 Average value of vorticity of freshwater into saltwater

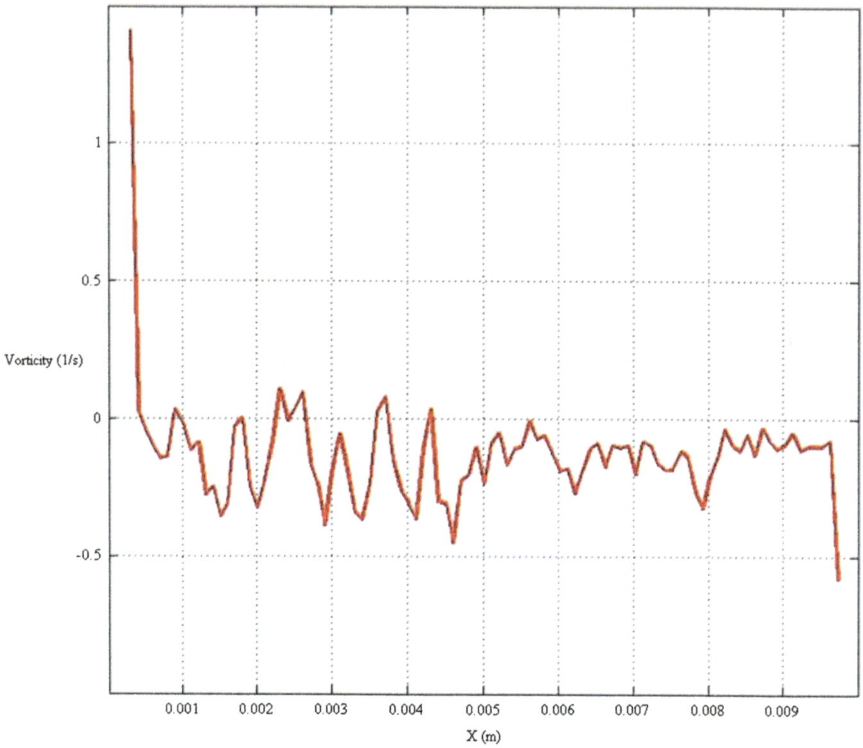

Fig. 24 Average value of vorticity of saltwater into freshwater

4 Conclusion

Selected parameters, which are Reynolds Stress, Turbulence Intensity, U for speed, V for velocity, and Vorticity, had been analyzed and discussed in order to study the mixing pattern of freshwater and saltwater with different flow rates that occurred.

It can be concluded that due to the difference in density between freshwater and saltwater, two layers of water had been established where saltwater had been observed to be at the bottom of the channel since the density is higher as compared to freshwater. Estuary can be classified through various characteristics and they are based on topography, mixing, and tidal influence. As for this research, the estuary is classified according to mixing characteristics and the type of estuary for this study is salt-wedge estuary.

Different values and patterns for all the parameters analyzed had been shown and obtained at the end of the analysis due to the difference in the value of the flow rate that had been introduced into the channel for two different conditions. The high value of parameters has been observed for introduction of saltwater since the flow rate used was higher as compared to freshwater. This is due to the higher forces and pressures created at the valve and the entrance area of saltwater.

Acknowledgment This study was partially funded by the Ministry of Higher Education (MOHE), Malaysia, and Universiti Teknologi MARA (UiTM), Malaysia, under the Research Acculturation Grant Scheme (RAGS).

References

1. R.H. Weisberg, L. Zheng, How estuaries work: A Charlotte Harbor example. J. Mar. Res. **61**, 635–657 (2003)
2. J. Cao, R. Li, Y. Zhu, Study on salt water intrusion in the Yangtze River Estuary by 3D numerical model, in *Education Technology and Training, International Workshop on Geoscience and Remote Sensing*, vol. 2 (IEEE, New York, 2008), pp. 81–84
3. F.H. Nuryazmeen, T. Wardah, Physical model of estuarine salinity intrusion into rivers: A review. Adv. Mater. Res. **905**, 348–352 (2014)
4. F.H. Nuryazmeen, T. Wardah, N.M. Irma, Laboratory investigations on estuary salinity mixing: preliminary analysis. Int. J. Sci. Basic Appl. Res. (IJSBAR) **13**(1), 36–41 (2014)
5. F.H. Nuryazmeen, T. Wardah, N.M. Irma, Potential of estuary transverse flow salinity intrusion due to extreme estuarine flooding, in *InCIEC 2013* (Springer, Singapore, 2014), pp. 343–352
6. C. Brossard, J.-C. Monnier, P. Barricau, F.-X. Vandernoot, Y.L. Sant, F. Champagnat, G.L. Besnerais, Principles and applications of particle image velocimetry. J. Aerospace Lab. (1), 1–11 (2009)

Part V
Weather and Climate

Tsunami Forecasting Due to Seismic Activity in Manila Trench of Malaysia Offshore Oil Blocks

N.H. Mardi, M.S. Liew, M.A. Malek, and M.N. Abdullah

Abstract Manila Trench is a potential generation source of tsunami in the South China Sea. Manila Trench is formed when Eurasian Plate is subducted under the Philippine Sea Plate. It has been classified as the most hazardous tsunami sources in USGS Tsunami Sources Workshop, 2006. The location of the trench begins at the north of Palawan, Philippines, continuing to the north along the west of Luzon, Philippine, and ends at Taiwan. The potential tsunami generated from Manila Trench affects Philippines and countries located in the vicinity of South China Sea including Malaysia. This chapter focuses on simulation of tsunami generation and tsunami wave propagation due to seismic activity in the Manila Trench. TUNA-M2 model is used to produce simulation results in terms of tsunami wave height and time of arrival of the first wave. The study area is Malaysian offshore reserves and operation where oil and gas platforms are located. There are three simulations performed at different moment magnitudes (M_w: 7.0, 7.5, and 8.0). The result shows the range of wave heights at the Malay basin to be between 0.002 and 0.122 m, the range of wave heights at Sabah basin to be between 0.004 and 0.168 m, and the range of wave heights at Sarawak basin to be between 0.004 and 0.230 m. These results are useful in the future design of offshore platform structure and operation

---

N.H. Mardi (✉) • M.A. Malek
Department of Civil Engineering, College of Engineering, Universiti Tenaga Nasional, Kajang, Malaysia
e-mail: nurulhanimardi@gmail.com; Marlinda@uniten.edu.my

M.S. Liew
Faculty of Geosciences and Petroleum Engineering, Universiti Teknologi Petronas, Tronoh, Malaysia
e-mail: shahir_liew@petronas.com.my

M.N. Abdullah
Department of Metocean Engineering, Petronas Carigali Sdn. Bhd., Kuala Lumpur, Malaysia
e-mail: mnasir_abdullah@petronas.com

S.H. Abu Bakar et al. (eds.), *ISFRAM 2014*, DOI 10.1007/978-981-287-365-1_20

Keywords Tsunami propagation • TUNA-M2 • Manila Trench • Malay • Sabah and Sarawak basin

1 Introduction

Tsunami is a combination of two Japanese words: *tsu* (wave) and *nami* (harbor). Tsunami is defined as waves that enter a harbor, with an amplified wave height and wave velocity [1]. Tsunami event is categorized as one of the most hazardous natural coastal disasters. It is caused by earthquakes that create a sudden movement on the tectonic plate. Other causes of tsunami include submarine landslides, volcanic eruption, or impact of large object on the sea surface.

Malaysia faced tsunami once on 26th December 2004, popularly known as the Boxing Day. The tsunami was occured due to earthquake measuring 9.3 on the Richter scale in the Indian Ocean. The earthquake is located off the coast of Aceh where the India Plate subducted under the Burma plate. The event created an estimated 800 km fault length, about 85 km fault width and initial vertical displacement of 11 m [2]. Almost 15 countries were affected. The tsunami caused massive loss of life and huge loss of property.

Since that event, Malaysia had embarked on many researches pertaining to tsunami hazards. As a result, TUNA-M2, an in-house tsunami numerical model, was developed by Universiti Sains Malaysia. Malaysian Meteorological Department (MMD) has also built a warning system to alert Malaysian community about tsunami occurrences. Various measures planned by Malaysian authorities to ensure the safety of Malaysian population, especially those who live near to the coastal areas, were implemented.

Recently, there are many research works on potential tsunami for South China Sea such as [1, 3–7]. GPS geodesy showed that the Manila Trench has about 8 cm/year convergence rate across the megathrust [4]. Therefore, the purpose of this study is to forecast tsunami occurrences due to earthquakes in the Manila Trench. The main focus of this study is to determine tsunami wave height at Malaysia offshore blocks of South China Sea where many oil and gas activities are located. The outcome of the study will provide the offshore professional the necessary preparations on possibilities of tsunami occurrences.

2 Tsunami Propagation Model

TUNA-M2 model is chosen to perform tsunami simulation. TUNA-M2 model is part of TUNA model developed by researchers from Universiti Sains Malaysia in 2005. TUNA could simulate tsunami events due to earthquake at various stages as generation, propagation, run-up, and inundation. For this study, TUNA-M2 is chosen since this study focuses on tsunami wave generation and propagation at offshore areas.

TUNA model has been used to simulate tsunami events in the Indian Ocean [2], Straits of Malacca, Andaman Sea [8], and South China Sea [1, 7]. TUNA has also been validated within other models. Simulation results from TUNA model show satisfactory performance when it is compared with simulation results from COMCOT (Cornell Multi-grid Coupled Tsunami Model) and on-site survey data [8].

The base equation used in TUNA-M2 model is linear Shallow Water Equation (SWE) in line with the Intergovernmental Oceanography Commission [9]. SWE is valid for the analysis of tsunami wave since the tsunami wave height is smaller in the deep ocean with large wavelength. TUNA-M2 is a 2D model that could simulate both tsunami generation and tsunami wave propagation. Equations (1)–(3) are hydrodynamic equations where the conservation of mass and momentum is at average depth since tsunami propagation occurs in the deep ocean [8].

$$\frac{\partial \eta}{\partial t} + \frac{\partial M}{\partial x} + \frac{\partial N}{\partial y} = 0 \tag{1}$$

$$\frac{\partial M}{\partial t} + \frac{\partial}{\partial x}\left(\frac{M^2}{D}\right) + \frac{\partial}{\partial y}\left(\frac{MN}{D}\right) + gD\frac{\partial \eta}{\partial x} + \frac{gn^2}{g^{7/3}} M\sqrt{M^2 + N^2} \tag{2}$$

$$\frac{\partial N}{\partial t} + \frac{\partial}{\partial x}\left(\frac{MN}{D}\right) + \frac{\partial}{\partial y}\left(\frac{N^2}{D}\right) + gD\frac{\partial \eta}{\partial x} + \frac{gn^2}{g^{7/3}} N\sqrt{M^2 + N^2} \tag{3}$$

From Eqs. (1)–(3), (M, N) in the x- and y-direction are related to u and v velocities as in Eqs. (4) and (5) where h is the sea depth and η is water elevation above the mean sea level.

$$M = u(h + \eta) = uD \tag{4}$$

$$N = v(h + \eta) = vD \tag{5}$$

The SWE is solved using explicit finite difference method with staggered grids. Figure 1 is the computational points for staggered scheme where η, u, and v are located. The partial derivatives had been replaced by finite difference as shown in Eq. (6). The time step Δt is restricted by the Courant criterion as shown in Eq. (7) [2]. This process is required in order to ensure that the numerical scheme is stable.

$$\eta_{i,j}^{k+1} = \eta_{i,j}^{k} - \frac{\Delta t}{\Delta x}\left[M_{i+0.5,j}^{k+0.5} - M_{i-0.5,j}^{k+0.5}\right] - \frac{\Delta t}{\Delta y}\left[N_{i,j+0.5}^{k+0.5} - N_{i,j-0.5}^{k+0.5}\right]$$

$$M_{i+0.5,j}^{k+0.5} = M_{i+0.5,j}^{k-0.5} - gD_{i+0.5,j}^{k}\frac{\Delta t}{\Delta x}\left[\eta_{i+1,j}^{k} - \eta_{i,j}^{k}\right]$$

$$D_{i+0.5,j}^{k} = h_{i+0.5,j} + 0.5\left[\eta_{i+1,j}^{k} - \eta_{i,j}^{k}\right]$$

$$N_{i,j+0.5}^{k+0.5} = M_{i,j+0.5}^{k-0.5} - gD_{i,j+0.5}^{k}\frac{\Delta t}{\Delta y}\left[\eta_{i,j+1}^{k} - \eta_{i,j}^{k}\right]$$

$$D_{i,J+0.5}^{k} = h_{i,J+0.5} + 0.5\left[\eta_{i,j+1}^{k} - \eta_{i,j}^{k}\right] \tag{6}$$

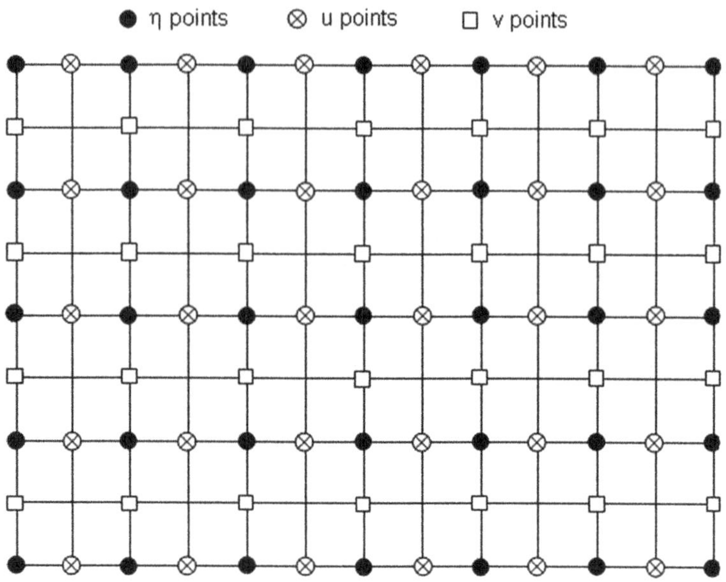

Fig. 1 Computational points for staggered scheme

$$\Delta t \leq \frac{\Delta x}{\sqrt{2gh}} \tag{7}$$

There are five main components in TUNA-M2 input file that control the tsunami source generation and results from wave propagation. These main components are (1) domain and output control, (2) sources control, (3) boundary control, (4) bathymetry control, and (5) location of point observation. The grid dimension of 1,851 × 1,851 is used in this study. It contains a study domain within a rectangle bounded by 100°E to 125° in longitude and 0°N to 25°N in latitude. The grid size chosen is 1,500 m. Bathymetry data used in the simulation were obtained from ETOPO1 using 1 arc-min.

3 Location of Study

The study area chosen is the South China Sea. The tsunami wave is generated in the Manila Trench due to earthquake phenomenon. It is estimated that the tsunami wave propagates toward Sabah, Sarawak, and Peninsular Malaysia at the offshore areas. Twelve observation points were selected as shown in Fig. 2. The study area is divided into another three locations: Peninsular Malaysia (six observation points), Sabah (three observations points), and Sarawak (three observations points) as shown in Table 1.

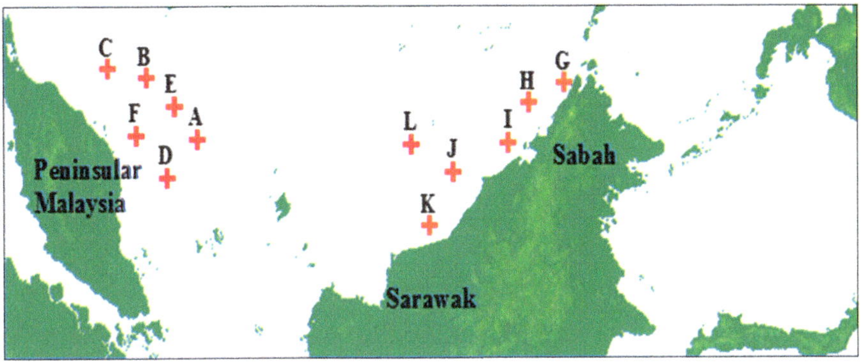

Fig. 2 Location of point observation

Table 1 Location of point observation

Location	Observation points
Peninsular Malaysia (MB)	A, B, C, D, E, F
Sabah (SB)	G, H, I
Sarawak (SWB)	J, K, L

4 Fault Parameter

The fault plane is the most important component in order to determine the source generation from the Manila Trench. In this study, the Manila Trench is divided into six hypothetical fault planes as identified [6]. Table 2 shows the hypothetical fault planes chosen along the Manila Trench. Each segment has its longitude and latitude of fault location, fault length, strike angle, dip angle, and rake angle.

Next is to determine the fault width parameter based on the length. There are many empirical equations available such as [10–12]. In this study, empirical equation developed by Papazachos et al. had been applied. Equations (8) and (9) show relationship of length and width of fault plane with moment magnitude for dip-slip fault in regions of lithospheric subduction. Both of these equations are applicable for events with moment magnitude ranging from 6.7 until 9.2.

$$\text{Log}\, L = 0.55M - 2.19 \tag{8}$$

$$\text{Log}\, W = 0.31M - 0.63 \tag{9}$$

where L: length of fault plane
W: width of fault plane
M: moment magnitude

Table 3 shows the rupture width calculated using Eqs. (8) and (9). Each segment has its own width ranging between 53 and 88 km. The average of the calculated width is chosen as one value for all fault widths. In this study, 71 km is used as the width of fault plane.

Table 2 Hypothetical fault planes along the Manila Trench, Philippines

Segment	Long. (°)	Lat. (°)	Length (km)	Strike (°)	Dip (°)	Rake (°)
1	120.5	20.2	160	10	10	90
2	119.8	18.7	180	35	20	90
3	119.3	17.0	240	359	28	90
4	119.2	15.1	170	3	20	90
5	119.6	13.7	140	320	22	90
6	120.5	12.9	100	293	26	90

Table 3 Estimated rupture width of Manila Trench

Segment	Length (km)	M (rupture length)	Width (km)
1	160	7.99	70.26
2	180	8.08	75.08
3	240	8.31	88.30
4	170	8.04	72.70
5	140	7.88	65.16
6	100	7.62	53.91

Table 4 Slip amount of fault parameters along the Manila Trench

Segment	M_w: 7.0	M_w: 7.5	M_w: 8.0
1	0.10	0.59	3.29
2	0.09	0.52	2.93
3	0.07	0.39	2.19
4	0.10	0.55	3.10
5	0.12	0.67	3.76
6	0.17	0.94	5.27

The next step is to determine the value of slip parameter. From USGS glossary, slip is defined as the relative displacement of formerly adjacent points on opposite sides of a fault measured on the fault surface. In this study, the slip analysis will be determined at three different scenarios of tsunami simulation with different moment magnitudes (M_w: 7.0, 7.5, and 8.0). The amount of slip motion will be calculated using Eqs. (10) and (11).

$$M_o = \mu DLW \quad (10)$$

$$M_w = (2/3)\log M_o - 10.7 \quad (11)$$

where μ: rigidity of earth mantle (3.0×10^{10} N/m)

D: amount of slip motion

L: length of fault plane

W: width of fault plane

M_o: scalar moment of an earthquake (dyn.cm)

M_w: moment magnitude of an earthquake

Table 4 shows the slip amount at each scenario of moment magnitude (M_w: 7.0, 7.5, and 8.0). All finalized fault parameters are included in TUNA-M2 model to perform the tsunami simulation. The fault parameters act as source control input to generate initial vertical displacement using Okada Source Model. TUNA-M2 model used Okada Source Model in order create initial displacement of tsunami wave. Okada model was developed in 1985 by Yoshimitsu Okada. This model is used as a function to calculate analytical solution for surface deformation due to shear and tensile faults in an elastic half-space. This model is widely used to simulate ground deformation produced by tectonic faults (earthquakes).

5 Result and Discussion

Table 5 exhibits results of tsunami wave height from TUNA-M2 simulation. It shows an increment of tsunami wave height as the moment magnitude increases from M_w: 7.0 to M_w: 8.0. Observation points located at Malay basin (MB) record the range of 0.002–0.122 m wave height. Among the six observation points, point E has the highest lead wave height. The range of wave height at Sabah basin (SB) is between 0.004 and 0.168 m, with point G producing the highest wave height, whereas in Sarawak basin (SWB), the range of wave height is between 0.004 and 0.230 m, with point K recording the highest wave height among other points.

Based on the results tabulated in Table 5, the values of wave height versus moment magnitude at observation points E, G, and K are graphically plotted as shown in Fig. 3. Observation point K at Sarawak basin shows the highest value of tsunami wave height. This is followed by point G at Sabah basin and point E at Malay basin.

Figure 4 shows the time arrival of tsunami wave height from the time sequenced pictures. Based on the observation, it can be summarized that tsunami wave height

Table 5 Result of tsunami wave height by different moment magnitude

Point observation	Location	Wave height (m)		
		M_w: 7.0	M_w: 7.5	M_w: 8.0
A	MB	0.003	0.016	0.090
B	MB	0.002	0.013	0.072
C	MB	0.002	0.011	0.062
D	MB	0.002	0.011	0.062
E	MB	0.004	0.022	0.122
F	MB	0.003	0.014	0.079
G	SB	0.005	0.030	0.168
H	SB	0.004	0.025	0.139
I	SB	0.005	0.029	0.161
J	SWB	0.005	0.028	0.157
K	SWB	0.007	0.041	0.230
L	SWB	0.004	0.020	0.112

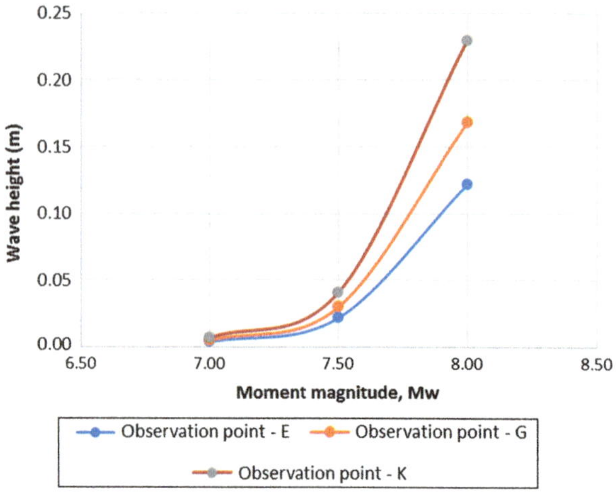

Fig. 3 Value of wave height vs. moment magnitude at observation points E, G, and K

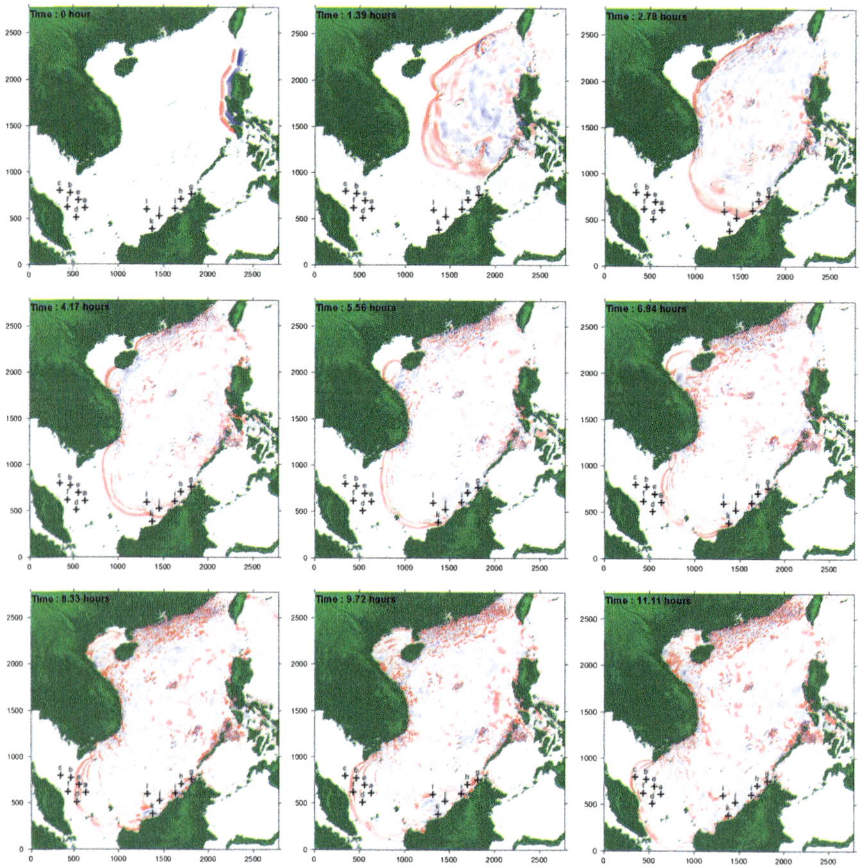

Fig. 4 Time sequence of tsunami wave propagation

from the Manila Trench will move through Sabah basin first, followed by Sarawak basin and Malay basin. The tsunami wave will arrive after 2 h at Sabah waters. It takes only 2 h for the tsunami wave to reach Sabah basin, 3 h to reach Sarawak basin, and 7 h to reach Malay basin.

Based on overall results above, it can be concluded that Sabah and Sarawak waters which are near to the Manila Trench generate higher wave heights as compared to Peninsular Malaysia water which was located much further from the source of earthquake. Therefore, out of the three locations in Malaysia offshore oil blocks, Sabah basin is the most hazardous followed by Sarawak basin and Malay basin.

6 Conclusion

Tsunami simulation has been performed using TUNA-M2 model due to earthquake in the Manila Trench. In this chapter, three simulation results have been presented using different moment magnitudes (M_w: 7.0, 7.5, and 8.0). There are 12 observation points located at three different locations in Malaysian waters: Malay basin, Sabah basin, and Sarawak basin. The tsunami wave height and time of arrival were generated from the simulation. It shows that Sabah basin has higher risk if there is tsunami generated in the Manila Trench. It is followed by Sarawak basin in the second place and Malay basin in the third place.

The tsunami wave height shows an increment in the moment magnitude. It was shown in wave height result (Table 5) on each observation point. The range of tsunami wave height result is as follows:

(a) Malay basin records the range of wave height between 0.002 and 0.122 m.
(b) Sabah basin records the range of wave height between 0.004 and 0.168 m.
(c) Sarawak basin records the range of wave height between 0.004 and 0.230 m.

Further analysis can be conducted to check the impact of tsunami wave at oil and gas offshore platform. It is important to determine whether current platform can resist the force from tsunami wave. This study is also useful for offshore community where oil and gas activities are ongoing in the area of South China Sea.

Acknowledgment The authors wish to express their sincere gratitude to Universiti Tenaga Nasional (UNITEN) and Universiti Teknologi PETRONAS (UTP). Further appreciation is due to Yayasan Universiti Teknologi PETRONAS (YUTP) and PETRONAS Carigali Sdn. Bhd. The authors also would like to thank Dr. Teh Su Yean from Universiti Sains Malaysia for TUNA-M2 model and providing technical support on using TUNA-M2.

References

1. S.Y. Teh, H.L. Koh, Tsunami simulation for capacity development, in *Proceedings of the International Multi Conference of Engineers and Computer Scientists 2011, vol. II, IMECS 2011, 16–18 March 2011*, Hong Kong, 2011

2. K.L. Cham, S.Y. Teh, H.L. Koh, A.I. Md Ismail, Numerical simulation of the Indian Ocean Tsunami by TUNA-M2, in *Proceeding of the 2nd IMT-GT 2006 Regional Conference on Mathematics, Statistic and Application*, Malaysia, 2006
3. A. Ruangrassamee, N. Saelem, Effect of Tsunami generated in the Manila trench on the Gulf of Thailand. J. Asian Earth Sci. **36**(1), 56–66 (2009)
4. K. Megawati, F. Shaw, K. Sieh, Z. Huang, T.-R. Wu, Y. Lin, S.K. Tan, T.-C. Pan, Tsunami hazard from the subduction megathrust of the South China Sea: Part I. Source characterization and the resulting tsunami. J. Asian Earth Sci. **36**(1), 13–20 (2009)
5. T.-R. Wu, H.-C. Huang, Modeling tsunami hazard from Manila Trench to Taiwan. J. Asian Earth Sci. **36**(1), 21–28 (2009)
6. P.L.-F. Liu, X. Wang, Tsunami hazard and early warning system in South China Sea. J. Asian Earth Sci. **36**(1), 2–12 (2009)
7. S.Y. Teh. H.L. Koh, Simulation of tsunami due to the megathrust in South China Sea, in *Proceedings of the 2013 3rd International Conference on Environmental and Computer Science (ICES 2010)*, 18–19 October 2010, Kunming, China, 2010
8. H.L. Koh, S.Y. Teh, P.L.-F. Liu, A.I. Md. Ismail, H.L. Lee, Simulation of Andaman 2004 tsunami for assessing impact on Malaysia. J. Asian Earth Sci. **36**, 74–83 (2009)
9. Intergovernmental Oceanographic Commission (IOC), *Numerical method of tsunami simulation with the Leap Frog Scheme, 1, shallow water theory and its difference scheme* (UNESCO, Paris, 1997), pp. 12–19
10. D.L. Wells, K.J. Coppersmith, New empirical relationships among magnitude, rupture length, rupture width, rupture area and surface displacement. Bull. Seismol. Soc. Am. **84**(4), 974–1002 (1994)
11. H. Tatehata, *The New Tsunami Warning System of the Japan Meteorological Agency. Perspectives on Tsunami Hazard Reduction* (Springer, Netherlands, 1997), pp. 175–188
12. B.C. Papazachos, E.M. Scordilis, D.G. Panagiotopoulos, C.B. Papazachos, G.F. Karakaisis, global relation between seismic fault parameters and moment magnitude of earthquake, in *Proceeding of the 10th International Congress, Thessaloniki*, 2004

Sustainable Trend Analysis of Annual Divisional Rainfall in Bangladesh

Muhammad Hassan Bin Afzal and Moniruzzaman Bhuiyan

Abstract The economy of Bangladesh predominantly depends on agriculture. As a result of that, agriculture is the particular prevalent earning sector of Bangladeshi economy since it competitively provides about 30 % of the national GDP. On top of that, more than 50 % of local labor forces are directly involved in agro-based work as the main source of earning. Despite being a riverine country blessed with so many rivers, the Bangladeshi farmers still largely dependent on rainfall. This chapter particularly analyzes the annual rainfall amount in different districts of Bangladesh based on last 10 years data. It also focuses on the month wise rainfall trends in seven different divisions such as Dhaka, Chittagong, Rajshahi, Khulna, Rangpur, Sylhet, and Barisal. This can be immensely useful information, because it can provide accurate informative suggestions to local farmers about their projected agricultural activities. Finally, this chapter also discusses about four-step cost-effective rainfall information sharing method which can be immensely helpful for the local farmers and can be hugely helpful for Bangladesh's agriculture.

Keywords Annual rainfall • Rainfall trend analysis • Divisional rainfall measurement • Agriculture • Irrigation management • Rainfall data distribution

1 Introduction

The unique geographical location of Bangladesh makes her more flood prone, as the Bay of Bengal is in south, the Himalaya is in north portion. The mean annual rainfall is as high as 2,320–6,000 mm. Due to lack of proper drainage system, heavy rainfall water clogged in one place which makes human life miserable and also other livelihood. Continuous 8 h 341 mm rainfall caused severe flooding in Dhaka on September 2004 and also in 28 July, 2009 due to 6 h 290 mm rainfall. In Chittagong 408 mm rainfall caused landslide and killed 124 people on 11 June, 2007 [1]. Therefore, every year proper rainfall analyses are being undertaken for precautionary measurements before heavy rainfall, improvement of drainage system, rainfall water

M.H.B. Afzal (✉) • M. Bhuiyan
Institute of Information Technology (IIT), University of Dhaka (DU), Dhaka, Bangladesh
e-mail: hasssans@ieee.org

S.H. Abu Bakar et al. (eds.), *ISFRAM 2014*, DOI 10.1007/978-981-287-365-1_21

Fig. 1 Six different divisions in Bangladesh

capture for irrigation in dry season, crop harvesting before rain, etc. Different types of rain gauge are used for rainfall analysis and satellite information is another reliable source for rainfall analysis. In Bangladesh climate change data are collected by Bangladesh Meteorological Department (BMD) operated in 34 stations [2].

The economy of Bangladesh predominantly depends on agriculture. As a result of that, agriculture is the particular prevalent earning sector of Bangladeshi economy since it competitively provides about 30 % of the national GDP. On top of that, more than 50 % of local labor forces are directly involved in agro-based work as the main source of earning. Despite being a riverine country blessed with so many rivers, still the Bangladeshi farmers largely reliant on rainfall. This chapter particularly analyzes the annual rainfall amount in different districts of Bangladesh based on last 10 years data. It also focuses on the month wise rainfall trends in six different divisions such as Dhaka, Chittagong, Khulna, Rangpur, Sylhet, and Barisal. This can be immensely useful information, because it can provide accurate informative suggestions to local farmers about their projected agricultural activities. Finally, this chapter also discusses about a sustainable cost-effective rainfall measurement method which can be immensely helpful for the local farmers and can be hugely helpful for Bangladesh's agriculture.

In the following section, rainfall analysis of past 10 years (2002–2011) is accomplished. Rainfall data are gathered from BMD. Here, rainfall data are analyzed division wise and countrywide with respect to predict that how much the rainfall deviates from mean in a particular district with the increase of years and determine the probability of uncertainties in terms of depth of rainfall. Probability of uncertainties in terms of rainfall duration is evaluated by deviation of predicted duration from actual duration [3]. The seven different divisions of Bangladesh are reflected in Fig. 1 [4, 5].

2 Statistical Methods for Rainfall Data Analysis

Analyzing statistical methods for rainfall data analysis of countrywide are mean, standard deviation, and coefficient of variation. In Bangladesh information on data quality is very poor. Due to improper data management system and old rain gauge, lots of data are being missing and this would affect the rainfall trend analysis. Different types of rain gauges are available for rainfall measurement, but they have some drawbacks such as cannot work under harsh environment, size of bucket, improper measurement of heavy rainfall due to speed of wind, clogging, duration of measurement, cannot sense low rate rainfall, etc. [6, 7].

Optical fiber rain sensor is more appropriate for effective annual rainfall measurement as it has many unique characteristics such as immune to harsh environments, small size, light weight, highly sensitive, and multiplexing capabilities. In countries like Bangladesh, with heavy rainfall, long-term flood, remote areas where communication is difficult and rain gauge is not available to collect data, many remote islands where electricity facility is hardly available, water clogged forests, and remote hilly tracks, fiber optic sensors are reliable enough to be used for data measurement. Along with fiber optic sensing technology and effective project management, it is possible to establish a dedicated information system to provide accurate forecasting and rainfall information to the local farmers for more planned farming [8–11].

Based on yearly weather condition in Bangladesh, rainfall data are generally categorized according to four different seasons, such as

- Pre-monsoon (March–May): before rainy season.
- Monsoon (June–September): during rainy season.
- Post-monsoon (October–November): after rainy season.
- Winter (December–February).

Table 1 Countrywide annual rainfall trend analysis

Year	Mean (mm/year)	Standard deviation (mm/year)	Coefficient of variation
2002	2,541.4	616.3	0.242
2003	2,077.7	739.5	0.355
2004	2,711.3	740.5	0.273
2005	2,494.0	714.8	0.286
2006	2,166.2	821.2	0.379
2007	2,789.1	905.4	0.324
2008	2,384.9	940.1	0.394
2009	2,200.7	759.3	0.345
2010	2,082.6	901.4	0.432
2011	2,477.5	966.7	0.39

3 Countrywide Annual Rainfall Trend Analysis

In this section, past 10 years (2002–2011), countrywide rainfall data are analyzed by mean, standard deviation, and coefficient of variation. After extensive analysis, it can be observed that standard deviation (S.D.) value largely deviate from mean value. All mean values are above 2,000 mm and S.D. values are below 1,000 mm. Large deviation of S.D. from mean value indicates that intensity of rain each year is not certain, sometime heavy rainfall and sometime very low rate.

From Table 1, it can be observed that standard deviation of rainfall is increasing with the increase of years but in winter it is different, as there is hardly any rainfall in this season. Other interpretation of this result is that rainfall intensity is not approximately same in different places of Bangladesh [8, 12]. After analyzing the last 10 years annual rainfall in Bangladesh, it can be seen that the highest rainfall average is 2,789.1 mm in 2007 and the second highest rainfall is 2,711.3 mm in 2004. After 2007, the rainfall gradually decreases and then in 2011 it increases again, but the difference between S.D. and mean decreases, which indicate the certainty of rainfall. Lowest rainfall is 2,077.7 mm in 2003.

4 Annual Divisional Rainfall Analysis

4.1 Dhaka Division

Rainfall in Dhaka fluctuates (Fig. 2) with respect to first few years 2002–2006, and then it increases about 2,405.2 mm in 2007, the highest rainfall among 10 years. From 2007, rainfall decreased gradually and then slightly increased in 2011 up to 1,776.8 mm. Here, both Figs. 2 and 3 represent the annual rainfall trend statistical format such as mean annual rainfall and standard deviation annual rainfall.

The values of standard deviation largely deviate (Fig. 3) from mean in 2004 and 2005 which indicate uncertainty of rainfall. It can be noticed that since 2004

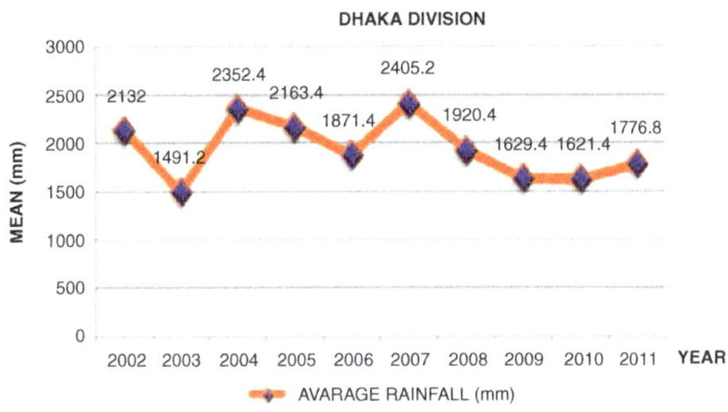

Fig. 2 Average (mean) rainfall of Dhaka division

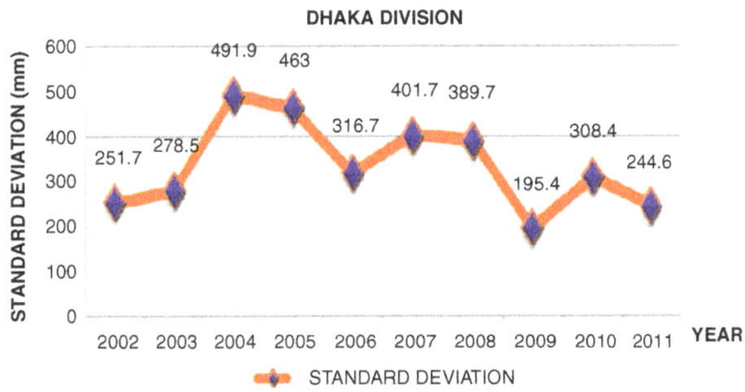

Fig. 3 Standard deviation (S.D.) variation of rainfall around 10 years in Dhaka division

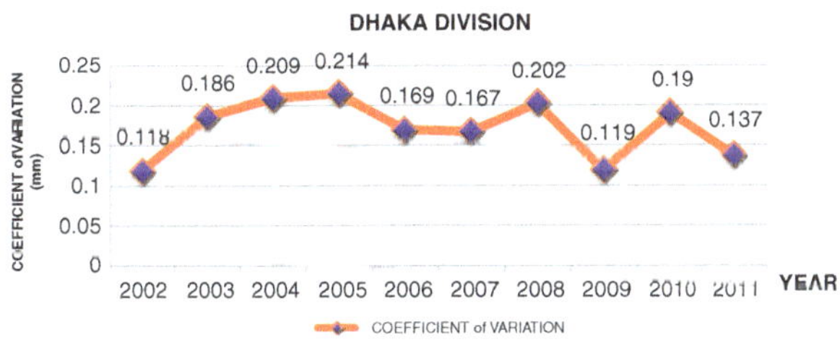

Fig. 4 Coefficient of variation of rainfall for last 10 years in Dhaka division

standard deviation decreases gradually and the rainfall becomes more predictable. Coefficient of variation in Fig. 4 gradually moves in upward direction till 2005, which indicates rainfall gradually deviates from mean rainfall and unpredictable rainfall. From 2005 to 2011, the trend is downward, which means that deviation of rainfall from mean value is low as the year advances. The rainfall occurrence is close to mean value and becomes more certain and predictable.

4.2 Chittagong Division

Most of the places of Chittagong division are hill tracks and forests, so the rainfall intensity is high, which is around 3,000 mm and above (Fig. 5). Fluctuation is also observed in mean values of rainfall. From 2002 to 2011, the rainfall means increase

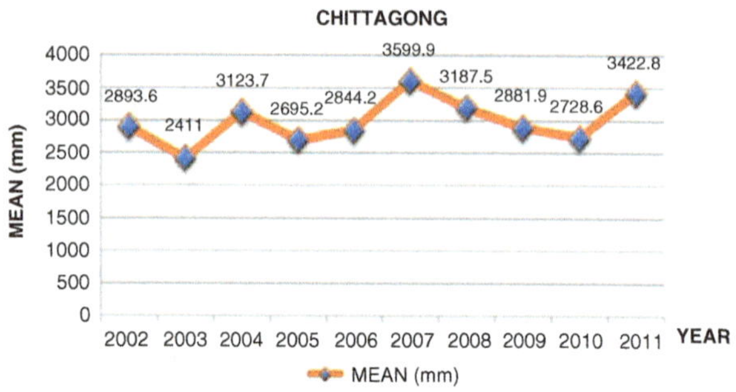

Fig. 5 Average (mean) rainfall of Chittagong division

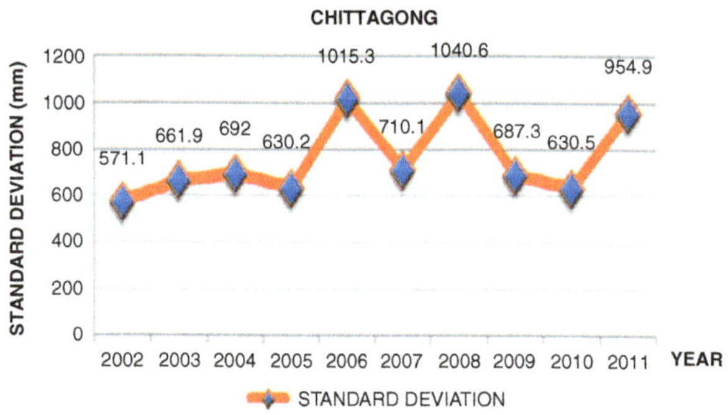

Fig. 6 Standard deviation (S.D.) variation of rainfall around 10 years in Chittagong

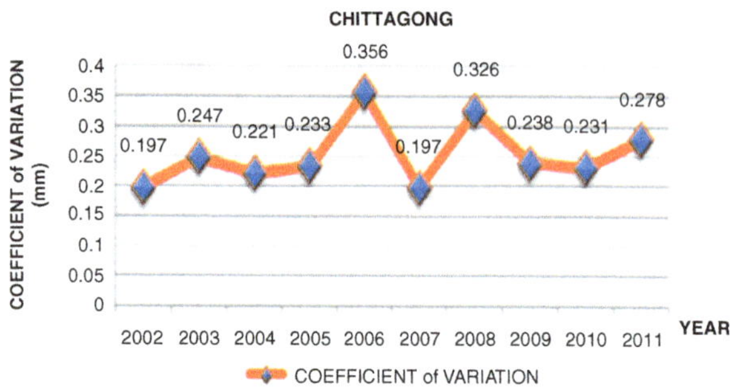

Fig. 7 Coefficient of variation of rainfall in 10 years in Chittagong

gradually. High deviation (Fig. 6) from mean values of 2006 and 2008 then in 2011 which is 1,015.3 mm, 1,040.6 mm, and 954.9 mm, respectively; these deviations indicate the uncertainties of rainfall. Upward trend is observed in standard deviation and mean values, which indicates that there is no consistency in rainfall in each year.

The coefficient of variation is lower (Fig. 7) from 2002 to 2005, which indicates lower deviation from the mean value of rainfall and predictability of rainfall, but in 2006 and 2008 the C.V. value is highest and uncertainty of rainfall is due to large deviation from mean value. From 2009, C.V. decreases and upward trend is observed from 2011 again, indicating uncertainty in rainfall predictability.

4.3 Rangpur Division

The average rainfall shows downward trend (Fig. 8) and the deviations are also low, which indicates the predictability of rainfall trend in Rangpur division. In 2006, a sharp deviation from mean value is observed and high coefficient of variation also indicates the uncertainty in rainfall. Average rainfall starts to decrease from 2006 and gradually increases from 2007 to 2011. Low deviation values indicate the predictability of rainfall. Lower the C.V. value lower the deviation from the mean or average rainfall.

4.4 Khulna Division

Figure 9 represents the average rainfall in Khulna division in past 10 years. The highest rainfall in 2002 is about 2,394 mm; then the trend of average rainfall goes downward till 2010 and then upward in 2011, which is about 1,859.8 mm. From this line graph, it is predictable that from last 10 years rainfall intensity decreases in Khulna division except the year 2011.

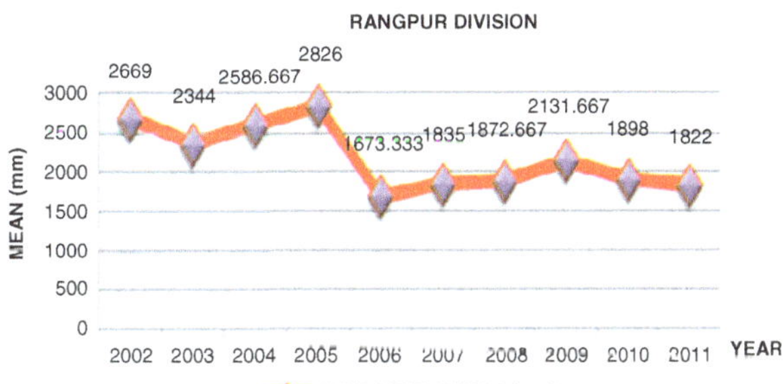

Fig. 8 Average (mean) rainfall of Rangpur division

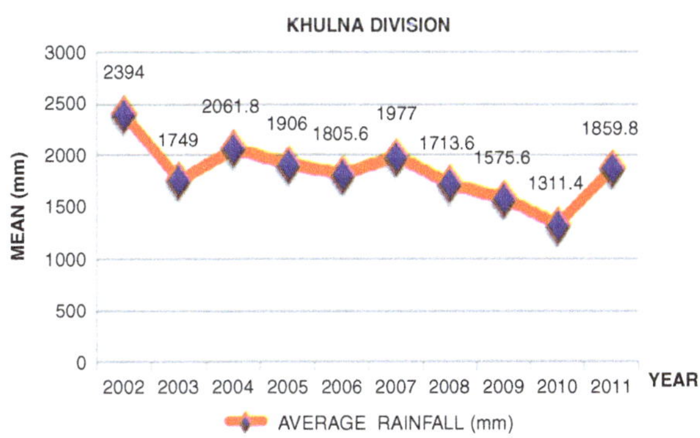

Fig. 9 Average rainfall (mm) in 10 years in Khulna division

4.5 Barisal Division

Rainfall average in Barisal district is approximately same, except small fluctuations, such as in 2003 the lowest and 2,962.75 mm the highest rainfall in Barisal in last 10 years. From 2007, the rainfall slightly decreased and then increased in 2011, but the mean value does not go below 2,000 mm. These mean values indicate the predictability of rainfall of upcoming years (Fig. 10).

4.6 Sylhet Division

The rainfall scenario in Sylhet division is different from other divisions, because the average rainfall decreases as the year increases. Interestingly, the rainfall intensity is higher than other divisions (Fig. 11); the range is about 4,000–3,500 mm. From the year 2002, the rainfall gradually increased up to 4,202.5 mm, the highest rainfall, and then sharply decreased to 2,746.5 mm in 2006. Fluctuations of mean value of rainfall started from 2006, and in 2011 the lowest rainfall in last 10 years is about 2,575.5 mm. According to this graph, the predictability of rainfall in Sylhet division is hardly possible.

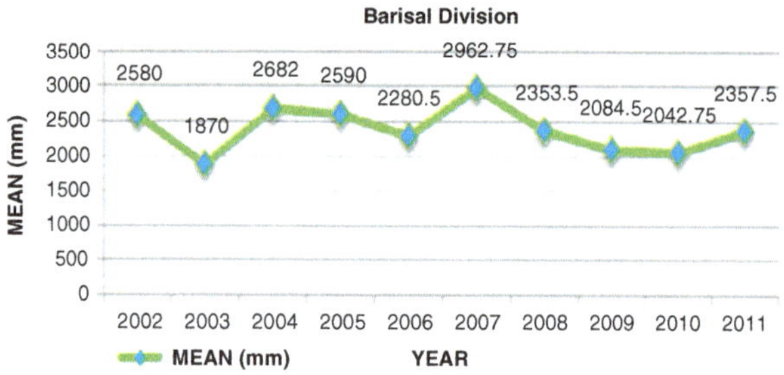

Fig. 10 Average rainfall (mm) in 10 years in Barisal division

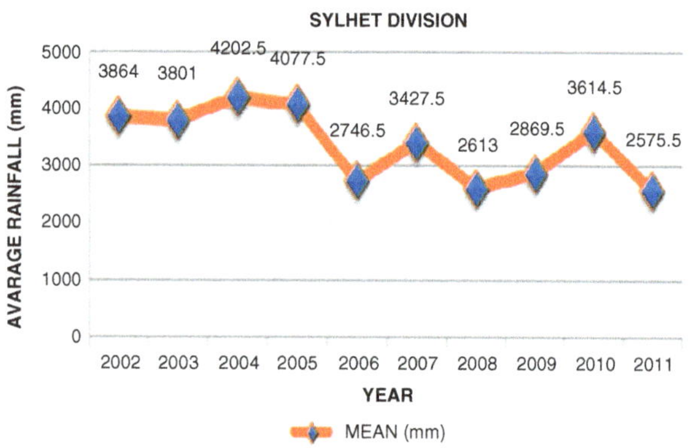

Fig. 11 Average rainfall (mm) in 10 years in Sylhet division

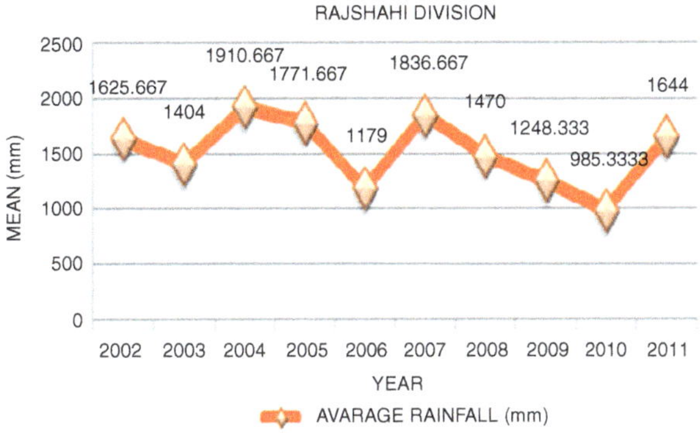

Fig. 12 Average rainfall (mm) in 10 years in Rajshahi division

4.7 Rajshahi Division

After analyzing the accumulated annual rainfall data for last 10 years for Rajshahi Division, highly fluctuated pattern of rainfall can be observed in this region. Rajshahi has the warmer weather than any other divisions in Bangladesh and rainfall intensity also lower. It can be documented from Fig. 12 that the highest rainfall is 1,910.6 mm in 2004, which is the lowest rainfall in other divisions, and the lowest rainfall is 985.3 mm in 2010. A sharp upward trend is observed in 2011 which is about 1,644 mm. Rainfall predictability is possible for Rajshahi division.

5 Rainfall Data Analysis and Findings

After analyzing the overall countrywide annual rainfall as well as divisional rainfall trend analysis, it can be clearly observed that Sylhet division is heavy rain area and Rajshahi is the lowest rainfall division and moderate rainfall division is Dhaka. Second and third heavy rainfall divisions are Chittagong and Barisal, respectively. Sylhet, Chittagong, and Barisal are forest rich divisions, and hill tracks are also observed in Sylhet and Chittagong. On the other hand, Rajshahi is most warm among other divisions and less rainfall is observed. From previous 10 years rainfall trend analysis in various divisions of Bangladesh, it has been found out that Rajshahi was the division with minimum rainfall, Dhaka with moderate rainfall, and Sylhet with heavy rainfall intensity compared to other divisions. Initially three divisions of Bangladesh were chosen into three categories such as high annual rainfall, medium annual rainfall, and low annual rainfall. Figure 13 represents the average highest and lowest annual rainfall in Bangladesh in last 10 (2001–2011) years.

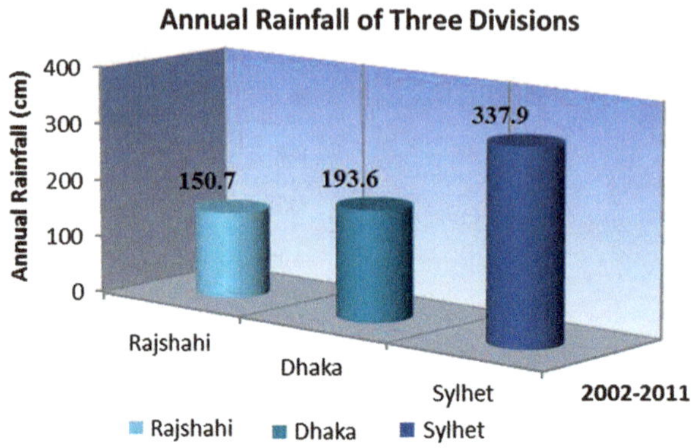

Fig. 13 Comparative total annual rainfall in three selected divisions of Bangladesh (2002–2011)

6 Future Works

The future steps are to create sustainable ways to disseminate this statistical analysis data in a personalized form among the rural people, who are primarily focusing on farming for a source of living in Bangladesh. It is one of the future scopes of this study to acquire and integrate more rainfall data from previous years to obtain more accuracy in rainfall trend analysis. With the help of appropriate planning and effective project management, there are four ways to promote and broadcast this information among the people to create awareness and provide live-in support in irrigation planning and cultivation procedure [8].

6.1 Basic SMS Service

The most basic service is to provide personalized SMS (Short Message Service) to the general people's mobile phone. It will integrate the network service provider's service to detect the user's location and provide the user personalized SMS explaining the rainfall trend in that specific area and also advice and/or alert them about upcoming rainfall prediction and also suggestion about irrigation planning for the farming. Although the ultimate goal is to distribute this information for free, still it might charge a nominal amount for providing the service.

6.2 Dedicated Website Development

The second step is to distribute these graphical chart and rainfall prediction analysis among the general people via developing, building, and promoting a dedicated website, whose sole purpose is to create awareness and provide experts' advice on rainfall trend in various areas of Bangladesh. It also provides personalized information on previous year's rainfall in a specific location via various charts and diagrams; user can interact online with the expert to seek advice and support for various rainfall-related issues. There are various platforms to promote and build a website such as Drupal, which is basically an open-source content management system to develop and design website with various contents that suits user's needs and requirements [8–10]. In advanced stage of website development, it is planned to include e-commerce module in website so that user can get advanced level of support by providing a nominal fee. This will dynamically help both users and developers and expert support providers.

6.3 Social Media Support

The third step is to create awareness and provide support to the local people of Bangladesh by designing and building a social media platform support station in Facebook and Twitter. The primary objective is to build a dynamic engaging Facebook page, which lets the users to read and study the rainfall trend analysis in Bangladesh and also real-time social interaction with other users in Bangladesh to share their views, thoughts, and problems and to obtain effective and sustainable solutions to their problems.

6.4 Mobile App Development

The final step is to develop a focused and personalized mobile application in android platform to broadcast this data, chart, prediction, and support service among the general people to create awareness and provide them the required service. In order to benefit from this service, user requires owning a smartphone which supports mobile apps and also certain level of knowledge to navigate a mobile app. A sample wireframe design is illustrated below in Fig. 14. This wireframe design for mobile app was created by using open-source wireframe software.

All these services are the future steps of this rainfall trend analysis project to create appropriate awareness among general people to make them realize and understand the proper utilization of rainfall and also take appropriate precaution to fight with excessive rainfall and also to support them in irrigation management and their farming works (Table 2).

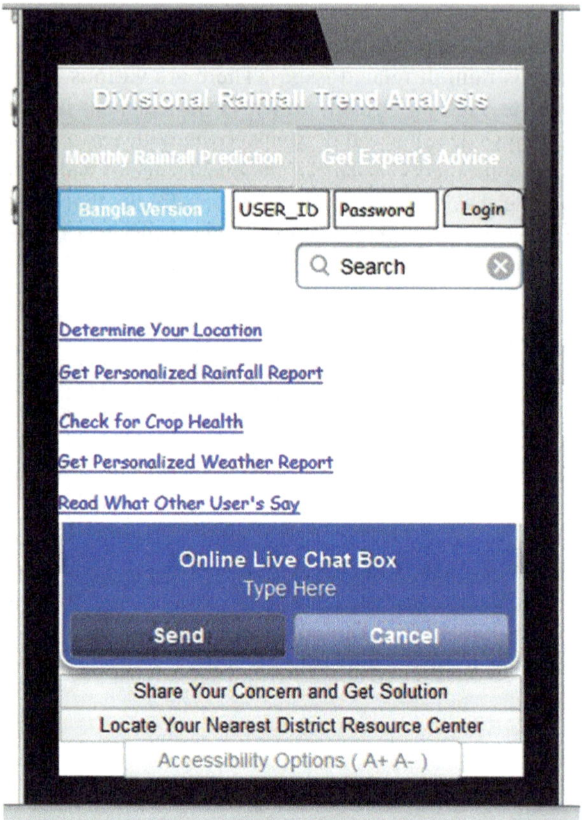

Fig. 14 This wireframe designed for mobile app

Table 2 Comparison of various services

Types of service	Scopes and requirements		
	Required platform (for developers)	User requirements	Intended support
Basic SMS service	Effective communication with mobile network providers to broadcast the required information (Bangla/English)	User requires having a basic mobile phone, which can send and receive SMS	To provide the user the required rainfall trend analysis information in return SMS
Website development	Using free web-development content management system (CMS) platform	User requires having internet access and device to access the website	More advance level of support for users to study, analysis the previous year's data, chart and diagram and also to gain the online expert's support

(continued)

Table 2 (continued)

Types of service	Scopes and requirements		
	Required platform (for developers)	User requirements	Intended support
Social media platform	Currently both Facebook and Twitter are trending in Bangladesh. The future plan is to develop a dedicated Facebook and Twitter page	User requires having internet access and also account in Twitter and Facebook (in order to interact with other users)	To provide dynamic interactive support, to communicate with other users and also share mutual concerns and to seek personalized expert's support
Mobile app development	To build and promote android-based mobile app, which specifically supports Ice Cream Sandwich (4.0–4.0.4) and later	User requires having android-based mobile phone which supports at-least Android version 4.0 and later in order to successfully navigate and execute the app	Total mobility support, dynamic and interactive chatting support, and dedicated web accessibility support (available both in Bangla and English)

7 Conclusion

The complete and exhaustive rainfall trend analysis for last 10 years (2001–2011) in Bangladesh has been carried out in the above sections to capture the inclusive rainfall measurement in this country [5, 6]. Annual rainfall data for few districts could not be able to extract because of remoteness, unavailability of effective rainfall measurement technology, and other relevant problematic scenarios. The future work is to overcome these difficulties so that the annual rainfall in overall Bangladesh can be effectively measured. These data sets are immensely useful for Bangladesh and can be implemented to prevent any future natural calamities caused by excessive rainfall. As well as, the local farmers can gain an accurate perspective in overall rainfall forecast which will help them to plan the harvest cycle and irrigation management. The goal is to omit these problems so that every part of Bangladesh can accumulate effective data collection of rainfall periodically to support and help the local residents as well as to improve the overall condition by providing the four aforementioned services effectively.

Finally, four types of information distribution services such as basic SMS service, Website Development, Social Media Platform, and Mobile App Development have been proposed, and a thorough analysis of each service and intended support systems has been discussed. The future goal and objective is to implement these services sequentially so that the local people of Bangladesh can be benefitted from these data analysis and interactive support system.

References

1. C. Zhao, L. Ye, J. Ge, J. Zou, X. Yu, Novel light-leaking optical fiber liquid-level sensor for aircraft fuel gauging. Opt. Eng. **52**(1), 014402 (2013)
2. A.F. Obaton, G. Laffont, C. Wang, A. Allard, P. Ferdinan, Tilted fibre Bragg gratings and phase sensitive-optical low coherence interferometry for refractometry and liquid level sensing. Sensor. Actuator. A **189**, 451–458 (2013). 2012 Elsevier B.V
3. S.B. Murshed, A.K.M. Saiful Islam, M. Shah Alam Khan, Impact of climate change on rainfall intensity in Bangladesh, in *3rd International Conference on Water and Flood Management, ICWFM*, 2011
4. A.U. Ahmed, S.R. Hassan, B. Etzold, S. Neelormi, "Where the Rain Falls" project. Case study: Bangladesh. Results from Kurigram District, Rangpur Division. Report No. 2. United Nations University, Institute for Environment and Human Security (UNU-EHS), Bonn, 2012
5. M.H.B. Afzal, Introduction to fibre-optic sensing system and practical applications in water quality management, in *2013 Fourth International Conference on Computing, Communications and Networking Technologies (ICCCNT)*, 4–6 July 2013, pp. 1–9
6. M.H.B. Afzal, Addition of polydimethylsiloxane in FOS technology for sustainable sensing, in *2013 Fourth International Conference on Computing, Communications and Networking Technologies (ICCCNT)*, 4–6 July 2013, pp. 1–6
7. X.D. Zhang, G.H. Huang, X.H. Nie, Optimal decision schemes for agricultural water quality management planning with imprecise objective. Agric. Water Manag. **96**, 1723–1731 (2009)
8. M.H.B. Afzal, Large scale IT projects: study and analysis of failures and winning factors. IETE Tech. Rev. **31**(3), 214–219 (2014)
9. A.F.M. Afzal Hossain, Management of irrigation and drainage systems using mathematical modelling. J. Appl. Water Eng. Res. **1**(2) (2013)
10. M.H.B. Afzal et al., Fiber optic sensor-based concrete structural health monitoring, in *2011 Saudi International Electronics, Communications and Photonics Conference (SIECPC)*, 24–26 Apr 2011, pp. 1, 5. doi:10.1109/SIECPC.2011.5876903
11. M.H.B. Afzal et al., An in-depth review: structural health monitoring using fiber optic sensor. IETE Tech. Rev. **29**(2), 105–113 (2012)
12. M. Styczen, R.N. Poulsen, A.K. Falk, G.H. Jørgensen, Management model for decision support when applying low quality water in irrigation. Agric. Water Manag. **98**, 472–481 (2010)

Flood Frequency Analysis Due to Climate Change Condition at the Upper Klang River Basin

Raksmey May, Zurina Zainol, and Muhamad Fuad Bin Shukor

Abstract The effects of flood in many countries in Asia and around the world are so tremendous, especially on economies and human lives. The climate change due to the global warming phenomena is projected to have serious impacts on the hydrological system such as severe droughts and floods. The mitigation measures to save lives and properties from the floods are related to the projection of the flood magnitude and return period of interest which is called design flood. The design flood is very important for construction and development plans along the river and flood plain. In this chapter, the study attempts to perform flood frequency analysis through the Log-Pearson type III distribution method for the historical and projected river discharges in the Upper Klang River Basin. The results show the flood frequency distribution estimated from the simulation results of the climate change data is lower than the distribution projected by the available historical recorded data of river discharges. The discharges from the projected results for the recurrence intervals of 100, 25, and 2 years are approximately 350, 182, and 61 m^3/s, respectively. The outcomes of the study provide the useful information associated with the flood magnitudes and recurrence periods for the sites along the river.

Keywords Upper Klang River Basin • Flood frequency analysis • Log-Pearson type III • Tank model • Climate change

1 Introduction

Many countries in Asia and around the world face one of the worst natural catastrophes, flood, which damages millions of dollars of properties and causes the loss of human lives. The cost of destruction from flood in Malaysia, for instance, reaches at one billion Malaysian Ringgit every year [1]. The recent floods in 2013 and 2014 caused the widespread human sufferings to many rural and urban

R. May (✉) • Z. Zainol • M.F.B. Shukor
Water Resources and Environmental Systems (WRES), Faculty of Civil Engineering, Universiti Teknologi MARA, 40450 Shah Alam, Selangor, Malaysia
e-mail: may_raksmey@salam.uitm.edu.my

S.H. Abu Bakar et al. (eds.), *ISFRAM 2014*, DOI 10.1007/978-981-287-365-1_22

residences in the country. The major causes of floods in the country are the unpredictable heavy downpour and high volume of surface runoff due to the rapid development within the catchment area and the deterioration of the river morphology [2]. Malaysia is predicted to confront with the serious difficulties associated with floods and drought when the temperature and precipitation change from 0.7 to 2.6 °C and −30 % to 30 %, respectively [3]. From the beginning of engineering hydrology [4], flood modeling has attracted more attention from the international research communities because of the serious impacts of floods in many countries. The climate change from the global warming has been ascribed to contribute to the recent devastating flood [5]. The extreme rainfall and a shift in the timing of the rainy season are predicted. Climate change is estimated to result in longer season [6–8].

Flood modeling is generally carried out to predict the floods that have the relationship with the return periods of interest. The estimation is called design flood which is so essential for various purposes such as design of hydraulic structures, management and development of the flood plain, flood mitigation and control plans, and flood insurance studies. Therefore, the design flood is very important to bring down the flood damage and save lives.

The flood frequency analysis is used to estimate the design floods, i.e., the return period related to a given flood magnitude, for sites along a river. To establish the flood frequency analysis, it needs longer time series data of historical peak flow discharge. The streamflow recorded at many gauged streams is often less than the return period of interest; therefore, the interpolation methods are usually applied to estimate the design flood. A probability distribution, one of the fundamental applications, is selected to establish the flood frequency analysis. Several statistical parameters such as mean values, standard deviations, skew coefficients, and recurrence intervals are determined to construct the frequency distributions consisting of graphs and tables that describe the different discharges associated with recurrence interval or exceedence probability. The forms of the flood frequency distributions are different based on the equations used to conduct the statistical analyses. The most common forms are the normal distribution, log-normal distribution, Gumbel distribution, and Log-Pearson type III distribution. The Log-Pearson type III distribution is the recommended method for flood frequency analysis [9].

To understand the hydrological system, it is so essential to predict the future streamflow, especially when the climate change data, e.g., rainfall distribution, surface temperature, or soil moisture, are available. The hydrological system comprises of groundwater flow, rainfall, evapotranspiration, and surface runoff. To evaluate each component of the system, it is hard without the comprehensive hydrological information and numerical simulation with fine grid cells [10]. Tank model is introduced in the simulation because it is easy to understand and flexible for land use and climate changes of the model input parameters [11].

In the present chapter, the flood frequency analysis for streamflow from 1976 to 2012 of the Upper Klang River Basin is estimated by using the Log-Pearson type III. Based on the climate change data of rainfall available from 2046 to 2065, the tank model is used to predict the future surface runoff contributing to the peak discharge of the river. The previous statistical technique to establish the flood frequency analysis is proposed for the simulation results of streamflow after calibration process is achieved. The estimated flood frequency distributions from historical and future data are compared to understand the changes of flood magnitude for any return period of interest.

2 Study Area

Klang River Basin is one of the four biggest basins in Selangor. The total catchment area of the basin is about 1,300 km^2, and the length of the longest river is about 102 km [12]. It originally flows from the Ulu Gombak reserved forest to Port Klang in the west of the country. Klang River Basin is the most built-up area compared to other basins in Selangor. Half of the basin area is currently developed and turned into industrial, commercial, residential, and institutional zones [12]. In the current study, the Upper Klang River Basin is selected for the flood frequency analysis because there are two dams and towns within the chosen area (Fig. 1). The total area of the selected site is approximately 466 km^2 with the longest stream around 30 km. Klang Gate Dam and Batu Dam are built for the water treatment plants downstream. The spatial extension of the reservoir for Batu Dam is around 2.5 and 2.4 km^2 for Klang Gate Dam. The topographical condition of the study area is in flat condition in the south and hilly and mountainous condition in the north (Fig. 2). The highest ground elevation is approximately 1,385 m above sea level (m a.s.l.). The lowest land is around 31 m a.s.l. and the average is about 245 m a.s.l.

3 Methods

3.1 Flood Frequency Analysis

The Log-Pearson type III distribution is used for the estimation of the flood frequency analysis in the present study. The benefit of the method is to extrapolate the design floods of the return period of interest beyond the available historical flood events. The general equation applied for the Log-Pearson type III distribution is as follows [13].

$$\log x = \overline{\log x} + K\sigma_{\log x} \tag{1}$$

where x is the flood discharge value of some specified probability, $\overline{\log x}$ is the average of the log x discharge values, K is a frequency factor that depends on the skew coefficient and return period, and σ is the standard deviation of the log

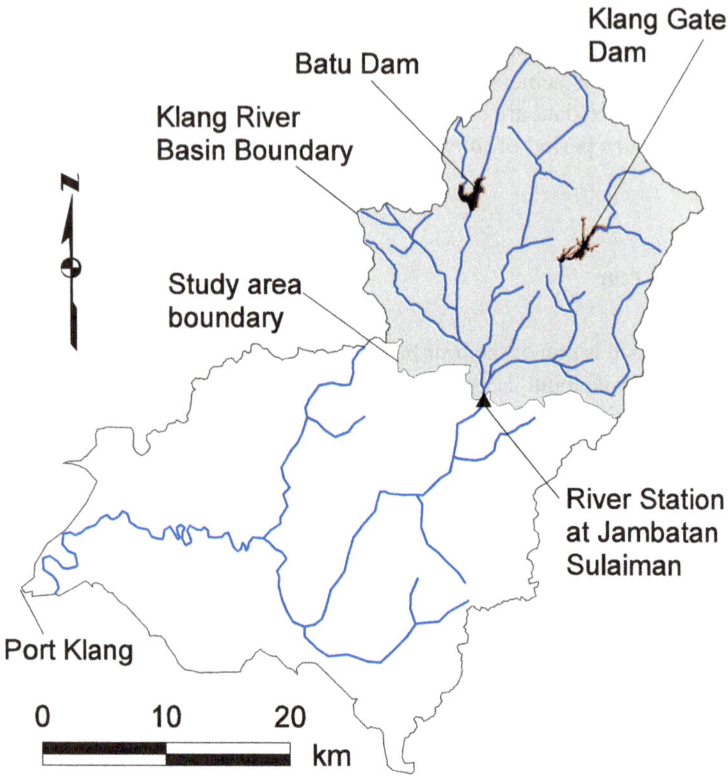

Fig. 1 Klang River Basin boundary and study area

x values. Equations (2), (3), and (4) are used to calculate the mean, standard deviation, and skew coefficient (C_s) of the data, respectively.

$$\overline{\log x} = \frac{\sum (\log x_i)}{n} \tag{2}$$

$$\sigma_{\log x} = \sqrt{\frac{\sum \left(\log x - \overline{\log x}\right)^2}{n-1}} \tag{3}$$

$$C_s = \frac{n \sum \left(\log x - \overline{\log x}\right)^3}{(n-1)(n-2)\left(\sigma_{\log x}\right)^3} \tag{4}$$

where n is the number of entries.

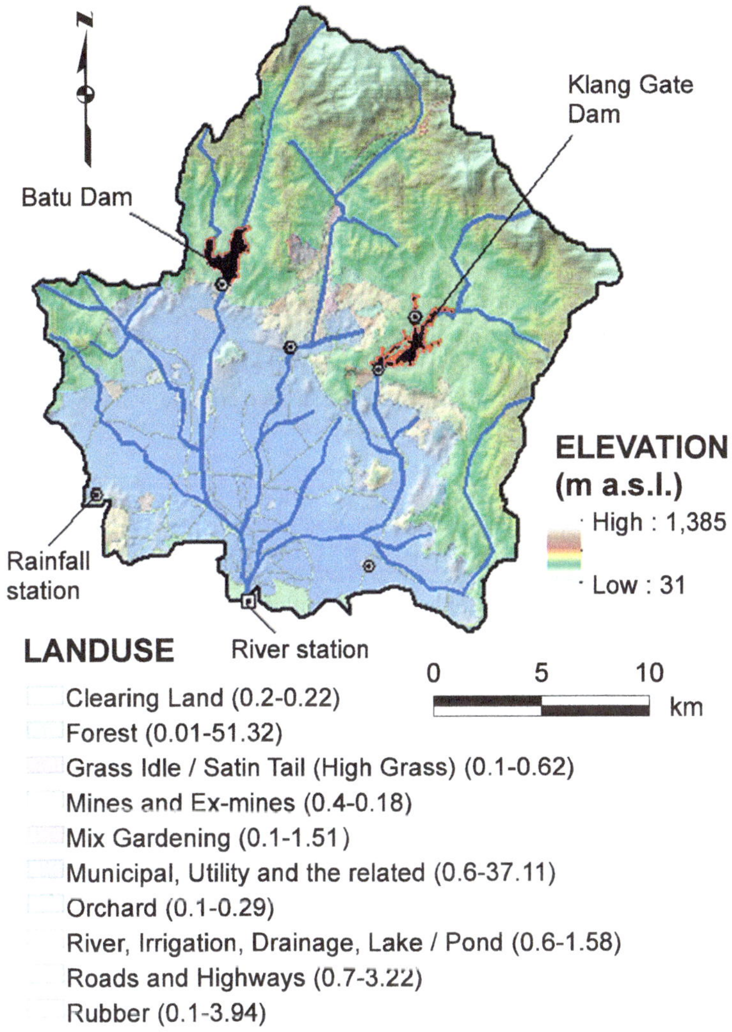

Fig. 2 Map of topography, river and rainfall stations, and land use types (the numbers in *brackets* represent the values of $F_{i\infty}$ and percentage of coverage area, respectively)

3.2 Tank Model

The conceptual hydrological model illustrated by [11] represents the hydrological system consisting of the rainfall, surface runoff, infiltration, and actual evapotranspiration. The component of surface runoff is estimated as the function of the rainfall and surface runoff coefficient, which depends also on the rainfall intensity. The flow rate, depth, and coefficient of surface runoff are given in (5), (6), and (7), respectively.

$$q_{is} = \frac{h_{is}A_i}{86,400} \tag{5}$$

$$h_{is} = r(t) \times F_i(r) \tag{6}$$

$$F_i(r) = \frac{r(t)}{r(t) + (r)_{1/2}} \times F_{i\infty} \tag{7}$$

where q_{is} is the flow rate of surface runoff (m^3/s), A_i is the area of the grid cell (m^2), h_{is} is the depth of surface runoff (mm) at the location i i.e., grid mesh of the specific dimension 50 m × 50 m, $r(t)$ is the rainfall intensity (mm), and $F_i(r)$ is the surface runoff coefficient at the location i. $(r)_{1/2}$ is the value of $r(t)$ when $F_i(r)$ is equal to $F_{i\infty}/2$. The value of $(r)_{1/2}$ is determined when the value of $F_{i\infty}$ is available in Table 1 of [11] based on the types of ground surface conditions.

3.3 *Input Data*

In the current study, the rainfall and streamflow data are collected from the Department of Irrigation and Drainage Malaysia (DID). The selected rainfall data are based on the spatial distribution of the stations within the study area. Figure 2 shows the locations of rainfall stations and streamflow station at Jambatan Sulaiman. Thiessen polygon method is used to average the rainfall distribution. The datasets of the rainfall and river discharge prepared for the model simulation and calibration from 1976 to 2012, respectively are presented in Fig. 3a. From the historical rainfall, the maximum daily rate is 179.5 mm in January 2008, and the average daily rainfall is around 6.91 mm. The maximum daily streamflow in the river is around 540.38 m^3/s in August 2010. The average river discharge rate is approximately 22.58 m^3/s.

The projected rainfall data that are used in this research are from the Global Climate Model (GCM) derived from data sets of the Data Integration and Analysis System (DIAS) from the University of Tokyo for the period 2046–2065 [14]. The projected daily rainfall data is presented in Fig. 3b. The projected maximum daily rainfall is around 151.67 mm and the average is about 7.09 mm.

The values of $F_i(r)$ relied on the values of $F_{i\infty}$ and rainfall rates. The $F_{i\infty}$ numbers are determined based on the trial-and-error processes during the simulation. The satisfied values are presented in Fig. 2. The highest $F_{i\infty}$ is around 0.7, which is the "Roads and Highways" land use type. The lowest $F_{i\infty}$ is approximately 0.01, which is the "Forest" land use cover.

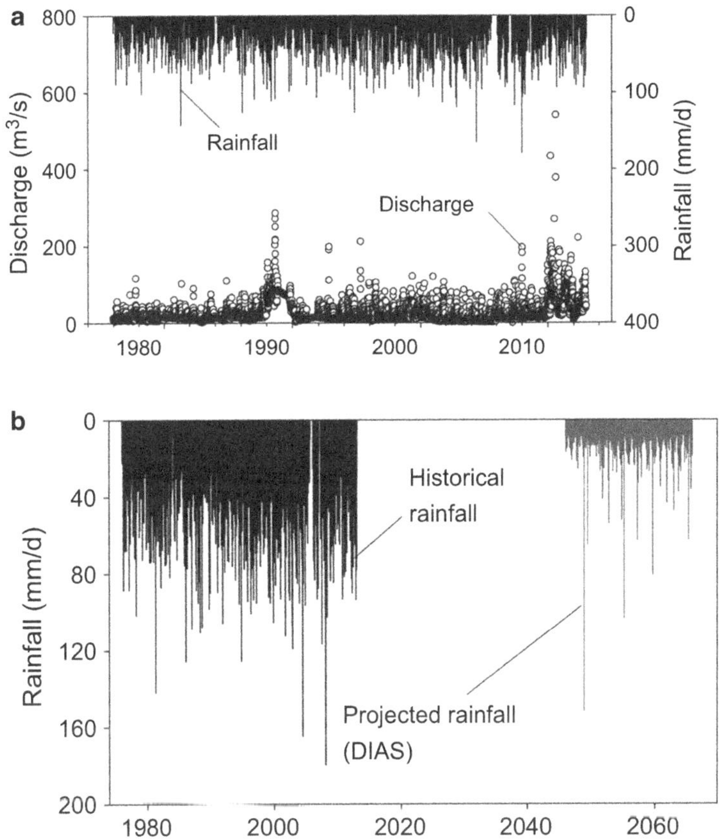

Fig. 3 Daily rainfall and river discharge for input data and calibration (**a**) historical rainfall and river discharge from 1976 to 2012 and (**b**) projected rainfall for simulation from 2046 to 2065

4 Results and Discussion

The flood frequency analysis statistically fit by the Log-Pearson type III distribution method tells the probable discharges in the river at various return periods of interests according to the available time series data. In the present study, eight recurrence intervals are selected to perform the analyses such as 1.25, 2, 5, 10, 25, 50, 100, and 200 years. The flood frequency analysis based on the historical record from 1976 to 2012 is presented in Fig. 4. The results from the Log-Pearson type III distribution show that streamflow varies from 60 to 667 m^3/s within the selected return periods. The maximum and minimum discharges of the upper and lower confidence intervals are 1,080 and 70 m^3/s and 477 and 50 m^3/s, respectively.

The tank model results are calibrated by applying the dataset from 1976 to 2012. The calibration results of the monthly discharges are shown in Fig. 5. The dataset

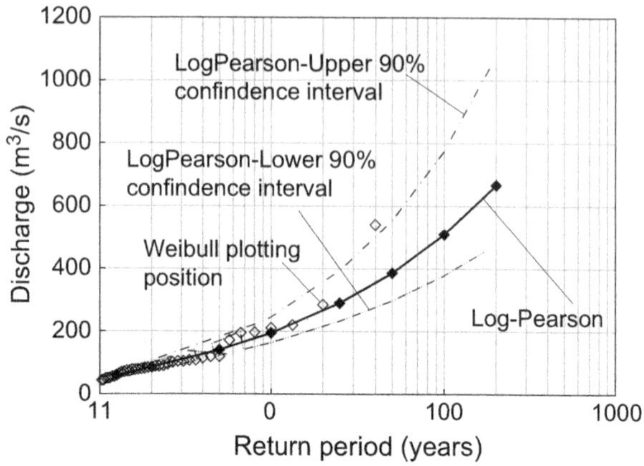

Fig. 4 Flood frequency analysis of the historical discharges consisting of the graph of Log-Pearson type III distribution, confidence interval (±90 %), and Weibull plotting positions

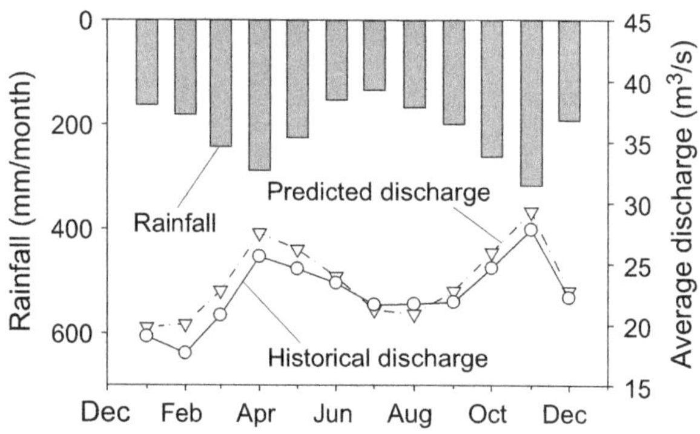

Fig. 5 Calibration results showing the average monthly river discharge and monthly rainfall from 1976 to 2012

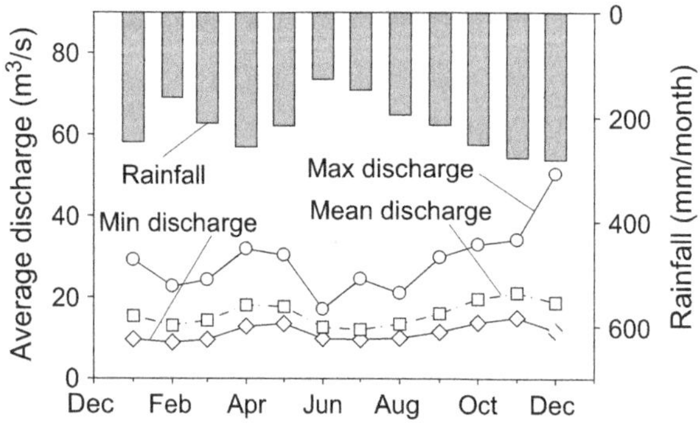

Fig. 6 Simulation results comprising of maximum, average, and minimum monthly river discharges and monthly rainfall from 2046 to 2065

from 2046 to 2065 is used as input for the calibrated model which results in the projected river discharges presented in Fig. 6. The average maximum estimated discharge is about 29 m^3/s; the average minimum discharge is approximately 11 m^3/s.

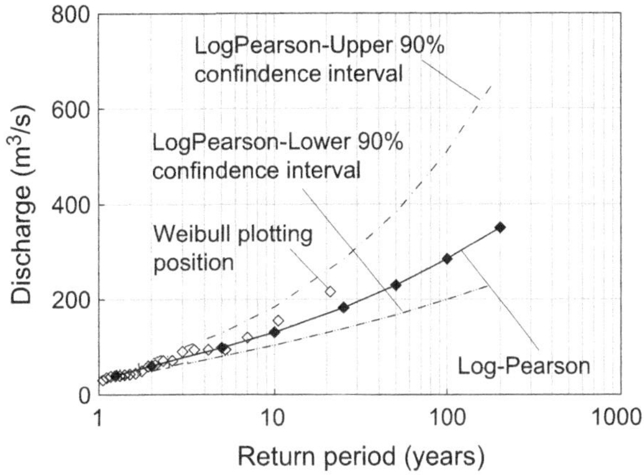

Fig. 7 Flood frequency analysis of the projected discharges, including the graph of Log-Pearson type III distribution, confidence interval (±90 %), and Weibull plotting positions

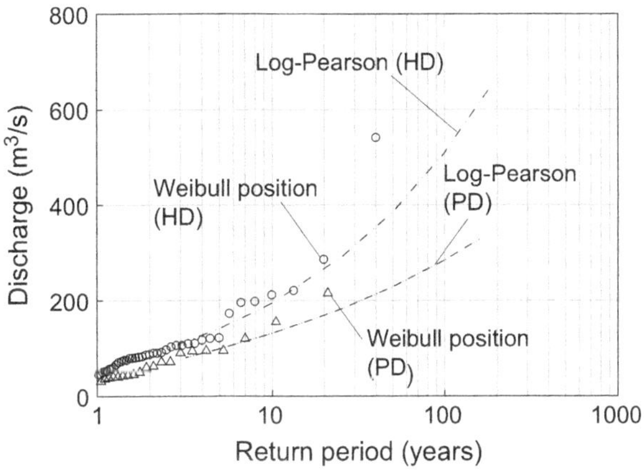

Fig. 8 Comparison between the flood frequency analyses of the historical discharges (HD) and projected discharges (PD)

The flood frequency analysis based on the climate change data from 2046 to 2065 is performed by using the projected discharges. Figure 7 presents the flood frequency distribution and confidence intervals. Within the selected recurrence intervals, the projected discharges vary from 41 to 350 m^3/s, while the lowest and highest discharges of the upper and lower confidence limits are 50–673 m^3/s and 31–235 m^3/s, respectively.

The comparison between the flood frequency distributions of both datasets (1976–2012 and 2046–2065) is made. The results are presented in Fig. 8, where the projected river discharges by the historical data are higher than the projected data. At the 100-year return period, the estimated discharge is around 667 m^3/s for the historical data and 350 m^3/s for the projected data. At the 25-year recurrence interval, the discharge is approximately 290 m^3/s for the historical data and 182 m^3/s for the projected data. At the 2-year return period, the projected discharge is about 85 and 61 m^3/s for the historical and projected data, respectively. From the flood frequency analysis results of the present study, the largest expected flood event is likely to be less than 350 m^3/s for the recurrence interval of 100 years. This is because the rainfall rate under the climate change condition of the Upper Klang River Basin is considered low, i.e., range between 127 and 281 mm/month and average about 216 mm/month.

5 Conclusions

The establishment of the flood frequency analyses at the Upper Klang River Basin is to estimate the flood magnitude associated with the return period of interest by using the Log-Pearson type III distribution. The flood frequency distributions are projected based on the available river discharge records and projected results from the climate change data. The tank model is used for the rainfall-runoff simulation. The historical discharge and rainfall data are available from 1976 to 2012. The flood frequency analysis of the historical dataset shows the variation of the discharges from 60 to 667 m^3/s within the recurrence intervals from 1.25 to 100 years. The climate change dataset of rainfall from 2046 to 2065 is used to project the future discharge in the river based on the calibrated model. The projected discharges are utilized to establish the flood frequency distribution. The results show that the projected flood magnitudes based on the climate change condition are lower than the results of the historical discharges. The discharges vary between 41 and 350 m^3/s within the same return periods. The expected flood event of the recurrence interval of 100 years is less than 350 m^3/s. This is because the low average rainfall in the future is approximately 216 mm/month. The results of the present study could inform the planners and developers at the Upper Klang River Basin the design flood for the constructions and sites along the river.

Acknowledgment The Department of Irrigation and Drainage (DID) Malaysia and National Hydraulic Research Institute of Malaysia (NAHRIM) are acknowledged for the hydrological data. The authors gratefully acknowledge the support from the Water Resources and Environmental System (WRES) Division, Faculty of Civil Engineering, UiTM, Shah Alam, and the Fundamental Research Grant Scheme [600-RMI/FRGS 5/3 (32/2012)] of the Ministry of Education (MOE), Malaysia. The authors would also like to thank the anonymous reviewers for their constructive comments and support during the review processes.

References

1. The Wall Street Journal (WSJ), Malaysia moves thousands to safety amid flash floods (2013), http://blogs.wsj.com/searealtime/2013/12/06/malaysia-moves-thousands-to-safety-amid-flash-floods/. Accessed June 2014
2. K.H. Ho, A.H. Ghazali, S.F. Chong, Calibration and evaluation of modified tank model (flood forecasting model) for Kelantan River Basin, in *Proceeding of Water Engineering Conference Malaysia*, 23–24 July 2002, UPM

3. M.K. Baharuddin, Climate change—Its effects on the agricultural sector in Malaysia (2007), http://www.met.gov.my/files/ClimateChange2007/session1b/8%20MustafaKamal_a.doc. Accessed June 2014
4. C. Cunnane, Statistical distributions for flood frequency analysis, operational hydrological Report No. 5/33, World Meteorological Organization (WMO), Geneva, Switzerland, 1989
5. Bureau of Meteorology (BOM), State of the climate report, Australian Bureau of Meteorology, Melbourne, 2012
6. Climate Impacts Group University of Washington (CIGUW), *The Washington Climate Change Impacts Assessment* (University of Washington, Washington, 2009)
7. Malaysian Meteorological Department (MMD), *Climate Change Scenarios for Malaysia 2001–2099* (Jabatan Meteorologi Malaysia, Petaling Jaya, 2009)
8. M.S. Kwan, F.T. Tangang, L. Juneng, Projection of climate change in Malaysia at the end of 21st century using SRES scenarios A1B by PRECIS model, in *Proceedings of Universiti Malaysia Terengganu 10th International Annual Symposium UMTAS*, 2011
9. U.S. Interagency Advisory Committee on Water Data, Guidelines for determining flood flow frequency, Bulletin 17-B of the Hydrology Subcommittee: Reston, Virginia, U.S. Geological Survey, Office of Water Data Coordination, 1982, p. 183
10. K. Beven, Towards a new paradigm in hydrology, in *Water for the Future: Hydrology in Perspective*, vol. 164. Proceeding of Rome Symposium, April 1987, IAHS, pp. 393–403
11. A. Tsutsumi, K. Jinno, R. Berndtsson, Surface and subsurface water balance estimation by the groundwater recharge model and a 3-D two-phase flow model. Hydrol. Sci. **49**(2), 205–226 (2004)
12. Department of Irrigation and Drainage (DID), Review of the national water resources study (2000–2050) and formulation of national water resources policy, Department of Drainage and Irrigation, 2011
13. P.B. Bedient, W.C. Huber, B.E. Vieux, *Hydrology and Floodplain Analysis* (Prentice-Hall, Upper Saddle River, 2002)
14. Data Integration and Analysis System (DIAS), A search and discovery system for DIAS datasets (2012), http://dias-dss.tkl.iis.u-tokyo.ac.jp/ddc/finder?lang=en. Accessed April 2014

Use of Numerical Weather Prediction Model and Visible Weather Satellite Images for Flood Forecasting at Kelantan River Basin

Intan Shafeenar Ahmad Mohtar, Wardah Tahir, Sahol Hamid Abu Bakar, and Ahmad Zikry Mohd Zuhari

Abstract This chapter presents the application of Numerical Weather Prediction (NWP) and weather satellite images for flood forecasting in Kelantan River Basin. Flood is known to be damaging due to its ability to cause losses to life and property. Many studies have indicated the adverse effects of flood to mankind. Therefore, it is crucial to develop an accurate flood forecasting system to provide sufficient time for a proper mitigation plan. This early flood warning can also provide decision makers early information on the flood event. The preliminary results have indicated the ability of NWP and weather satellite images to produce a weather forecast. It is believed that there is a bright future direction to adapt NWP for flood forecasting as a tool to produce forecasts. It is also believed that NWP will be used in an effective way to forecast any changes to the flood even though there are still many uncertainties, where it involves the complexity of the NWP model itself to serve as a precise forecasting method.

Keywords Flood • Flood forecasting • Numerical Weather Prediction (NWP) • MM5 • WRF

1 Introduction

Flood is one of the most hazardous natural disasters affecting many parts of the world causing deaths and damages to properties [1, 2, 18]. It is estimated that flooding contributes 40 % from all kinds of disasters to the total economic loss [3]. According to [3], flood can generally be characterised into several types which are fluvial (river) floods, pluvial (or overland) floods, coastal floods and groundwater floods, and the causes can vary from heavy downpour to sea level rise. Flood in Malaysia is a common problem especially during monsoon season. Flood in

I.S. Ahmad Mohtar (✉) • W. Tahir • S.H. Abu Bakar • A.Z. Mohd Zuhari
Flood Control Research Center, Faculty of Civil Engineering, Universiti Teknologi MARA, Shah Alam, Malaysia
e-mail: intanshafeenar@gmail.com

S.H. Abu Bakar et al. (eds.), *ISFRAM 2014*, DOI 10.1007/978-981-287-365-1_23

Malaysia is often associated with the monsoonal rainfall events during the northeast monsoon season where it is common to occur at the eastern coast of Peninsular Malaysia. The monsoonal flood prevails heavy rain during the months of October to March.

The flood episodes have indicated several events where thousands of people need to be evacuated especially in extreme cases of flooding that are caused by monsoonal factors. Some of the recorded flood episodes in the country occurred in 1926, 1931, 1947, 1954, 1957, 1963, 1965, 1967, 1969, 1971, 1973, 1983, 1988, 1993, 1998 and 2001 [5]. The occurrence of flash flood has always been reported by [17] in areas of big cities such as Kuala Lumpur and Shah Alam. Frequent flash flood in big city like Kuala Lumpur can cause so much distress if proper mitigation is not implemented.

The natural factors such as heavy monsoon rain and intensive convective rain have led to flooding where uneven distribution of rainfall occurs throughout the year. Around 60 % of annual rainfalls fall in the month of November and January which have led to flood occurrences to some areas in the country especially to low lying areas due to the seasonal distribution and variation of rainfall throughout the year [7]. The other cause is due to human factors such as improper drainage systems [8], where soil is covered by impervious surfaces in most of the big cities.

Study by [19] had indicated that flood episodes that occur during monsoon periods can cause severe loss of public infrastructure, crop yield and also loss of lives especially in low lying areas in the eastern coast of Peninsular Malaysia. It is also estimated that the area that is prone to flooding is about 29,000 km^2 or 9 % of the total area in the country which affects 2.7 million people [10]. In between December 2006 and January 2007, extreme flood events that had occurred in the southern Peninsular Malaysia had led to the evacuation of more than 200,000 people, 16 deaths and economic losses of US$500 million [11].

A precise method to forecast flood is crucial as it will provide early warning of the disaster, and this will give decision makers and planners adequate time to implement evacuation plan. It is also helpful to prevent the occurrence of the disaster in the future. Previously, many researchers have created a model or graph by using relevant historical data, and this has lead to pattern recognition [12]. Previous work on flood forecasting for Malaysian tropical condition can be found from [8, 13–15], who used infrared satellite images and Numerical Weather Prediction (NWP) data input in Artificial Neural Network (ANN)-based model. NWP, which consists of Fifth Penn State/NCAR Mesoscale (MM5) and Weather Research and Forecasting (WRF), has been used variously in different forecasting. There are some studies that have shown the application of hydro-meteorological study that uses radar images to extrapolate and to improve lead time of flow forecasts [16].

The challenges in flood forecasting are to provide precise and reliable prediction. NWP has been used worldwide for weather forecasting. However, it has limited use for flood forecasting. Therefore, comprehensive study on NWP in the application for flood forecasting should be increased, especially in tropical condition in which the weather pattern is more chaotic and random. Providing an early warning system, especially to catchment with short response time and dynamic runoff regimes, will help to enhance any flood event management. This type of catchment requires more advanced technique rather than the traditional way which uses observed precipitation data for hydrological simulations, especially for flash floods.

2 Background of Study

2.1 Flood Forecasting Model

Many flood forecasting systems rely on the input initially from observation networks, which is mainly from rain gauges and radar [3]. The observation networks provide precipitation input. There are many variables involved in order to provide precise flood forecasting. It is vital to utilize the most suitable technique because the technique that will be used will affect the accuracy of the forecasting.

2.2 Flood Forecasting Model Using Numerical Weather Prediction

Numerical Weather Prediction (NWP) was initially developed in the early 1900s by a Norwegian meteorologist named Bjerknes. Then, it was discovered that the process for forecasting was too difficult because it counters a huge amount of calculation. This has resulted in the development of limited area models (LAMs) in 1970s, where this model uses higher resolution over smaller area. The LAMs also took boundary conditions from a larger hemispheric or global model. Later on, LAMs have been developed to models such as MM4 and then later the new WRF model. Many studies have indicated the possibility of NWP to serve as an alternative input for flood forecasting.

NWP can be used for medium term forccast (~2–15 days ahead) and when equipment or transmission data fail especially in extreme flood cases. NWP usually provides forecasting over a period of several hours up to 1 or 2 weeks ahead. If the time intervals can be shortened, there will be adequate time for mitigation method to be implemented. Weather forecasting in Malaysia is conducted by the Malaysian Meteorological Department (MMD), where they currently use the Fifth Penn State/NCAR Mesoscale (MM5) and the Weather Research and Forecasting (WRF). Any minor deviation in NWP can have crucial impact on flood forecasting.

2.3 Fifth Penn State/NCAR Mesoscale

MM5 model is a non-hydrostatic primitive equation model with versatility to choose the domain region of interest. It has various options to choose parameterisation schemes for convection, planetary boundary layer, explicit moisture, radiation and soil processes.

2.4 Weather Research and Forecasting

The Advanced Research WRF system (WRF-ARW 3.1) can be used as an alternative meteorological driver for MM5 in the forecasting model. It is considered by NCAR as the successor of MM5, since further development of MM5 has come to an end in favour of WRF. The WRF-ARW system is a non-hydrostatic model (with a hydrostatic option) using terrain-following vertical coordinate based on hydrostatic pressure. Weather Research and Forecasting (WRF) model is a mesoscale NWP model which is suitable for research and operational forecasting.

3 Methodology

3.1 Case Study of Kelantan River Basin

Kelantan River basin is located on the northern part of Peninsular Malaysia (Fig. 1). It is located in the district of Kelantan Darul Naim. The case study period is for the year of 2009. Kelantan River basin received annual rainfall of about 2,700 mm during the northeast monsoon between October and January. The river is about 248 km long and the main tributaries are Lebir River (2,430 km^2) and Galas River (7,770 km^2). Kelantan River system flows northward passing through such major towns as Kuala Krai, Tanah Merah, Pasir Mas and Kota Bharu, before finally discharging into the South China Sea.

3.2 Datasets Used

The study involves two main sources of data, which are visible weather satellite images and NWP model products. The NWP will generate quantitative rainfall estimates as an output for flood forecasting model. Figure 2 illustrates the methodology followed to achieve research objectives. The method begins with the data acquisition from NWP and visible images from the geostationary meteorological satellite model. The concept of using the weather satellite images can be referred

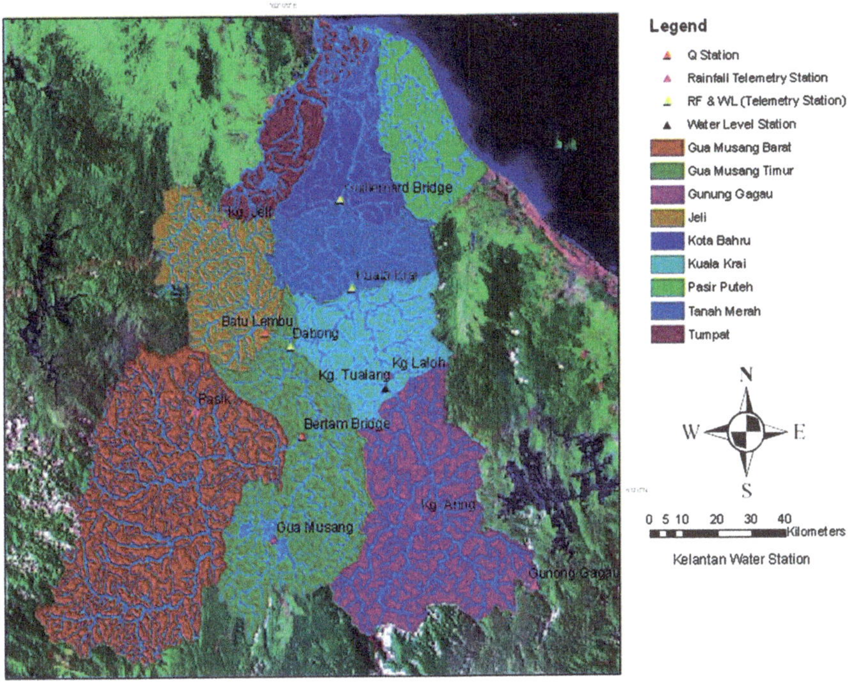

Fig. 1 Kelantan River basin (*Source: Pradhan, 2009*)

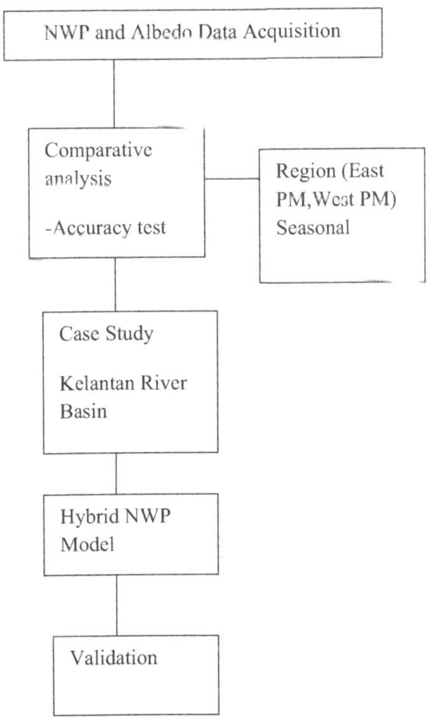

Fig. 2 Flowchart on research methodology

from [8, 13–15]. The current study used visible images in addition to the infrared images. The albedo values from the visible images provide information of the thickness of the cloud. When the cumulonimbus raining cloud gets thicker and taller, it indicates the impending rainfall events [8]. The correlation of the thickness of the cloud and rainfall will aid in the rainfall forecast model development. The output from this model will be used to forecast flood. In order to assess the preciseness of NWP model products and albedo, correlation and regression analysis with the gauged rainfall will be done to validate the data.

4 Preliminary Result and Analysis

4.1 Correlation Between the Albedo of the Visible Satellite Images with Rain-Gauged Data

The albedo values of the selected locations at Kelantan are determined by using McIDAS-V software processing tools. Using the grayscale satellite images, hourly albedo data have been determined for each coordinate of rainfall stations in Kelantan areas, which have been selected based on the rainfall events during the year 2009.

The albedo (%) was plotted as a function of the rainfall depth (mm) versus time (UTC) to show the trend line correlation, while the power regression was derived based on the correlation between both data (rainfall and albedo). Two methods have been used to generate the correlation between these data: by considering the average amount of rainfall and albedo based on the number of rainfall stations involved and hourly data of rainfall and albedo for selected rainfall station. Figures 3, 4 and 5 show the plotted graphs of relationship between albedo and the rainfall depth for total average rainfall event on 3rd January 2009, 15th April 2009 and 28th June 2009 for Kelantan River basin and 8th September 2009, 7th November 2009 and 26th December 2009 for Klang River basin. Figures 6, 7, 8 and 9 show the plotted graph for correlation of the albedo with the rainfall depth for Station 5820006—Bendang Nyior, Station 6019004—Rumah Kastam Rantau Panjang, Station 6120014—Kuala Jambu, 6021013—Rumah Kerajaan Jps. Meranti, Station 3116074—Leboh Pasar, 3217004—Kg. Kuala Sleh and Station 3216001—Kg. Sg. Tua. All correspond graphs of rainfall declines at the point where the high values of albedo occur. The albedo is high at highest rainfall depth recorded by DID rain gauge.

Kelantan River Basin

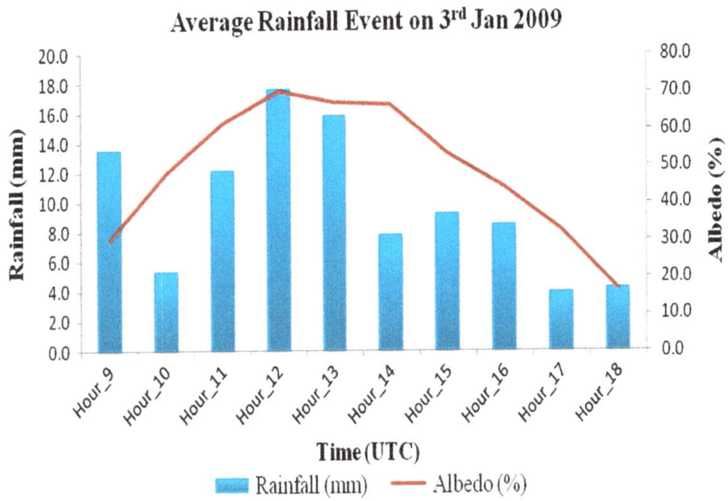

Fig. 3 Albedo plotted as function of the rainfall depth versus time on 3rd January 2009—total average of rainfall and albedo

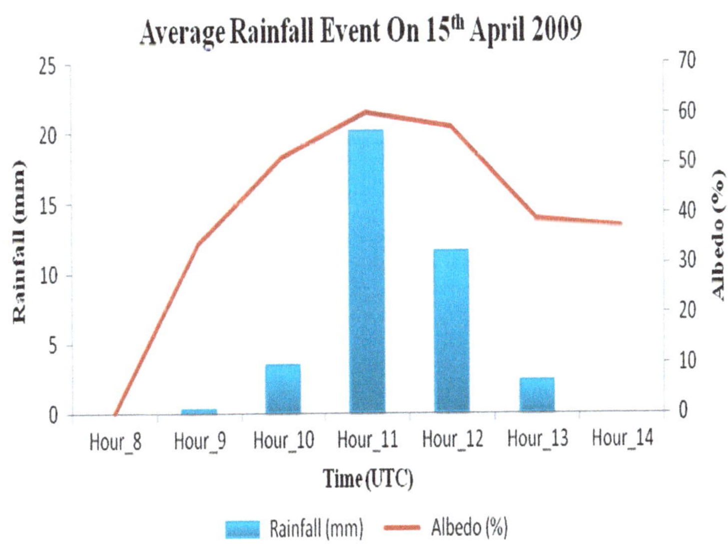

Fig. 4 Albedo plotted as function of the rainfall depth versus time on 15th April 2009—total average of rainfall and albedo

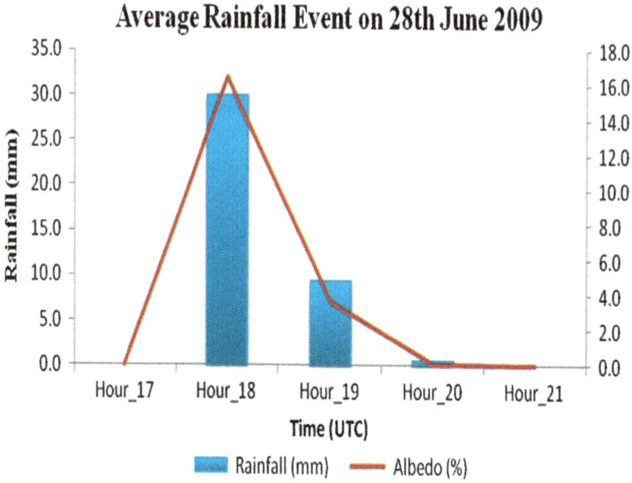

Fig. 5 Albedo plotted as function of the rainfall depth versus time on 28th June 2009—total average of rainfall and albedo

Rainfall Station at Kelantan

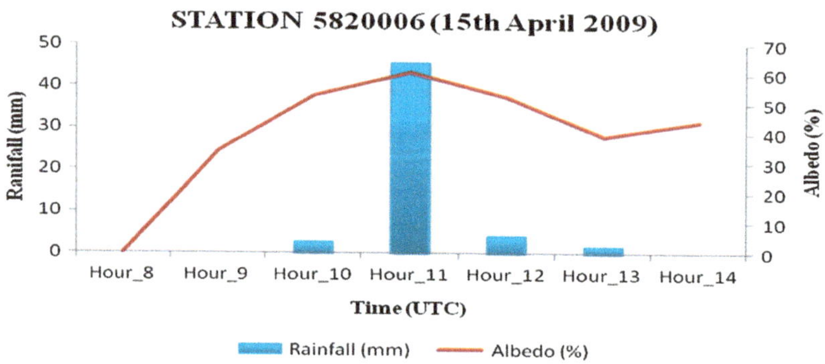

Fig. 6 Albedo plotted as function of the rainfall depth versus time at station 5820006—Bendang Nyior

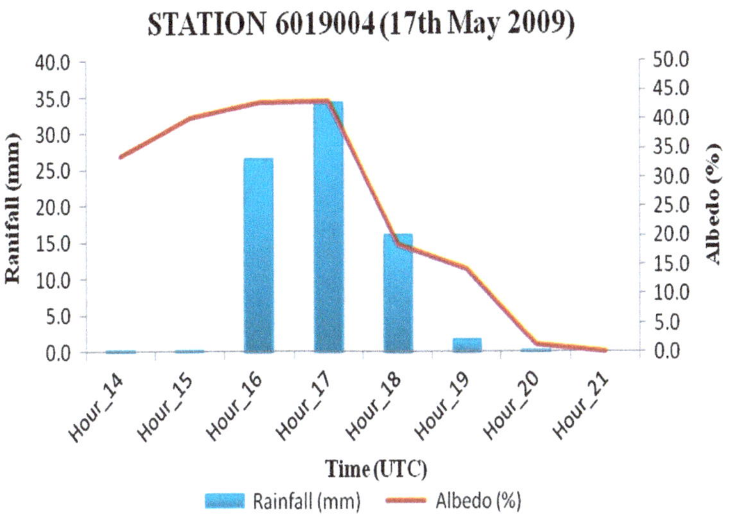

Fig. 7 Albedo plotted as function of the rainfall depth versus time at station 6019004—Rumah Kastam Rantau Panjang

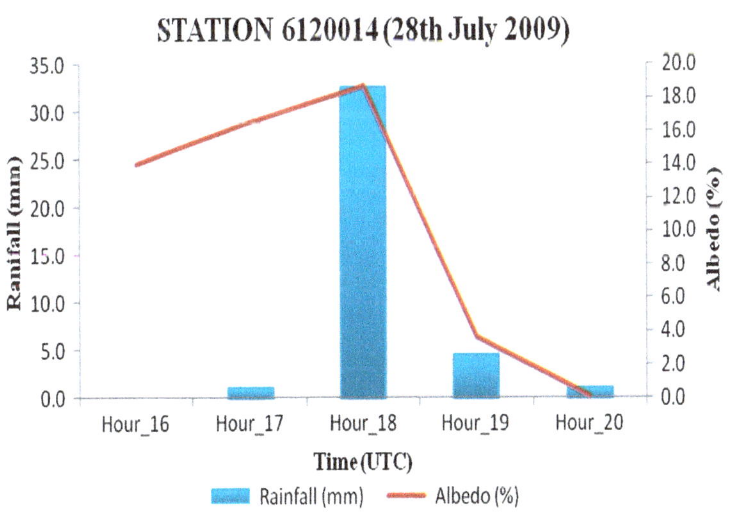

Fig. 8 Albedo plotted as function of the rainfall depth versus time at station 6120014—Kuala Jambu

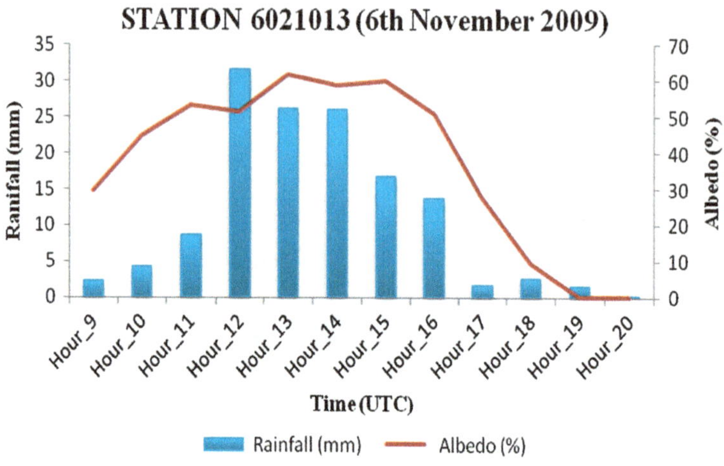

Fig. 9 Albedo plotted as function of the rainfall depth versus time at station 6021013—Rumah Kerajaan Jps. Meranti

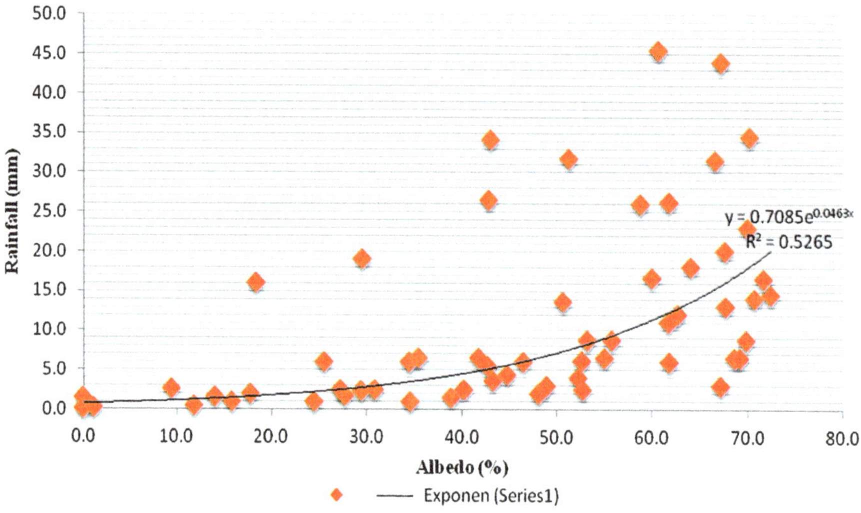

Fig. 10 Correlation between albedo (%) and rainfall depth (mm) for Kelantan River basin

Exponential law regression and linear law regression were derived to show the correlation between the albedo and rainfall depth from nine stations at Kelantan River basin. The plotted data represent the average rainfall and average albedo for all rainfall events selected at both areas. From Fig. 10, the graph shows the correlation between albedo and rainfall depth which is equal to 0.5265 or 52.7 % for Kelantan River basin.

5 Conclusion and Future Work

In this chapter, the preliminary results indicate that there is a good correlation between albedo and rainfall at the Kelantan River basin. There is a bright future direction to integrate the albedo values in NWP model products for flood forecasting. The future work will investigate the effectiveness of the NWP for rainfall and flood forecasting for selected areas in Malaysia. It is believed that NWP can be used in an effective way to forecast the flood even though there are still many uncertainties due to the complexity of the NWP model itself to serve as a precise forecasting method. Combining the NWP model products with other tools such as weather satellite images may enhance its capability and potential for flood forecasting in Malaysia.

Acknowledgement The author would like to express gratitude to the Malaysian Meteorological Department and Faculty of Civil Engineering, UiTM, for the research support and guidance. Appreciation also goes to the Research Management Institute of Universiti Teknologi Mara for DANA Grants (600-RMI/DANA 5/3/CIFI (127/2013).

References

1. B. Pradhan, Flood Susceptible mapping and risk area delineation using logistic regression, GIS and remote sensing. J. Spatial Hydrol. **9**(2), 1–18 (2009)
2. B. Pradhan, A.M. Youssef, A 100-year maximum flood susceptibility mapping using integrated hydrological and hydrodynamic models: Kelantan River Corridor, Malaysia. J. Flood Risk Manag. **4**, 189–202 (2011)
3. H.L. Cloke, F. Pappenberger, Ensemble flood forecasting: A review. J. Hydrol. **375**(3–4), 613–626 (2009), doi:10.1016/j.jhydrol.2009.06.005
4. K.J. Abhas, R. Bloch, J. Lamond, *Cities and Flooding—A Guide to Integrated Urban Flood Risk Management for the 21st Century* (World Bank, Washington, DC, 2012)
5. A.J. Hassan, A. Ab. Ghani, R. Abdullah, Development of flood risk map using GIS for Sg. Selangor Basin, in *Proceeding 6th International Conference on ASIA GIS, 9-10 Mac, UTM*, retrieved from redac.eng.usm.my/html/publish/2006_11.pdf
6. Department of Irrigation and Drainage, 2005 2006 Annual Report, Ministry of Natural Resources and Environment, Malaysia (ISBN 1823-6277, 2007)
7. N.W. Chan, A comparative study of water resources usage by households in Georgetown-Malaysia and Pattaya-Thailand, Iran. J. Environ. Health **3**(4), 223–228 (2006)
8. T. Wardah, S.H. Abu Bakar, A. Bardossy, M. Maznorizan, Use of geostationary meteorological satellite images in convective rain estimation for flash-flood forecasting. J. Hydrol. **356**(3), 283–298 (2008)
9. L. Juneng, F.T. Tangang, H. Kang, W.J. Lee, K.S. Yap, Statistical downscaling forecasts for winter monsoon precipitation in using multimodel output variables. J. Climate **23**(11), 17–22 (2010)
10. G. Mohd Azhar, Managing Malaysian water resources development. Buletin Kesihatan Masyarakat Isu Khas **200**, 40–58 (2000)
11. F.T. Tangang, L. Juneng, E. Salimun, P.N. Vinayachandran, Y.K. Seng, C.J.C. Reason, S.K. Behera, T. Yasunari, On the roles of the northeast cold surge, the Borneo vortex, the Madden-Jullan Oscillation, and the Indian Ocean Dipole during the extreme 2006/2007 flood

in southern Peninsular Malaysia. Geophys. Res. Lett. **35**, L14S07 (2008). doi:10.1029/2008GL033429

12. U.R. Acharya, P.S. Bhat, Classification of heart rate data using artificial neural network and fuzzy equivalence relation. Pattern Recogn. **36**, 61–68 (2003)
13. T. Wardah, I. Zaidah, R. Suzana, Geostationary meteorological satellite-based quantitative rainfall estimation (GMS-Rain) for flood forecasting. Malaysian J. Civ. Eng. **21**(1), 1–16 (2009)
14. T. Wardah, S. H. Abu Bakar, M. Mohamad, Intense Convective rain estimation using geostationary meteorological satellite, in *Advances in Geosciences*. Hydrological Science (HS), vol. 11, 2009, p. 133
15. T. Wardah, R. Suzana, S.Y.S.N. Huda, A.A. Kamil, Multi-sensor data inputs rainfall estimation for flood simulation and forecasting, in *2012 I.E. Colloquium on Humanities, Science and Engineering (CHUSER)*, pp. 374–379
16. J. Joregeson, P. Julien, Peak flow forecasting with radar precipitation and the distributed model CASC2D. IWRA Water Int. **30**(1), 40–49 (2005)
17. L.H. Feng, J. Lu, The practical research on flood forecasting based on artificial neural networks. Expert. Syst. Appl. **37**(4), 2974–2977 (2010)
18. A. Suliman, N. Nazri, M. Othman, M. Abdul, K.R. Ku-mahamud, Artificial neural network and support vector machine in flood forecasting: A review, in *Proceedings of the 4th International Conference on Computing and Informatics, ICOCI 2013*, 28–30 August 2013 Sarawak, Malaysia. Universiti Utara Malaysia (030), pp. 327–332
19. T. Wardah, A.A. Kamil, A.B.S. Hamid, W.W.I Maisarah, Statistical verification of numerical weather prediction models for quantitative precipitation forecast, in *IEEE Colloquium on Humanities, Science and Engineering*, 2011, pp. 88–92

Radar Rainfall for Quantitative Precipitation Estimates at Johor River Watershed

Sharifah Nurul Huda Syed Yahya, Wardah Tahir, Suzana Ramli, Asnor Muizan Ishak, and Haji Ghazali Omar

Abstract The frequent occurrence of floods in many low-lying areas throughout Malaysia demands for improved flood forecasting system. An advanced flood forecasting system named AMRFF (Atmospheric Model Based Rainfall and Flood Forecasting), which uses numerical globe scale atmospheric model, is being developed by the Department of Irrigation and Drainage (DID) to provide early flood warning. One component of AMRFF is rainfall input from weather radar. Use of radar rainfall will enable wider spatial coverage including ungauged areas. The contribution of Constant Plan Position Indicator (CAPPI) data from Doppler weather radar helps to give input data for the area without rain gauge station. This chapter presents a technique to automate the radar rainfall data into AMRFF system for the purpose of flood forecasting. It was shown that the weather radar input can satisfactorily provide the information on rainfall intensity for ungauged location.

Keywords Flood forecasting • AMRFF • Radar • Atmospheric model and CAPPI

1 Introduction

The applicability of weather radar data in hydrological and flood modelling has been studied extensively by many researchers. The main function of the radar in Malaysia is to provide the most important element of meteorological observation for aircraft safety measures, such as wind shear, micro-burst detection, temperature, rainfall, direction and speed of wind and pressure. It also provides rainfall intensity or rainfall accumulation maps over a large area at short time intervals. Malaysia has been using the conventional weather radar WSR 3D Rapic Radar since about 50 years ago. In 1998, the Malaysian Meteorological Department (MMD) had

S.N.H.S. Yahya (✉) • W. Tahir • S. Ramli
Flood Control Research Center, Faculty of Civil Engineering, UiTM, Shah Alam, Malaysia
e-mail: sha_huda2@yahoo.com

A.M. Ishak
Department of Irrigation and Drainage, Ampang Kuala Lumpur, Malaysia

H.G. Omar
S.I Protech Sdn Bhd, Kuala Lumpur, Malaysia

S.H. Abu Bakar et al. (eds.), *ISFRAM 2014*, DOI 10.1007/978-981-287-365-1_24

replaced the conventional weather radar with Doppler Weather Radar [1]. The difference between a conventional weather radar and Doppler weather radar is that the conventional radar can only detect the characteristic, size, direction and distance of precipitations, while Doppler radar can also measure micro-burst, radial wind speed and shear [2, 3].

The MMD is using Interactive Radar Information System (IRIS) software to process the raw radar signal, produce radar display and rainfall estimates. Radar rainfall estimates have the advantages of enhanced spatial and temporal resolution despite its limitation and inherent errors. Factors contributing to radar errors include the conversion from reflectivity to rainfall (Z–R relation), beam blockage, ground clutter and hardware calibration errors [4, 5]. The accuracy of the radar rainfall product needs to be examined due to the high degree of variation [6]. Rain gauge is an instrument that is considered as one that truthfully measures the actual amount of rain that falls over a small area and has become the long-established measurement of rainfall. Practically being the 'ground truth' of rainfall values, rain gauge data are considered as the most viable means of verification of rainfall estimate products using radar [8, 9]. This chapter describes the work on radar rainfall as potential input to AMRFF using the current raw CAPPI radar data transmitted from the MMD.

This chapter describes the ability of radar rainfall data as the input to AMRFF server by using the current raw radar data that was transmitted from the MMD. The use of radar data will help to monitor a rainfall pattern for over a large area. It will give real time information through the *InfoBanjir*, which is monitored and developed by the DID. The type of weather radar product received by the DID was CAPPI data. It is 1 km high from the ground and gives the reflectivity value for each of (720 × 720) pixel. The process of producing rainfall estimates from the raw CAPPI data involved some programming using LINUX and Microsoft Visual Studio. The primary step before the automation of the radar rainfall input is the investigation of the accuracy of the weather radar data to ensure that the rainfall data produced is reliable.

2 Study Area

This study focused on Johor River watershed as the case study. The watershed is located at the east of Peninsular Malaysia. The watershed covers an area of approximately 2,690 km^2. The Johor River watershed can be considered as having flat terrain except on the northern and the eastern region with an elevation of up to 1,010 m [7].

The main tributaries of Johor River are Linggiu, Sayong, Pengeli, Semangar, Lebak, Telor, Panti, Temboyah and Permandi River. There are two dams available, which are Linggiu Dam and Lebam Dam. Figure 1 shows the image of Doppler Weather Radar for Kluang Station. The type of Doppler radar used is S band with

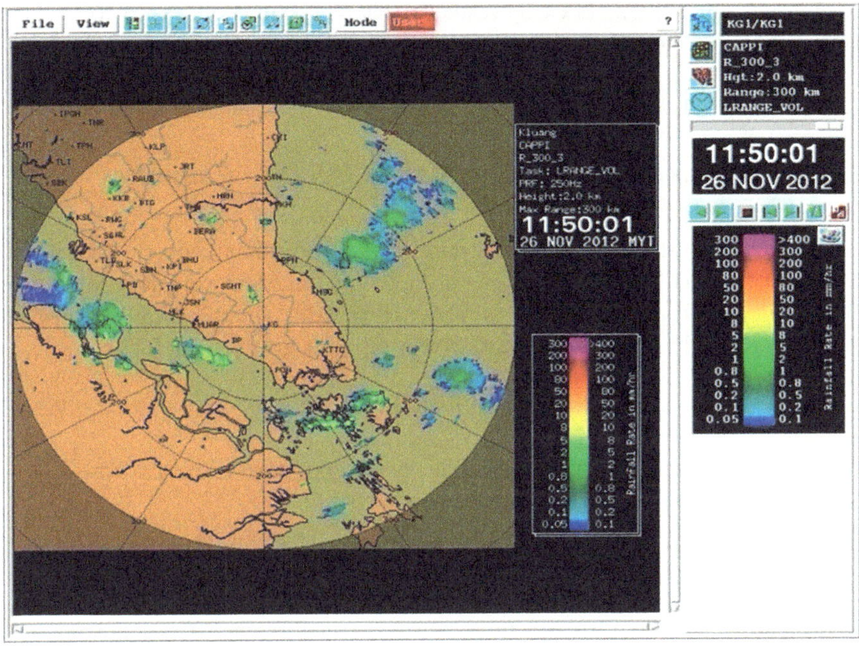

Fig. 1 Kluang radar display image from IRIS software (*Sources*: *MMD*)

the wavelength of around 10.7 cm. It may cover areas up to 300 km range from the initial point of 0.9′N 103 19.0′E.

The weather radar generates the data for every 10 min interval while being monitored by the MMD. Figure 1 shows a sample of weather radar image taken on 26th November 2012. The image will give the information on rainfall at each pixel for the size of 720 × 720. The presented values of rainfall on the map are in terms of reflectivity. MMD is adopting the Marshall–Palmer equation to convert the reflectivity data (dBZ) into rain rate (mm/h).

3 Data Analysis

3.1 Raw CAPPI Data Conversion to Readable Reflectivity Values (dBZ)

The first scope of work was to decipher the raw radar data into readable file format. This step involved LINUX programming to retrieve and process the raw radar data. The radar data that had been acquired were from Kluang radar station for record dated 25th July, 2012 until 4th December, 2012. All the data were in raw binary code forms which were not easily readable. The first step done was to convert the binary code into the reflectivity (dBZ).

3.2 Conversion of Reflectivity Values (dBZ) to Rainfall

The reflectivity values in dBZ at 10 min interval were then converted to rainfall using the Marshall–Palmer equation as used by the Meteorological Malaysian Department [8, 9]. Reflectivity, Z is measured in dBZ. The dB is called a decibel and was originally devised to express power ratios as given in the following:

$$\mathrm{dB} = 10\log[P1/P2] \tag{1}$$

where $P1$ and $P2$ are the two power levels being compared and log is base 10.

Mathematically, reflectivity is defined as

$$Z(\mathrm{dBZ}) = 10\log\left[z/\left[1\ mm^6/m^3\right]\right] \tag{2}$$

where z is the measured backscattered power received by the radar.

Using the Marshall–Palmer equation adopted by the MMD

$$z = 200\,R^{1.6} \tag{3}$$

Hence to calculate the rainfall intensity, R for $Z = 40$ dBZ

$$\begin{aligned} z &= \mathrm{antilog}(40/10) \\ R &= (z/200\}^{(1/1.6)} \\ &= 11.53\ \mathrm{mm/h} \end{aligned}$$

3.3 Radar Rainfall Verification

There are four rainfall stations near Kluang which will be considered in testing the radar rainfall accuracy. The rainfall stations are:

(a) Ulu Sebol at Kota Tinggi
(b) Kota Tinggi
(c) Felda Air Hitam at Kluang
(d) Emp Macap at Kluang

The comparison was made for selected good events which had occurred between 2nd and 4th November 2012. 72 hourly radar rainfall data were matched with hourly gauged rain for each of the rainfall station. Figure 2a–d shows the correlation between radar rain and gauged rain for the four stations.

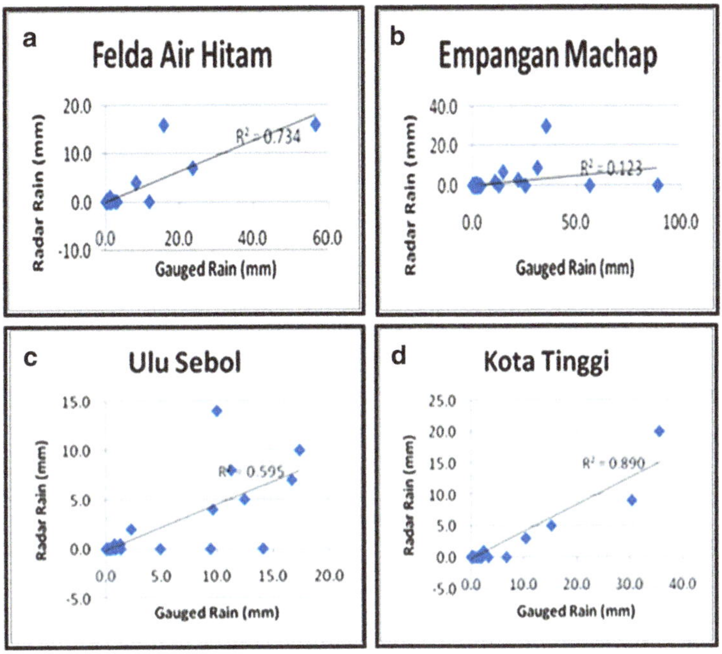

Fig. 2 Comparison between radar rain and gauged rain for different stations: (**a**) Felda Air Hitam, (**b**) Empangan Machap, (**c**) Ulu Sebol, (**d**) Kota Tinggi

The highest r value was for Station Kota Tinggi ($r = 0.890$) and the lowest was for Empangan Machap ($r = 0.123$). The data at Empangan Machap show that there were several rainfall signals detected by the radar but not captured by the rain gauge.

The details of the rainfall events can be observed by the hyetographs given by Figs. 3, 4, 5 and 6.

3.4 *Imaginary Rainfall Stations using Radar Rainfall*

The effectiveness of using radar rainfall as complementary to gauged rain to be input to AMRFF is going to be investigated through the creation of six imaginary rainfall stations with radar rainfall values. The six additional rainfall station values are to be read from the locations as in Table 1 (from E to J).

Figure 7 shows the location of these stations on Johor map and the watershed. Point A was for Felda Air Hitam station, Point B for Empangan Machap station, point C for Ulu Sebol station and point D for Kota Tinggi station. Other points indicated by E, F, G, H, I and J were meant to represent the imaginary rainfall stations. These locations were identified to be crucial in providing rainfall input to

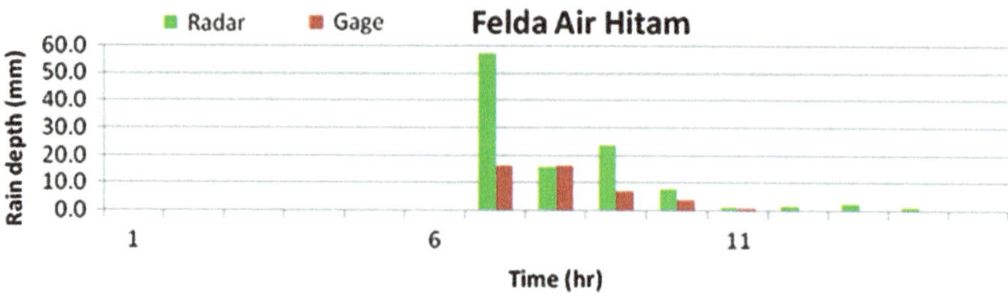

Fig. 3 Felda Air Hitam

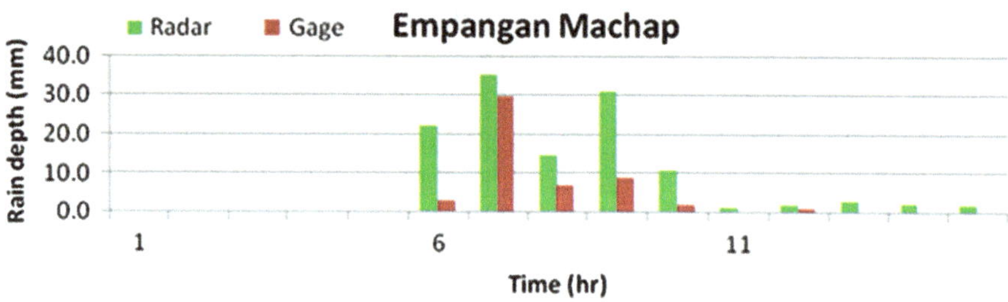

Fig. 4 Empangan Machap

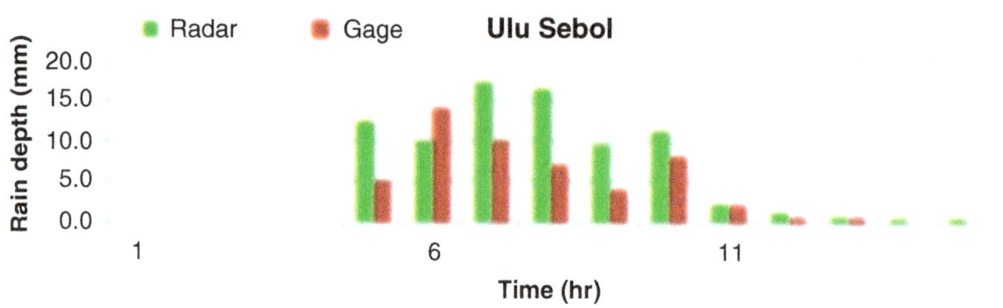

Fig. 5 Ulu Sebol

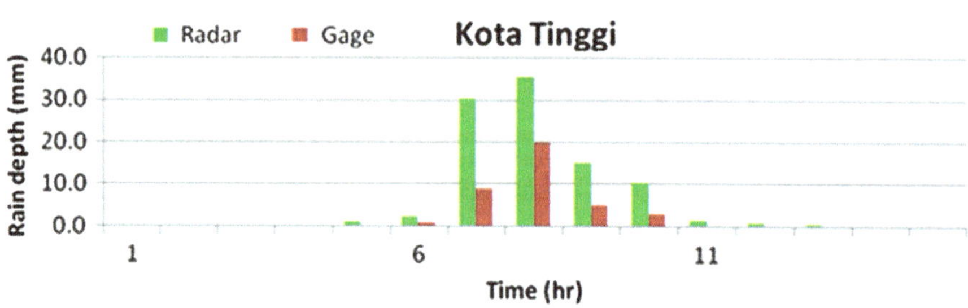

Fig. 6 Kota Tinggi

Table 1 The location of radar rainfall values

Station	Longitude	Latitude
A (Felda Air Hitam)	103.2340	1.9548
B (Empangan Machap)	103.3315	2.0598
C (Ulu Sebol)	103.6389	1.8799
D (Kota Tinggi)	103.8938	1.7299
E	103.3315	2.0523
F	103.6239	1.6775
G	103.7139	1.7674
H	103.5265	1.9473
I	103.6840	2.0147
J	103.8264	1.8273

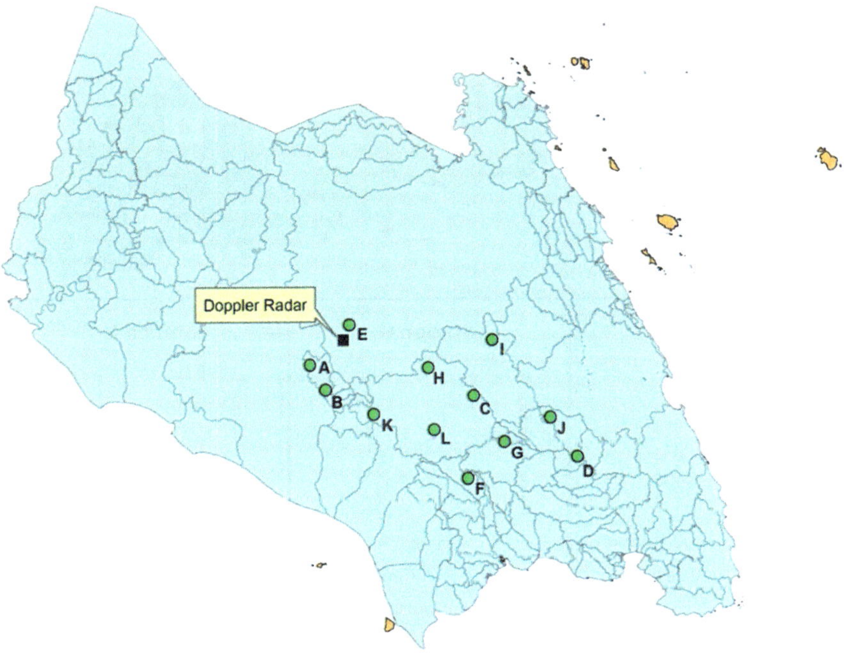

Fig. 7 The location of the rainfall stations and Kluang radar on Johor map

the AMRFF system. The unavailability of rainfall stations was to be overcome by the use of radar rainfall. The graphs in Fig. 8 show the rainfall hyetograph for the imaginary stations for event dated 3rd November, 2012.

The accuracy of raw radar data has to be determined. A comparison was made between the radar rainfall and gauged rainfall. Table 1 shows the values of coefficient correlation, *R*, and the root mean square error, RMSE.

Table 2 shows that Kota Tinggi station produced results of radar rainfall with good accuracy. The location of Kota Tinggi, which is closed to the Doppler

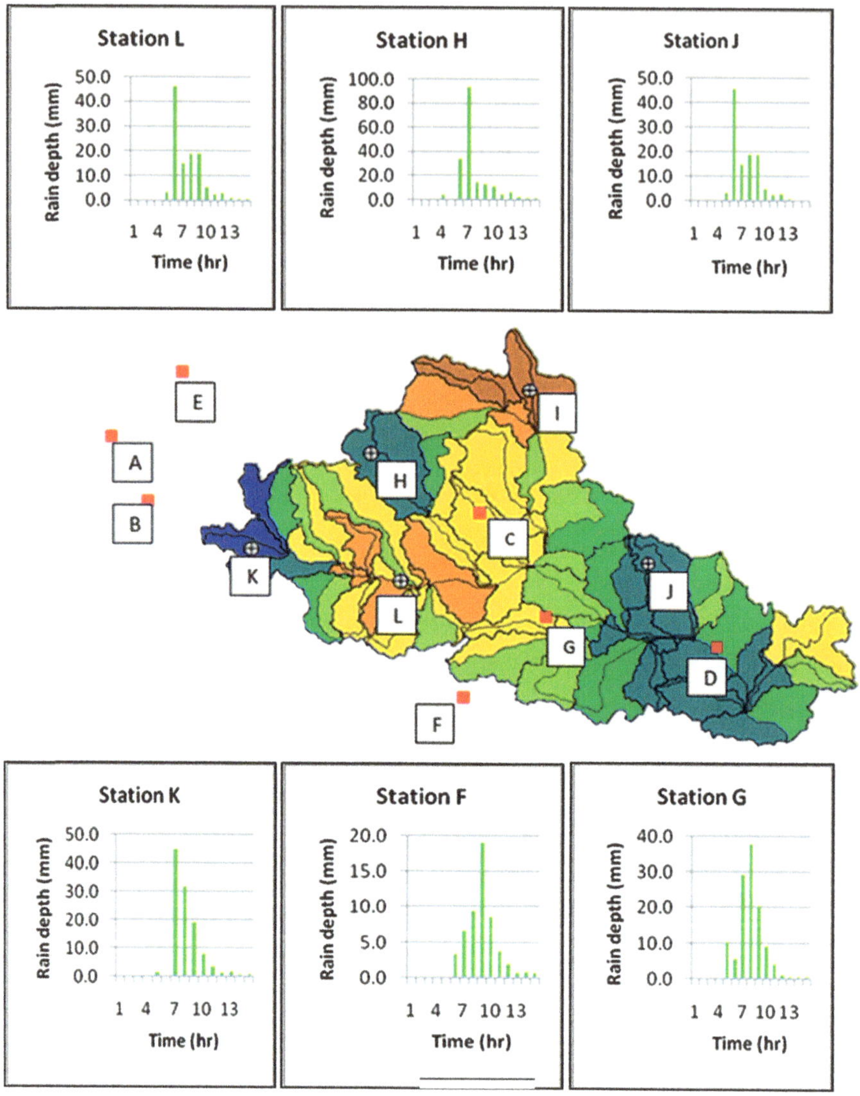

Fig. 8 Radar rainfall for event dated 3rd November 2012

Table 2 Accuracy measures of radar rainfall

Rainfall station	R	RMSE
Felda Air Hitam	0.734	5.476
Empangan Machap	0.123	13.956
Ulu Sebol	0.595	3.735
Kota Tinggi	0.89	1.933

Weather Radar Station, may contribute to the positive finding. The least accurate result was found at Empangan Machap station. The reason could be due to its location which is far from Kluang Weather Station. The use of inadequate data together with unpredictable weather would also produce unreliable results [10, 11]. The results will be improved with the use of longer record of weather radar data.

The radar rainfall assimilation may enhance the QPF input to the AMRFF system. The high accuracy of weather radar input plays an important role to give information for a larger area in a short time. It is recommended that the raw radar data be filtered and monitored in improving its capability and effectiveness [11, 12].

Acknowledgments The authors gratefully acknowledge the contributions of Department of Irrigation and Drainage (DID) and Malaysian Meteorological Department (MMD) for providing the data that have been used in this study. Also for the contributions of DANA Grants (600-RMI/DANA 5/3/PSI (262/2013) from the Research Management Institute of Universiti Teknologi Mara, Faculty of Civil Engineering, Faculty of Science, Computer and Mathematics, Uitm Shah Alam, and special thanks for the cooperation from the S.I Protect Sdn. Bhd. and Servis EDA Sdn. Bhd. in the process of implementing this chapter.

References

1. A.R.M. Waled, M.S.M. Amin, G.A. Halim, A.R.M. Sharif, W. Aimrun, Calibrated radar-derived rainfall data for rainfall-runoff modeling. Eur. J. Sci. Res. **30**(4), 608–619 (2009). ISSN 1450-216X
2. T. Wardah, S.Y. Sharifah Nurul Huda, S.M. Deni, B. Azwa, Radar rainfall estimates comparison with Kriging interpolation of gauged rain, in *Colloquium on Humanities, Science and Engineering, CHUSER 2011*, art. no. 6163877, pp. 93–97
3. T. Wardah, S.H. Abu Bakar, A. Bardossy, M. Maznorizan, Use of geostationary meteorological satellite images in convective rain estimation for flash-flood forecasting. J. Hydrol. **356**(3), 283–298 (2008)
4. T. Wardah, I. Zaidah, R. Suzana, Geostationary meteorological satellite-based quantitative rainfall estimation (GMS-rain) for flood forecasting. Malaysian J. Civ. Eng. **21**(1), 1–16 (2009)
5. T. Wardah, S.H. Abu Bakar, M. Mohamad, Intense convective rain estimation using geostationary meteorological satellite, in *Advances in Geosciences: Hydrological Science (HS)*, vol. 11, ed. by N. Park, et al. (World Scientific Publishing Company, New Jersey, NJ, 2009), p. 133
6. R. Suzana, T. Wardah, A.S. Hamid, Radar hydrology: New Z/R relationships for Klang River Basin Malaysia based on rainfall classification. World Acad. Sci. Eng. Tech. **59** (2011)
7. M. Jafri, Z. Hashim, M.L. Kavvas, Z.Q. Chen, N. Ohara, Development of atmospheric based flood forecasting and warning system for selected river basins in Malaysia, http://www.met.gov.my/images/pdf/national_papers/nsm2010/pn.%20hjh.%20zainab%20hashim%20(presentation).pdf. Accessed 2 Dec 2013
8. R. Suzana, T. Wardah, Radar hydrology: New Z/R relationships for Klang River basin, Malaysia, in *International Conference on Environment Science and Engineering. IPCBEE*, vol. 8, IACSIT, Singapore, 2011
9. R. Suzana, T. Wardah, Radar hydrology: New Z/R relationships for quantitative precipitation estimation in Klang River basin, Malaysia. Int. J. Environ. Sci. Dev. Monit. **2**(3), 2011
10. T. Wardah, C. Hamid, H. Aimme, Flood forecasting using tank model and weather surveillance radar (WSR) input for Sg Gombak catchment. Int. J. Civ. Environ. Eng. **13**(2) (2013)

11. S.Y. Sharifah Nurul Huda, T. Wardah, R. Suzana, A. Hamzah, M.S. Muhammad Faiz, Improved estimation of radar rainfall bias over Klang River basin using a kalman filtering approach, in *Business, Engineering and Industrial Applications (ISBEIA), 2012 I.E. Symposium*, art. no. 13285358, pp. 368–373
12. T. Wardah, R. Suzana, S.Y. Sharifah Nurul Huda, Multi-sensor data inputs rainfall estimation for flood simulation and forecasting, in *2012 I.E. Colloquium on Humanities, Science and Engineering (CHUSER)*, IEEE, December 2012, pp. 374–379

Enhanced Flood Forecasting Based on Land-Use Change Model and Radar-Based Quantitative Precipitation Estimation

Salwa Ramly, Wardah Tahir, and Sharifah Nurul Huda Syed Yahya

Abstract For a country located in the equatorial region, flooding in the event of heavy rain is something that is inevitable. Malaysia is located in the equatorial region and experiences tropical climate. Kuala Lumpur, the capital city of Malaysia located in the Klang River Basin, is prone to flooding in the event of heavy rains in the catchment. Therefore, flood forecasting is a necessity, as the system helps in planning for flood events and helps to prevent loss of lives and minimise damages. Increased development in Klang Valley has resulted in the need of commercial space and housing demand in the Klang Valley. Therefore, many new areas have been developed for commercial land residential purposes. This has resulted in significant changes in land use due to the development since the past 10 years. The overall objective of the study is to enhance flood forecasting system for the provision of flood warning and emergency response with a convenient lead time. Expected outputs from the study are the incorporation of the effects of land-use change in flood model development and enhancement of the flood forecasting by using radar-based quantitative precipitation estimation (QPE).

Keywords Flood • Hydro-meteorological • Radar • Land use • Rainfall-runoff

1 Introduction

Flooding is a natural phenomenon in the world. Flooding occurs when the normally dry areas are inundated with water. Flooding can happen at any time depending on the input factors that cause flooding. Floods also vary in terms of size of the area affected by flooding, flood duration and depth. There are many factors causing floods. Among the factors that could cause flood are excessively heavy and prolonged rainfall, urbanisation, river erosion, deforestation and drainage systems

The authors would like to thank Research Management Institute, Universiti Teknologi Mara, ERGS grants 600-RMI/ERGS 5/3 (21/2012).

S. Ramly (✉) • W. Tahir • S.N.H. Syed Yahya
Flood Control Research Center, Faculty of Civil Engineering, Universiti Teknologi MARA, Shah Alam, Selangor, Malaysia
e-mail: salwaramly@yahoo.com

S.H. Abu Bakar et al. (eds.), *ISFRAM 2014*, DOI 10.1007/978-981-287-365-1_25

that are not organised. According to [1], flood prone area in Malaysia is estimated to be approximately 29,800 km^2, which is 9 % of total land and 23 % from this flood prone area is urban area. There are 2.7 million people who are exposed to the flooding problem. In the National Register of River Basin Study [2], it was estimated that the average flood damage in 1980 was RM100 million, and this value increased to RM915 million in 2000. Therefore, flood preparedness is important to reduce damages and protect lives from the hazard. One of the options for flood preparedness is early flood warning. Flood forecasting and warning system can be used to alert and warn the public to better prepare for incoming flood. Flood forecasting in terms of quantitative measurement and forecasting of precipitation is vital for preparation of flood issues. Good understanding on meteorological and hydrological behaviour of the catchment is essential for flood forecasting. Flood forecasting uses rainfall data as an input in a rainfall-runoff model to forecast flow and water level in the river. From years to years, many works on flood forecasting development have been done to improve the quality of flood forecasting. One of the methods is utilising the weather radar as quantitative precipitation estimation (QPE) in the flood forecasting system in an attempt to extend the lead time.

Precipitation is one important element in the hydrological cycle. Precipitation is formed from atmospheric water and influenced by other elements such as wind, temperature and atmospheric pressure. Precipitation is in the form of rain, snow and hail. Rainfall could be classified into three types: convective rainfall, frontal or cyclonic rainfall and orographic rainfall [3]. This classification is based on the conditions that generate vertical air motion. Convectional rainfall is typical to tropics and is brought about by heating of the air at the interface with the ground, and locally strong vertical air motions occur [4]. When the air is thermally unstable, the cloud continues to rise and creates towering cloud and may lead to locally intense rainfall but sometimes in the limited duration. Malaysia is located in tropic country and experiences convectional rainfall and risk to flash flood due to locally intense rainfall. Therefore, understanding on convectional rainfall is essential since this type of rain varies in terms of temporal and spatial pattern [5]. It is a challenge to develop QPE in the situation of convectional rainfall in this region. Therefore, comprehensive study on QPE in this region is needed to establish better method for QPE. This study will focus on convectional rainfall.

Land-use change also plays an important role in hydrological cycle. Changes in land use and land cover over time may interrupt the hydrological processes and affect the ecosystem, river system and also the drainage system. Rapid development and continuous population growth give pressure on undeveloped area to be converted into urban area. These may lead to less impervious area for rainfall to intercept into the ground and more runoff into the river system. If the land use is continuously changing from time to time, the existing river system and drainage system would fail to cater the increase of surface runoff in the river catchment. This situation will contribute to flash flood in the urban area, and more damages would happen to the infrastructures and people. Analysis of the land use and land cover change (LUCC) plays important role in flood forecasting to get accurate flow forecast and better prediction in the future.

2 Background of Study

2.1 Climate of Malaysia

Malaysia is a humid tropical climate country. The characteristics of this climate are uniform temperature, between 21 and 32 °C. Malaysia also experiences a high rate of humidity and high rainfall. Malaysia's average annual rainfall is between 2500 to 3000 mm. On average, the wind speed in Malaysia is relatively low with wind speed around 3 m/s. Malaysia is located near the Equator between latitudes 1° and 7° North and between longitudes 99° and 120° East. It is surrounded by the sea, causing frequent cloudy conditions. Total area of Peninsular Malaysia is 131,587 km^2. As a maritime country, Malaysia receives plenty of rainfall from very high convective rainfall and localised thunderstorms [6]. The rainfall pattern in Peninsular Malaysia is characterised by two rainy seasons associated with the Southwest Monsoon (SWM) starting from the end of May to September and the Northeast Monsoon (NEM) from November to March. Heavy rainfall also occurs in the inter-monsoon season in April and October, especially in west peninsular Malaysia [7–9]. Kuala Lumpur is located in the west coast of Peninsular Malaysia and is the capital and largest city in Malaysia. Kuala Lumpur is located in the Klang Valley. This valley is bordered by the Titiwangsa Range in the east, several minor ranges in the north and south and the Straits of Malacca in the west. Klang Valley is the most densely populated area of the country with 1.7 million populations. The river system flowing through the Klang Valley is Sungai Klang and its tributaries, which together form the Klang River Basin. The total length of Klang River is 120 km, while the total area of the Klang River Basin is approximately 1,290 km^2. Kuala Lumpur city centre is located at the confluence of two major rivers, namely Sungai Klang and Sungai Gombak. This area often suffers from the threat of flooding in the event of heavy rain in the catchment area of the river. The rapid development in this area causes buildings and infrastructure facilities to be crowded in the river corridor. Many studies have been done to determine spatial and temporal pattern of rainfall for Klang Valley catchment [5, 10]. Study by [5] shows the storm movement in Klang catchment is not in specific direction, and propagation of the storm is in chaotic direction. Furthermore, [10] concluded that there is significant variation of storm in spatial and temporal pattern for this catchment based on existing gauge network.

2.2 SMART

SMART or Stormwater Management and Road Tunnel is a Klang River's flood control system for the Kuala Lumpur City Centre, which was built in 2003 and completed and started operating in July 2007. SMART is a dual-function tunnel, with the objective to solve the problem of flood in Kuala Lumpur City Centre. Apart from that, SMART also functions as highways for 3 km of Sungai Besi/Jalan Tun Razak's roads to solve the problem of traffic congestion on the route. SMART components are divided into two parts: storm water component and traffic component.

Storm water component consists of two main ponds. The holding pond sited at Kampung Berembang and the storage reservoir located at Taman Desa, near to Sungai Besi. Storm water components are controlled by a system named SMART Flood Detection System (FDS). This system is comprised of hydrological and hydrodynamic models and a database and scheduler. The hydrologic rainfall-runoff models provide warning time for tunnel opening using real-time rainfall information from the upper catchment to predict stream flows. The Australian Water Balance Model (AWBM) hydrologic process rules were used to convert the rain into runoff [11]. There are 22 rainfall stations and 16 discharge measurement and water level stations in the SMART catchment area that had been installed for the purpose of flood forecasting and the operation of the SMART [12]. Design's criterion for SMART is to mitigate floods up to 100-year event level. The specific areas targeted for flood protection by SMART are the flood prone areas on either side of the Klang River immediately above its confluence with the Gombak River.

2.3 *Upper Klang Ampang River Catchment*

In this study, Upper Klang Ampang River catchment has been selected to be a case study on the flood forecasting considering impact of land-use change using radar-based QPE. This Upper Klang Ampang River catchment is the most important area for SMART catchment area. The hydrologic model in the FDS uses this catchment area to measure and forecast the flow of Klang River for operation of the SMART specifically at the confluence of Klang/Ampang River.

Catchment area for Upper Klang Ampang River is approximately 160 km^2. Topography of this catchment varies from vertical slopes of hills, 1,430 m mean sea level at upper part of the catchment, and low lying area in the downstream of the river near the city centre. The annual mean rainfall in this catchment is 2,600 mm at the hilly areas [13].

Upstream of the Klang River, approximately 10 km from the confluence of Klang/Ampang River, there is a Klang Gate Dam. This dam is surrounded by forest as water catchment for the dam. This dam is used for water supply purposes and also as flood mitigation for Klang Valley. Immediately after the dam, there are many residential areas being developed and caused upper Klang river watershed to experience land-use change. There are high demands of residential area on the hillside of Upper Klang.

A water treatment intake plant is also located at the upstream of the Ampang River, approximately 8 km from confluence of Klang/Ampang River. Significant land-use changes have occurred in the upper Ampang River catchment due to the opening of the residential and commercial areas. The confluence of Klang/Ampang River will be the study point of flood forecasting for this study.

3 Background Information

3.1 Quantitative Precipitation Estimation in Flood Forecasting

Flood forecasting uses rainfall data as an input in a rainfall-runoff model to forecast flow and water level in the river. QPE is a method to estimate the distribution of rainfall amount that has fallen at a location or a region. Accurate QPE is important to produce high-quality flow forecast with longer lead time in flood forecasting. Rainfall data from rain gauge data or radar rainfall data would be the input into the flood forecasting model as QPE. QPE could be derived by using different data sources including field observation such as rain gauge, weather radar [14–16] and geostationary meteorological images-based rainfall model [17, 18]. Rain gauge is a field observation's equipment which is installed on the ground to measure the accumulated rain that falls on the ground over a set period of time. Rain gauge measures point rainfall data in the catchment, and the accuracy of the data is satisfactory [19]. However, to implement rain gauge data into hydrological modelling for flood forecasting, this rain gauge data need to be derived into an areal rainfall data for the target catchment. Weather radar gives information on radar reflectivity of precipitation, dBZ, and has been used to estimate a rain rate at certain distance above the ground level. Radar technology is capable to observe large area with high spatial resolution compared to the rain gauge network. However, the measurement of rainfall from radar may be different from ground-based rain gauge as radar measures the reflectivity at higher level in the atmosphere and does not measure the rainfall directly. Therefore, bias in rainfall data between rain gauge data and radar should not be ignored. Several works for Malaysian case studies have shown that radar rainfall can improve its accuracy by filtering its noises [20, 21] and optimising the Z–R relationship [22, 23].

3.2 Weather Radar

Weather radar is a remote sensing instrument that emits electromagnetic pulses into the surroundings. Doppler weather radar provides information about speed and direction of the object. Radar not only shows the location but more importantly, the strength of a storm. It uses radio waves or microwaves to determine the range, movement's direction, speed of objects and also altitude [24]. Weather radar system has four main components. The first component is a transmitter, which converts power supplied and generates high frequency signals and transmits it to an antenna. The second component is an antenna. A single antenna reflector can be rotated freely and sends the energy waves or signal out to atmosphere at specific direction, and it receives the echoes returned. The third component is a receiver that processes the returned signal, and the fourth component is a data display system [25]. Frequency for radio waves used in weather radar is in the range of 100 MHz to 100 GHz.

Weather radar in Malaysia is operated by the Malaysian Meteorological Department (MMD). There are two integrated radar networks, one in Peninsular Malaysia and the other in Sabah and Sarawak. In 2011, the exisiting radar network has been enhanced by installing six systems of S-Band Doppler radars over Peninsular Malaysia. In addition, four systems of C-band and one S-band (Miri) Doppler radar were also installed in Sabah and Sarawak [9, 26].

3.3 *Radar Rainfall Measurement*

Estimation of radar rainfall could be made by several measuring methods. The common method is by using radar equation known as power law Z–R relationship developed by Marshall and Palmer in 1948. This method assumes radar-measured volume is filled homogeneously with a known raindrop size distribution (DSD) and is falling vertically in still air [27]. Marshall and Palmer equation, $Z = 200R^{1.6}$, has been widely used by many researchers to estimate radar rainfall. Another method to estimate rainfall radar is by probability matching method [27].

Many studies have been conducted to improve flood forecasting, especially flash flood forecasting in small spatial and short time scales. Sun et al. [28] discussed an accurate rainfall input into hydrological application, which is a key factor for accurate flood estimation. In their study, four different rainfall inputs were used to estimate catchment rainfall: direct use of rain gauge data, using Kriging of Darwin network rain gauge data, direct use of radar rainfall data by using window probability matching method (WPMM) and using cokriging of rain gauge and radar rainfall data. The findings of the study showed that the rainfall estimated by cokriging of rain gauge and radar rainfall data considered improved flood estimation because it optimally combines both the rain gauge and radar data to improve the estimate of sub-catchment rainfall.

Smith et al. [29] studied on flash flood forecasting by using radar rainfall estimation in small urban watershed. Radar rainfall estimation was derived based on power law method (Z–R relationship) and used rain gauge network to examine the bias estimates of radar rainfall. The study found that as spatial and time scales decrease, the predictability of flash flood response from radar rainfall decreases significantly. Possibly, bias correction should be considered to improve radar rainfall estimation and flood forecast response.

Z–R relationships also could be defined by using Z–R matching approach. Many researchers had conducted studies in this approach and proposed several techniques such as traditional matching method (TMM), probability matching method (PMM), window probability matching method (WPMM) and window correlation matching method (WCMM).

3.4 Land Use and Land Cover Change (LUCC)

Changes in land use and land cover influence a watershed's response to precipitation, increase in volumes of surface water runoff, increase in incidence of flooding, erosion or deposition in downstream river channel geometry and also disturbance on aquatic habitat of fish and other aquatic life [30]. Rainfall that falls on developed catchment areas will increase runoff and rapidly flow into the river system causing increase in peak discharge and erosion to the river bank. Many researchers applied numerous methods to investigate and analyse the hydrological effects of LUCC in watershed. The combination of LUCC model with hydrologic model can determine the impact of land-use change on regional or local level [31].

Land-use model could be classified into three classes: empirical model, dynamic model and integrated model [32]. Empirical model uses a best-fit model coefficient to express a statistical relationship between a dependent variable (land use) and a series of independent variables (driving factors of LUCC). The model output produces a transition probability, based on independent variables. Examples of empirical and statistical models are Markov Chain model, regression model, conversion of land use and its effects (CLUE) model and many more. CLUE-S model (conversion of land use and its effects at small region extent) is a modified modelling approach from CLUE model specific for regional scale. It is suitable for fine resolution data with less than 1×1 km grid. Land-use data could be derived from maps or remote sensing images [31]. Dynamic model in land-use model is a spatially explicit method to determine how rapid the land is converted into other uses and where the change will take place. Examples of dynamic model are cellular automata model, agent-based model and system dynamic model. Integrated model is based on multidisciplinary and combined elements of different modelling technique. The model is being developed with empirical and dynamic approaches to produce and simulate land-use change processes. It is also incorporated with geographic information system (GIS) to define initial conditions of the model. Examples of integrated model are IDRISI and GEOMOD.

Cellular automata (CA) model is a dynamic model which can simulate spatial pattern change with local rules and taking into account of neighbourhood configuration and transition maps [32–34]. CA model consists of an arrangement of two-dimensional cells, a characterisation of the neighbourhood of cell, a set of transition rules and a set of cell states [35]. In CA simulation, the output of the previous iteration plays an important role and gives effects to the output for the next iteration [33]. Application of CA in land-use change model had been introduced and demonstrated by [36] in his paper 'Cellular Geography' in the year 1979. Later, this method has been explored and tested by many researchers and integrated with other methods such as with Markov model, artificial neural network (ANN) and fuzzy logic method to suit with the objectives of the land-use change studies.

ANN is useful to quantify complex behaviour and patterns of models by using learning approach method. The application of ANN in the land-use model has been adopted to calculate conversion probabilities for multiple land uses. Studies by [33, 34] on the integration of ANN and CA model in the land-use change model give satisfactory results on the flexibility and freedom of the model to simulate the land-use changes.

3.5 Rainfall-Runoff Modelling

The Hydrologic Engineering Centre-Hydrologic Modelling System HEC-HMS will be used to develop a hydrologic model for the Upper Klang Ampang River Catchment in a distributed modelling scheme. The HEC-HMS is a mathematical model to simulate precipitation-runoff and routing processes in natural or controlled watershed. The model shall predict flow, stage and timing. The HEC-HMS is a distributed runoff model and uses separate sub-models to represent each component of the runoff process. The separate components include models that compute rainfall volume, models of direct runoff (overland flow and interflow), models of base flow and models of channel flow [37].

HEC-HMS has been widely used as rainfall-runoff model for different objectives of the studies including flood forecasting as researched by [29, 34, 38, 39]. Besides land-use change model outcome, land-use maps could be utilised in the HEC-HMS model to support the spatial data for sub-basin input. [40] highlighted that flood forecasting in urban area requires distributed precipitation estimates and distributed hydrologic model. HEC-HMS is capable to simulate the flood forecast with these conditions by using Mod Clark model for direct runoff model. The soil moisture accounting model allows continuous modelling of water movements between the surface and in the ground layers and suitable for runoff-volume model in HEC-HMS [37].

4 Method

Based on the research objectives, the research methodology has been developed to carry out this study. Research methodology is divided into three main parts to achieve the objectives of the research as shown in Fig. 1, the conceptual framework for research methodology.

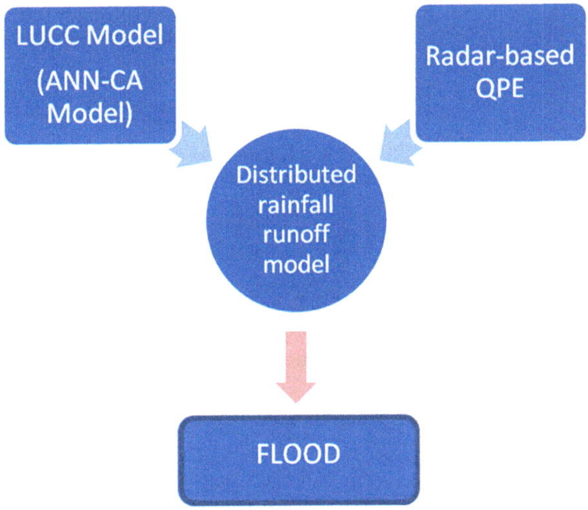

Fig. 1 Conceptual framework for research methodology

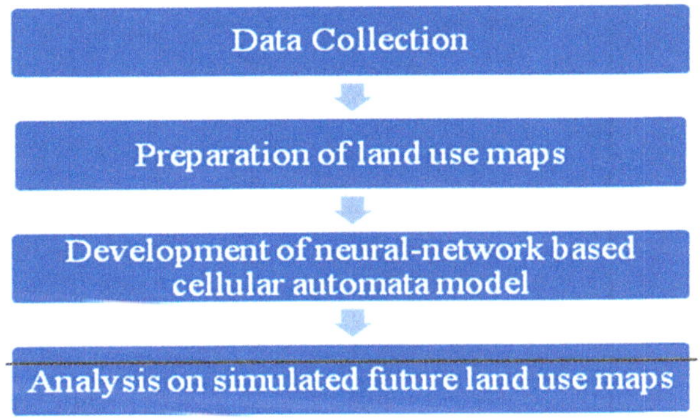

Fig. 2 Methodology for land-use change analysis

4.1 Land-Use Change Analysis Methodology

The objective of the investigation on the existing and future condition of land use at upper Klang Ampang River catchment is to identify the changes of land use which generate a significant contribution to the downstream flood peak (Fig. 2).

4.2 Radar-Based Quantitative Precipitation Estimation Methodology

The method will involve the development of radar-based QPE by using PMM, as Z–R relationship. Bias adjustment method will be used to adjust radar rainfall over the study area before applying the radar rainfall in the rainfall-runoff modelling as given in Fig. 3.

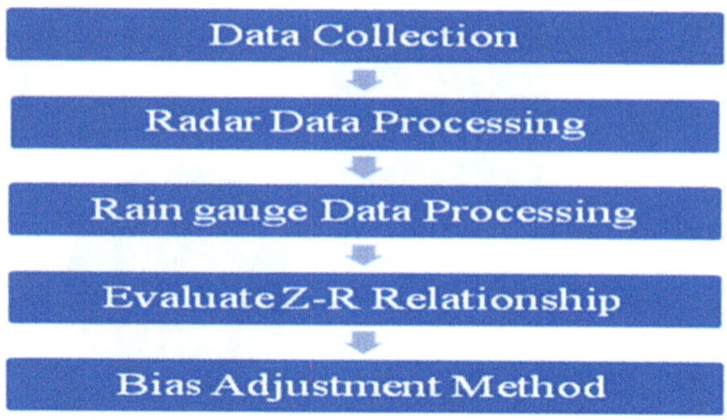

Fig. 3 Methodology for radar-based quantitative precipitation estimation (QPE)

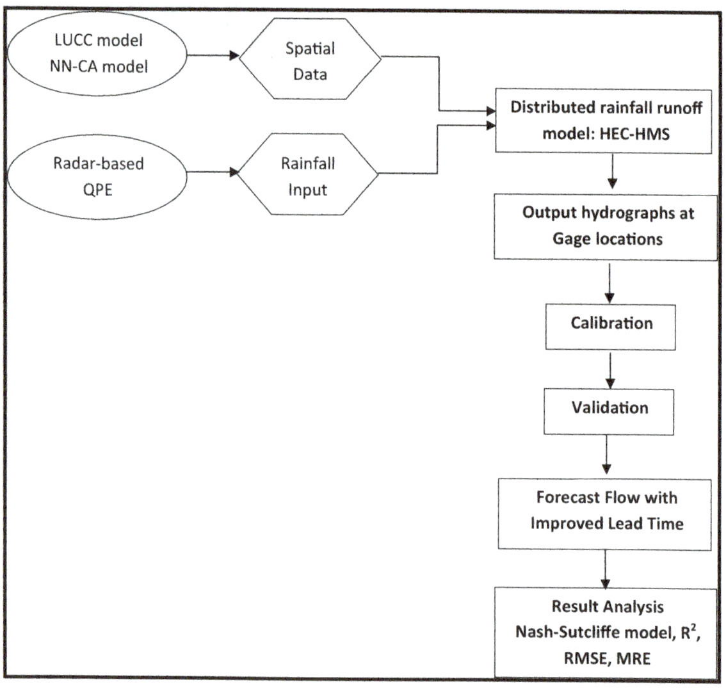

Fig. 4 Methodology for flood forecasting coupled with radar-based QPE and land use input

4.3 Flood Forecasting Coupled with Radar-Based QPE and Land-Use Input

The study will also develop distributed rainfall-runoff model by using the HEC-HMS. The hydrologic model will then be coupled with radar-based QPE integrated with ground rain gauge data to improve accuracy, reliability and lead time for flood forecasting (Fig. 4).

5 Conclusion and Future Research Recommendations

In this study, two new approaches will be introduced for flood forecasting in urban area. The first approach is hydrologic model development using neural network–cellular automata (NN-CA) to analyse the existing land-use and future land-use change for the study area. The focus of the analysis is the effect of downstream flow caused by deforestation and urbanisation in the area of Upper Klang/Ampang River catchment. Results from this analysis will be utilised in the flood forecasting modelling in order to enhance the quality of flow forecast.

The second approach is the development of radar-based QPE by using PMM for Z–R relationship. The QPE will be adopted into the distributed rainfall-runoff model. The aim of this approach is to enhance the lead time and accuracy of flood forecasting. Longer lead time and accurate flow forecast are essential in flood forecasting system and will help relevant authorities, such as SMART, for flood operation management. Thus, the enhanced flood forecasting is expected to be a useful tool for future planning of water resources.

Acknowledgements The authors gratefully acknowledge the contributions of the Department of Irrigation and Drainage (DID) and Malaysian Meteorological Department (MMD) in providing the data that have been used in this study. Appreciation also goes to the Research Management Institute, Universiti Teknologi MARA, for Exploratory Research Grant Scheme (ERGS) Grant (600-RMI/ERGS 5/3 (21/2012)) funded by the Ministry of Higher Education.

References

1. Department of Irrigation and Drainage, 2005–2006 Annual Report. Ministry of Natural Resources and Environment, Malaysia, 2007. ISBN 1823 6277
2. Department of Irrigation and Drainage, National Register of River Basin Study, 2003
3. J. Warren Viessman, G.L. Lewis, *Introduction to Hydrology*, 4th edn. (HarperCollins College Publishers, New York, 1996)
4. R.C. Ward, M. Robinson, in *Principle of Hydrology*, 4th edn., ed. by E. Robinson (Alfred Waller, Glasgow, 2000)
5. M.M.N. Desa, J. Niemczynowicz, Spatial variability of rainfall in Kuala Lumpur, Malaysia: Long and short term characteristics spatial variability of rainfall in Kuala Lumpur, Malaysia. Hydrol. Sci. J. **41**(3), 345–362 (1996)
6. K.S. Sow, L. Juneng, F.T. Tangang, A.G. Hussin, M. Mahmud, Numerical simulation of a severe late afternoon thunderstorm over Peninsular Malaysia. Atmos. Res. **99**(2), 248–262 (2011). doi:10.1016/j.atmosres.2010.10.014
7. C.L. Wong, R. Venneker, S. Uhlenbrook, Variability of rainfall in Peninsular Malaysia. Hydrol. Earth Syst. Sci. Discuss. **6**(4), 5471–5503 (2009). doi:10.5194/hessd-6-5471-2009
8. S. Jamaludin, A.A. Jemain, Fitting daily rainfall amount Malaysia. J. Appl. Sci. **7**(14), 1880–1886 (2007)
9. T. Wardah, R. Suzana, S.Y. Huda, Multi-sensor data inputs rainfall estimation for flood simulation and forecasting, in *2012 I.E. Colloquium on Humanities, Science and Engineering (CHUSER)*, 2012, pp. 374–379

10. A. Akbari, F. Othman, A. Abu Samah, Probing on suitability of TRMM data to explain spatio-temporal pattern of severe storms in tropic region. Hydrol. Earth Syst. Sci. Discuss. **8**(5), 9435–9468 (2011). doi:10.5194/hessd-8-9435-2011
11. C. Ludlow, J. Peterson, G. Smith, Stormwater Management and Road Tunnel (SMART) Project, Functional Specification—FDS Modelling System, 2008
12. Setia Sepakat Perunding, SMART Tunnel: Designer's Operation and Maintenance Manual, 2008
13. K. Abdullah, in *Kuala Lumpur: Re-engineering a Flooded Confluence*. Fourteenth Professor Chin Fung Kee Memorial Lecture, Kuala Lumpur, 2 Oct 2004
14. R. Suzana, T. Wardah, Radar hydrology: New Z/R relationships for quantitative precipitation estimation in Klang River Basin, Malaysia. Int. J. Environ. Sci. Dev. **2**(3), 248–251 (2011)
15. T. Wardah, C. Hamid, H. Aimme, Flood forecasting using tank model and weather surveillance radar (WSR) input for Sg Gombak catchment. Int. J. Civ. Environ. Eng. **13**(2), 42–46 (2013)
16. T. Wardah, S.H. Abu Bakar, A. Bardossy, M. Maznorizan, Use of geostationary meteorological satellite images in convective rain estimation for flash-flood forecasting. J. Hydrol. **356**(3), 283–298 (2008)
17. T. Wardah, I. Zaidah, R. Suzana, Geostationary meteorological satellite-based quantitative rainfall estimation (GMS-Rain) for flood forecasting. Malays. J. Civ. Eng. **21**(1), 1–16 (2009)
18. T. Wardah, S.H. Abu Bakar, M. Mohamad, Intense convective rain estimation using geostationary meteorological satellite, in *Advances in Geosciences, Volume 11: Hydrological Science (HS)*, vol. 11 (2009), pp. 133–147
19. K. Nagata, *Quantitative Precipitation Estimation and Quantitative Precipitation Forecasting by the Japan Meteorological Agency*. Technical review no. 13, RMSC Tokyo—Typhoon Center (2011), pp. 37–50
20. T. Wardah, S.Y. Sharifah Nurul Huda, S.M. Deni, B. Azwa, Radar rainfall estimates comparison with kriging interpolation of gauged rain, in *Colloquium on Humanities, Science and Engineering, CHUSER 2011*, 2011, art. no. 6163877, pp. 93–97
21. T. Wardah, S.Y. Sharifah Nurul Huda, R. Suzana, A. Hamzah, W.W.I. Maisarah, WRF model input for improved radar rainfall estimates using Kalman filter, in *2014 International Symposium on Technology Management and Emerging Technologies (ISTMET)*, IEEE, 2014, pp. 322–327
22. R. Suzana, T. Wardah, A.S. Hamid, Radar hydrology: New Z/R relationships for Klang River Basin Malaysia based on rainfall classification. World Acad. Sci. Eng. Technol. **59** (2011), pp. 248–251
23. R. Suzana, T. Wardah, Radar hydrology: New Z/R relationships for Klang River Basin, Malaysia, in *International Conference on Environment Science and Engineering, IPCBEE*, vol. 8 (IACSIT Press, Singapore, 2011)
24. S.N.H.S. Yahya, W. Tahir, S. Ramli, S.M. Deni, H. Arof, M.F.M. Saaid, Improved estimation of radar rainfall bias over Klang River Basin using a Kalman filtering approach, in *2012 I.E. Symposium on Business, Engineering and Industrial Applications (ISBEIA)*, IEEE, September 2012, pp. 368–373
25. R.E. Rinehart, *Radar for Meteorologists* (Rinehart Publications, Grand Forks, ND, 1997)
26. K. Subramaniam, L.K. Ling, W.A. Wan Hassan, Variational data assimilation of Doppler radar reflectivity data into WRFV3.2, 2013
27. D. Rosenfeld, D.B. Wolff, E. Amital, The window probability matching method for rainfall measurements with radar. J. Appl. Meteorol. **33**, 682–693 (1994)
28. X. Sun, R.G. Mein, T.D. Keenan, J.F. Elliott, Flood estimation using radar and raingauge data. J. Hydrol. **239**(1–4), 4–18 (2000). doi:10.1016/S0022-1694(00)00350-4
29. J.A. Smith, M.L. Baeck, K.L. Meierdiercks, A.J. Miller, W.F. Krajewski, Radar rainfall estimation for flash flood forecasting in small urban watersheds. Adv. Water Resour. **30**(10), 2087–2097 (2007). doi:10.1016/j.advwatres.2006.09.007

30. C. McColl, G. Aggett, Land-use forecasting and hydrologic model integration for improved land-use decision support. J. Environ. Manag. **84**(4), 494–512 (2007). doi:10.1016/j.jenvman.2006.06.023
31. P.H. Verburg, W. Soepboer, A. Veldkamp, R. Limpiada, V. Espaldon, S.S.A. Mastura, Modeling the spatial dynamics of regional land use: The CLUE-S model. Environ. Manag. **30**(3), 391–405 (2002). doi:10.1007/s00267-002-2630-x
32. D. Guan, H. Li, T. Inohae, W. Su, T. Nagaie, K. Hokao, Modeling urban land use change by the integration of cellular automaton and Markov model. Ecol. Model. **222**(20–22), 3761–3772 (2011). doi:10.1016/j.ecolmodel.2011.09.009
33. X. Li, A.G.-O. Yeh, Neural-network-based cellular automata for simulating multiple land use changes using GIS. Int. J. Geogr. Inf. Sci. **16**(4), 323–343 (2002). doi:10.1080/13658810210137004
34. X. Xu, J. Zhang, X. Zhou, Modeling urban land use changes in Lanzhou based on artificial neuralnetwork and cellular automata, Geoinformatics 2008 and Joint Conference on GIS and Built Environment: Geo-Simulation and Virtual GIS Environments, Lin Liu, Xia Li, Kai Liu, Xinchang Zhang, Aijun Chen, Eds., Proc. of SPIE Vol. 7143 71431A · © 2008 SPIE · CCC code: 0277-786X/08/$18 · doi: 10.1117/12.812574
35. R. White, G. Engelen, Cellular automata and fractal urban form: A cellular modelling approach to the evolution of urban land-use patterns. Environ. Plann. A **25**(8), 1175–1199 (1993)
36. W.R. Tobler, *Cellular Geography: Philosophy in Geography* (D. Reidel, Dordrecht, 1979), pp. 379–386
37. USACE-HEC, Hydrologic modeling system HEC-HMS technical reference manual (2000), retrieved from http://www.hec.usace.army.mil/software/hec-hms/documentation/HEC-HMS_TechnicalReferenceManual_(CPD-74B)
38. H.A.P. Hapuarachchi, Q.J. Wang, T.C. Pagano, A review of advances in flash flood forecasting. Hydrol. Process. **25**(18), 2771–2784 (2011). doi:10.1002/hyp.8040
39. M.R. Knebl, Z.-L. Yang, K. Hutchison, D.R. Maidment, Regional scale flood modeling using NEXRAD rainfall, GIS, and HEC-HMS/RAS: A case study for the San Antonio River Basin Summer 2002 storm event. J. Environ. Manag. **75**(4), 325–336 (2005). doi:10.1016/j.jenvman.2004.11.024
40. H.O. Sharif, D. Yates, R. Robert, C. Mueller, The use of an automated nowcasting system to forecast flash floods in an urban watershed. J. Hydrometeorol. **7**, 190–202 (2006)

CPSIA information can be obtained at www.ICGtesting.com
Printed in the USA
BVOW10*2212230415

397560BV00001B/9/P

9 789812 873644